AI with Integrity

Dimitrios Sargiotis

AI with Integrity

Shaping Ethical, Transparent, and Equitable Urban Futures

Springer

Dimitrios Sargiotis
School of Civil Engineering
National Technical University of Athens
Athens, Greece

ISBN 978-3-032-31132-0 ISBN 978-3-032-31133-7 (eBook)
https://doi.org/10.1007/978-3-032-31133-7

This Springer imprint is published by the registered company Springer Nature Switzerland AG
The registered company address is: Gewerbestrasse 11, 6330 Cham, Switzerland

Preface

In an age where algorithms have begun to govern the very tissues of our urban lives allocating resources, shaping mobility, predicting crime, and distributing public services the question is no longer whether Artificial Intelligence (AI) belongs in cities, but how it must be deployed with conscience, transparency, and justice. This volume, the fourth in a series dedicated to AI-driven urban transformation, takes an unflinching look at the ethical, legal, and philosophical imperatives of integrating AI into the daily structures of governance, infrastructure, and civil existence.

What does it mean for a city to think? And when it does, who is it thinking for? These are not rhetorical provocations they are practical, computational, and political design problems that demand both rigorous mathematics and ethical clarity. The premise of this book is profound in its simplicity: intelligent systems must not merely be intelligent they must be just. Their architectures must embed accountability. Their optimizations must center equity. Their predictions must respect human dignity, not abstract it away.

This book unpacks these imperatives across a sweeping intellectual arc, introducing a constellation of AI frameworks including Adversarial Debiasing, Fairness-Constrained Reinforcement Learning, Explainability-Augmented PPO, Governance-Driven Multi-Agent Models, and Game-Theoretic Penalty Structures not as technical flourishes, but as instruments of urban ethics. Each equation is a covenant with the future; each model, a deliberation on power, bias, and justice.

We dive deep into the infrastructural logics of fairness in city planning, the political economy of algorithmic surveillance, the fragile contracts between privacy and public interest, and the mechanisms through which explainable AI becomes not only a regulatory requirement, but a democratic act. Through novel architectures such as Participatory XAI, Federated Transparency Constraints, and Dispute Resolution Functions, we reframe the AI-city nexus from a technological opportunity to a moral responsibility.

This volume insists that ethical AI in urban systems is not a layer to be added after the fact it must be foundational, mathematically encoded, and publicly verifiable. To achieve this, it traverses interdisciplinary terrain: weaving civil engineering, law, philosophy, environmental science, and AI research into a unified framework for ethical urban intelligence.

For researchers, this book provides operationally reproducible frameworks, robust theoretical models, and rigorous proofs of ethical AI design. For policymakers, it offers governance blueprints to ensure AI accountability and transparency. For citizens, it provides the language and logic to interrogate the systems that increasingly govern their lives. And for cities themselves, it offers a vision: not merely to be "smart," but to be fair, inclusive, and legible to the people they serve.

Let this book be not a manual, but a manifesto. A call for AI that does not automate inequality but interrogates it. A vision where cities do not become code-saturated fortresses of efficiency, but algorithmically augmented sanctuaries of justice. In the silence between datasets and decisions, we must hear the voice of the citizens.

Athens, Greece Dimitrios Sargiotis

AI with Integrity: Shaping Ethical, Transparent, and Equitable Urban Futures

This book presents a powerful ethical manifesto for the integration of Artificial Intelligence (AI) into urban environments, challenging technocratic assumptions by grounding smart city innovation in the principles of fairness, transparency, accountability, sustainability, and privacy. It establishes that intelligent urban systems cannot merely optimize performance; they must be morally legible, socially just, and legally compliant. The book argues that the city of the future must be governed not just by data but by dignity, not just by efficiency but by equity.

The discussion begins by tackling the structural problem of algorithmic bias in urban AI systems, particularly in predictive policing, public service allocation, and transport optimization. Through a mathematically rigorous framework that includes Human-Centered AI, Fairness-Constrained Optimization, and Fairness-Aware Reinforcement Learning, the book introduces adaptive models that correct historical inequalities without compromising performance. These are not mere technical appendices; they represent an evolving ethical infrastructure that underpins responsible AI deployment.

Moving beyond fairness, the book develops a comprehensive theory of AI accountability in urban governance. It introduces a series of interlinked models including Explainability-Optimized PPO updates, Policy-Justification Functions, and Dispute Resolution Functions within Multi-Agent Reinforcement Learning (MARL) systems that make algorithmic decisions auditable, contestable, and legally actionable. Accountability here is framed not only as a technical imperative but as a democratic necessity.

Transparency and explainability emerge as central pillars, developed through a suite of self-interpreting AI models. The book introduces Interpretable Deep Reinforcement Learning (I-DRL), Self-Interpretable Policy Optimization, Explainability-Augmented Reward Functions, and the SHAP-based Decision Attribution Framework. These ensure that AI-driven decisions in public domains such as housing, transportation, or energy allocation are both intelligible and reviewable by stakeholders across disciplines and literacy levels. The book further explores Participatory

Explainable AI (P-XAI) and Federated Explainability Constraints, positioning community-informed decision-making as a new standard of civic-algorithmic governance.

The book also addresses the political economy of AI deployment examining how corporate incentives, regulatory gaps, and economic pressures can corrupt ethical intentions. In response, it proposes a Game-Theoretic Reinforcement Learning (GTRL) framework that explicitly encodes public-interest optimization alongside profit constraints, ensuring that AI models in urban planning cannot prioritize economic gain at the expense of social justice. This framework is supported by the Ethically Constrained Nash Equilibrium Function and the Regulated AI Economic Penalty Function, ensuring balance between innovation and responsibility.

The book then expands its ethical analysis into AI-driven urban resource management, introducing the Fairness-Constrained Multi-Agent Reinforcement Learning (FCRL-MARL) and Sustainability-Constrained Reinforcement Learning (SC-RL) frameworks for equitable and human-centric energy, water, and waste governance. These frameworks integrate comfort-aware AI adjustments, green federated learning, and multi-objective environmental optimization, ensuring that smart city infrastructures are simultaneously sustainable and socially inclusive.

Finally, the book develops an extensive architecture for privacy and data governance in smart urban systems. Through Federated Learning (FL), Differential Privacy, Blockchain-Based Data Integrity Models, and Consent-Aware Federated Learning (CA-FL), it constructs a decentralized ethical infrastructure that safeguards citizen autonomy, prevents surveillance, and complies with legal frameworks such as GDPR. These are operationalized with equations for privacy-preserving optimization, SHAP-based fairness-aware attribution, and auditable AI transparency scoring grounded in information theory.

The frameworks developed in this volume should be read as a layered response to the central governance problem of contemporary urban AI: the difficulty of aligning adaptive, data-intensive, and often opaque computational systems with democratic accountability, distributive justice, privacy protection, and ecological responsibility. Rather than treating fairness, transparency, sustainability, and human oversight as isolated compliance requirements, the book organizes them as interdependent design constraints that must be embedded across the full AI life cycle, from data collection and representation to model optimization, explanation, auditing, appeal, and institutional review. The proposed frameworks therefore move beyond conventional "smart city" optimization by integrating fairness-constrained learning, explainable and participatory AI, privacy-preserving federated architectures, sustainability-aware reinforcement learning, adversarial debiasing, auditability mechanisms, and legally contestable governance functions into a unified urban intelligence paradigm. In this sense, the following table does not merely summarize technical models; it establishes a systematic bridge between urban AI use cases, the risks they generate, the stakeholders they affect, and the governance architectures required to make intelligent cities accountable, equitable, and publicly defensible.

Urban AI Challenges, Governance Risks, and Proposed Technical Responses

Chapter	Urban domain/use case	Ethical risk	Affected stakeholders	Governance requirement	Representative proposed frameworks
Chapter 1	Urban governance and public-service allocation	Algorithmic bias, inequitable prioritization, weak accountability, incentive capture	Citizens, municipalities, underserved districts, public agencies, regulated vendors	Bias auditing, fairness constraints, explainability, contestability, public-interest alignment	UABMF, FAI-AOUD, RL-XAI-US, MARL-AI-AG, GTRL-EAI-US
Chapter 2	Transparent and participatory urban decision-making	Black-box optimization, low public trust, limited stakeholder understanding, superficial explanations	Citizens, planners, policymakers, technical auditors, community representatives	Interpretability, participatory validation, post-hoc attribution, fairness-constrained explainability	X-DRL, I-DRL, SI-DRL, XG-DRL, P-XAI, SHAPE-SCA, FCEM
Chapter 3	Energy, water, waste, and urban resource allocation	Resource inequity, sustainability trade-offs, efficiency-over-welfare optimization, limited accountability in adaptive allocation	Households, vulnerable groups, municipalities, utilities, infrastructure operators	Fairness-sustainability balancing, transparent resource allocation, accountable optimization, privacy-preserving coordination	FCRL-MARL, SC-RL, XRL-URM, AC-RL, FC-FL
Chapter 4	Urban data governance	Privacy erosion, re-identification, consent failure, opaque data processing, unfair data-driven decisions	Citizens, data controllers, municipalities, service providers, regulatory bodies	Privacy preservation, consent governance, anonymization, accountability, transparency reporting	FL-PPAM, CA-FL, AIDAG, AI-AG, FEA-UG, TEA-UG
Chapter 5	Civil engineering, urban development, and infrastructure governance	Unequal infrastructure investment, unclear responsibility, explainability-performance tensions, sustainability burdens, privacy and compliance risks	Engineers, residents, public authorities, peripheral communities, infrastructure owners, future urban populations	Sociotechnical fairness, responsibility attribution, compliance auditing, human-centric oversight, sustainable optimization	FEA-UG, FEA-ERD, AI-GUS, SFAI-EUD, AI-TGUS, AI-TGUS II, AIAGF, AIRAF, AIAF, HCEOA, AID-USOS, FPSUF, FCDAI-GM, FLDM
Chapter 6	Inclusive urban planning	Spatial exclusion, biased datasets, unequal participation, opaque planning decisions, weak trust in algorithmic recommendations	Low-income residents, marginalized groups, planners, community organizations, municipal authorities	Fairness-constrained planning, participatory review, explainability, dynamic auditing, privacy-preserving compliance	FCAI-UP, FairGAN, FCDRL-UP, MO-DRL-UP, TA-FC-DRL-UP, FC-BDRM, FC-ADM, FA-XRL, XAI-UP, XAI-TC-UP, AI-PEUP

(continued)

Chapter	Urban domain/use case	Ethical risk	Affected stakeholders	Governance requirement	Representative proposed frameworks
Chapter 7	AI surveillance and responsible security governance	Privacy violation, discriminatory targeting, chilling effects, authoritarian misuse, human-rights risks, weak legal accountability	Citizens, activists, minority communities, regulators, law-enforcement bodies, public-interest organizations	Proportionality, privacy, anti-bias safeguards, legal compliance, independent audit, stakeholder-centric explainability	AI-PSS, PPAISF, PPAISF II, PCASIF, SEAISF, FCPPA, FC-AISF, RL-XAM, SVB-XAI, AAAM, SCEF, AISLECF, AISLEG

Acknowledgments

This book is born not merely of research, but of necessity a necessity to question, reimagine, and ultimately re-engineer the very moral substratum of the algorithmic city. As machine systems advance toward omnipresence in urban governance, a new civic ontology demands articulation: one in which justice is not an afterthought, but the primal code.

What unfolds in these pages is not simply a synthesis of AI models, governance theories, and mathematical formulations, but a deliberate excavation of the ethical contracts that must govern our technological futures. Each chapter is a quiet rebellion against the illusion that urban intelligence can be reduced to performance metrics and predictive analytics alone. Instead, it argues firmly, mathematically, ethically that the city of the future must not only calculate, but contemplate; it must not only optimize, but recognize; and above all, it must never forget that beneath every dataset lies a human life.

Though I stand as the sole author of this volume, its lineage is indebted to an intellectual multitude: the ethicists who have refused to separate logic from justice; the engineers who have infused their code with care; the urban theorists who remind us that a city is more than infrastructure it is a living narrative of belonging, exclusion, power, and hope. Their influence, whether direct or ambient, resonates through every layer of this work.

I am profoundly grateful to Springer's editorial and production teams for their unwavering commitment to scholarly excellence. Their exacting standards transformed abstract ideals into publishable architecture, ensuring that the argument advanced here reaches an international audience equipped to act upon it.

To the institutions that advocate for algorithmic fairness, data sovereignty, human rights, and ecological stewardship you are the ethical scaffolding of this volume. Your vision of inclusive, transparent, and accountable systems has lent direction and urgency to every model proposed and every framework constructed.

Finally, to those who have walked with me in silence my family, my colleagues, my inner circle of confidants your presence, patience, and belief have made this endeavor not only possible, but meaningful. In the long hours between proof and prose, it was your quiet companionship that gave this manuscript its final form.

This book is offered as both a theoretical framework and a civic invocation. It insists that ethical AI is not a technical feature to be appended, but a foundational architecture to be designed, debated, and defended. It is my conviction that the urban future belongs not to the most intelligent systems, but to the most principled ones.

Let this work stand, then, as a scholarly act of conscience a refusal to automate injustice and a commitment to reprogramming the city in the image of dignity.

Competing Interests The author affirms unequivocally that no competing financial, institutional, or proprietary interests have influenced the conception, authorship, or dissemination of this volume. *AI with Integrity: Shaping Ethical, Transparent, and Equitable Urban Futures* was developed as an entirely autonomous scholarly undertaking absent of external funding, consultancy affiliations, corporate sponsorship, or institutional directive.

No financial compensation neither overt nor tacit was received from industry partners, government agencies, or private enterprises throughout any phase of this book's ideation, research, or editorial production. The theoretical frameworks, ethical postulates, and AI architecture articulated herein are the sole products of independent inquiry, crafted through long-standing interdisciplinary engagement and guided by an unwavering commitment to academic freedom, scientific reproducibility, and civic responsibility.

This work is grounded in the pursuit of epistemic integrity and the ethical application of artificial intelligence within the public domain. Every algorithmic framework, case context, and governance critique presented in this volume has been selected strictly for its scholarly relevance, ethical salience, and conceptual robustness without promoting, endorsing, or depending upon any commercial platform, proprietary software, or vendor-specific methodology.

This declaration confirms the intellectual and ethical independence of the present volume, which stands as a contribution to the global academic commons crafted not for commercial gain, but for the advancement of critical knowledge, democratic accountability, and urban justice in the age of intelligent systems. Readers may engage with confidence, knowing that the content has not been shaped by economic incentives or institutional pressures, but by a principled commitment to truth, equity, and the public good.

About the Book

As Artificial Intelligence (AI) becomes integral to urban governance, infrastructure, and decision-making, the question is no longer whether AI will shape the urban future—but whether it will do so justly, transparently, and inclusively. The book *AI with Integrity: Shaping Ethical, Transparent, and Equitable Urban Futures* part of the book series "AI and the Intelligent City. Data-Driven Innovations for Urban Transformation" presents a groundbreaking framework that integrates ethical AI at the heart of smart city development. Reframing fairness, transparency, accountability, privacy, and sustainability as essential design principles, the book demonstrates how these values can be mathematically embedded and democratically governed, rather than viewed as mere constraints on technological progress.

Addressing the challenge of algorithmic bias in AI-driven systems for urban services, the book introduces advanced frameworks such as Fairness-Constrained Reinforcement Learning, Adversarial Debiasing Models, and Causal Inference for Bias Mitigation. These models enable AI to adapt in real-time, correcting inequitable decision-making while maintaining both accuracy and social equity.

To ensure transparency and auditability, the book proposes architectures including Interpretable Deep Reinforcement Learning (I-DRL) and Explainability-Augmented Reward Models, which make AI decisions intelligible and justifiable. Furthermore, Participatory Explainable AI (P-XAI) and Federated Explainability Protocols enable community-informed oversight, empowering stakeholders to participate actively in governance.

The book introduces innovative governance frameworks like the Ethically Constrained Nash Equilibrium and the Risk-Aware Urban Regulation Framework (RAURF), which ensure AI deployment aligns with democratic values, regulatory frameworks, and environmental responsibility.

Privacy concerns are addressed through Federated Learning, Differential Privacy, and Blockchain-Based Audit Trails, providing a decentralized, legally compliant infrastructure that safeguards citizen autonomy and trust.

This book proposes a unified vision for ethical AI in urban life—one where technology serves justice, predictions are bounded by explainability, and innovation is grounded in human dignity. It calls for a moral reimagining of AI governance, where smart cities are not only more efficient but more equitable, accountable, and collectively governed.

Contents

About the Author

Dimitrios Sargiotis is a Faculty Dr. and Deputy Director of the PC-Laboratories within the Digital Technology Laboratory for Technical Projects at the School of Civil Engineering, National Technical University of Athens. With more than three decades of academic and professional experience, he stands at the intersection of Artificial Intelligence, Computer Science, Data Governance, and Civil Engineering, pioneering the digital transformation of infrastructures, cities, and societies.

He has authored 17 scholarly books, 137 book chapters, and over 50 scientific publications, articles, research reports, and other scholarly contributions disseminated through Springer and other prestigious international outlets, establishing himself as one of the leading voices in AI for Smart Cities, Digital Twins, Intelligent Transportation Systems, Green and Sustainable Infrastructure, and Human-Centric AI. His frameworks emphasize ethics, explainability, and governance, aligning technological innovation with public value and societal well-being.

Dr. Sargiotis provides leadership in European Commission-funded research projects, serves as a reviewer and program committee member for international journals and conferences, and collaborates with esteemed organizations, including Springer, the Programming Research Group of Paris, and the Union of Computer Engineers and Communications of Greece.

Beyond academia, he is an advocate of life-long learning, physical well-being, and cultural engagement, embodying a vision where scientific rigor, ethical responsibility, and human creativity converge. His career demonstrates how Artificial Intelligence can act as a catalyst for sustainable development, social justice, and the flourishing of digitally governed societies worldwide.

Abbreviations

AI	Artificial Intelligence: machine-based systems that simulate human cognitive functions such as perception, reasoning, and decision-making; in this volume, AI is examined as a force shaping governance, ethics, and civic life.
DP	Differential Privacy: a formal method of injecting statistical noise into data queries to protect individual identity, supporting ethical data use in AI training across public datasets.
DRL	Deep Reinforcement Learning: a method that combines deep neural networks with reinforcement learning to handle high-dimensional state spaces and optimize complex civic decisions.
FAIR	Findable, Accessible, Interoperable, Reusable: a set of principles guiding ethical data stewardship, enhancing the transparency, reusability, and accountability of civic datasets.
FC	Fairness Constraint: a learning or optimization restriction applied to AI models to ensure nondiscriminatory outcomes across protected demographic groups.
FCEM	Fairness-Constrained Ethical Model: a conceptual framework in which ethical parameters (e.g., equity, transparency) are explicitly modeled in AI learning objectives.
FCRL	Fairness-Constrained Reinforcement Learning: a reinforcement learning approach that integrates fairness objectives into the reward function to ensure equitable decision-making.
FL	Federated Learning: a decentralized training paradigm where models are trained across distributed nodes without centralizing sensitive data, preserving privacy while enabling collaborative intelligence.
GAN	Generative Adversarial Network: a deep learning framework where two networks (generator and discriminator) compete to produce realistic synthetic data, discussed in bias detection.

GDPR General Data Protection Regulation: the European Union's legal framework for personal data protection and privacy, frequently referenced in AI ethics and urban data governance.

GNN Graph Neural Network: a deep learning architecture suited for graph-structured data, applied in modeling interconnected urban systems, infrastructures, and social topologies.

KL Kullback-Leibler divergence: a statistical measure of difference between two probability distributions, often used in policy regularization within reinforcement learning.

LIME Local Interpretable Model-Agnostic Explanations: an interpretability method that approximates the behavior of black-box models using interpretable local surrogates.

MARL Multi-Agent Reinforcement Learning: a framework involving multiple agents learning concurrently through interaction, coordination, or competition, essential for simulating city-wide systems.

PIA Privacy Impact Assessment: a procedural tool for evaluating the potential risks to personal privacy posed by AI implementations and data systems.

PPO Proximal Policy Optimization: a reinforcement learning algorithm employed to maintain stability and efficiency in policy updates, central to fairness-aware learning in dynamic civic environments.

RL Reinforcement Learning: an AI paradigm in which agents learn optimal behavior through interaction with an environment via reward-maximizing policies.

SHAP SHapley Additive exPlanations: a model-agnostic approach based on cooperative game theory to assign feature importance, used for post-hoc transparency in decision-making systems.

UP Urban Policy: used as an abbreviation in context to refer to AI-modulated urban policy instruments or strategic design parameters.

XAI Explainable Artificial Intelligence: a class of AI methods designed to make models' outputs transparent and interpretable, fostering public trust and ethical compliance in urban deployments.

Annotations

Accountability-Aware Decision Function A reinforcement learning mechanism that integrates legal and ethical oversight into the value function, enabling traceable, justifiable decisions in urban governance.

Adversarial Debiasing A dual-network training approach in which a predictive model is penalized for encoding sensitive attributes, with an adversary learning to detect unfair biases and reduce discriminatory outcomes.

AI Surveillance A critical lens examining how AI systems are deployed in urban environments for monitoring behavior, often raising concerns about privacy, bias, and the need for democratic oversight.

Bias Mitigation	A suite of techniques for detecting, quantifying, and reducing bias in AI systems, applied at the data, algorithmic, or decision-output level to ensure equitable treatment.
Bias-Aware Reward Adjustment	A technique in which AI reward functions are penalized in proportion to the bias detected, guiding learning toward equitable and nondiscriminatory outcomes.
Comfort-Aware AI Adjustments	An adaptive AI mechanism that preserves human well-being (e.g., temperature, lighting) while optimizing for sustainability, applied in smart buildings and public infrastructure.
Data Governance	The normative and procedural framework that regulates how urban data is collected, processed, shared, and preserved ensuring legal compliance, ethical integrity, and algorithmic transparency.
Deep Reinforcement Learning	A class of AI algorithms that combine deep learning with reward-driven exploration, enabling decision-making in high-dimensional, uncertain civic environments such as traffic control or resource allocation.
Demographic Parity	A fairness constraint requiring that positive prediction rates be statistically equal across demographic groups, helping to detect and correct structural discrimination in algorithmic outputs.
Differential Privacy	A mathematical framework that injects calibrated noise into data analyses, protecting individual-level information while preserving the utility of aggregated datasets for urban AI development.
Dispute Resolution Function	An AI module that evaluates contested decisions based on fairness, legal thresholds, and citizen appeals operationalizing redress mechanisms in algorithmic governance.
Equalized Odds	A fairness criterion requiring that prediction accuracy be statistically independent of protected attributes such as gender or race, ensuring parity in true positive and false negative rates.
Error Attribution Function	An explainability mechanism that identifies which features contribute to undesirable or biased outcomes, supporting ethical debugging and system refinement.
Ethical Reinforcement Learning Penalty	A regulatory constraint that penalizes AI systems when economic incentives override social fairness, preventing ethically compromised optimization.

Explainable AI (XAI)	A methodological paradigm ensuring that AI decisions can be understood, interrogated, and contested by human users, foundational for public trust and regulatory compliance in civic applications.
Fair Representation Learning	A machine learning technique that transforms input data into latent representations that obfuscate sensitive attributes, promoting fairness in downstream decisions.
Fairness-Constrained Reinforcement Learning	An adaptive learning paradigm that embeds fairness constraints directly into the reward structure of reinforcement learning, ensuring ethical policy development in dynamic environments.
Federated Explainability Constraint	A validation condition in federated systems requiring that model decisions remain interpretable and reviewable across all decentralized nodes and stakeholder groups.
Federated Learning (FL)	A decentralized machine learning architecture in which models are trained across multiple nodes without transferring raw data preserving data privacy and regulatory compliance.
Governance Oversight Model	A parallel AI system tasked with evaluating primary decision outputs for fairness, legality, and error, ensuring institutional accountability in multi-agent settings.
Human-in-the-Loop	An interactive AI framework where human feedback, judgment, or correction is integrated into the training or deployment process to ensure alignment with social norms.
Layered AI Architecture	A hierarchical design paradigm in which AI systems are structured into modular, interpretable layers each governed by distinct ethical, functional, and regulatory constraints to ensure clarity, accountability, and alignment in complex urban applications.
Multi-Agent Reinforcement Learning	A framework in which multiple AI agents simultaneously learn through interaction in shared environments, enabling distributed governance and cooperative civic optimization.
Participatory Explainable AI (P-XAI)	A governance-aligned AI model in which explainability is evaluated and validated through the lens of diverse stakeholder input, supporting inclusive and democratic decision-making.

Policy-Justification Function	A mathematical function embedded in reinforcement learning models that generates human-readable explanations for policy actions, enhancing interpretability and accountability.
Privacy-Preserving Federated Learning	An extension of FL that incorporates differential privacy and encryption to protect citizen data while enabling decentralized learning in urban AI deployments.
Proximal Policy Optimization (PPO)	A reinforcement learning algorithm designed to ensure stable and incremental updates to policy, applied in fairness-aware and transparency-constrained urban systems.
Reinforcement Learning	An adaptive AI paradigm where agents learn optimal behavior by interacting with an environment and receiving feedback, foundational to the book's treatment of ethical policy modeling.
SHAP-Based Feature Attribution	A post-hoc interpretability technique using Shapley values from cooperative game theory to assign causal significance to input features in model outputs.
Sustainability-Constrained Optimization	A multi-objective reinforcement learning formulation that balances environmental targets with operational performance, safeguarding long-term ecological impact.
Transparency-Enhanced State Representation	An AI design method that encodes state variables in interpretable formats, ensuring that decision rationales remain legible to human evaluators.
Urban Planning	An integrative discipline referenced in the context of AI-enhanced governance, wherein intelligent models inform long-term spatial, social, and infrastructural transformations in urban landscapes.

Chapter 1
Ensuring Fairness, Transparency, and Accountability in AI-Driven Urban Governance

Artificial Intelligence (AI) now forms the cognitive backbone of modern urban governance, orchestrating the complex dynamics of transportation, energy management, healthcare, security, and public administration. Yet, as these systems gain power and autonomy, they also inherit and amplify the biases, inequities, and asymmetries embedded within the data and institutional structures that sustain them. This chapter, *Ensuring Fairness, Transparency, and Accountability in AI-Driven Urban Governance*, confronts the urgent question of how cities can deploy AI systems that are not only efficient but also ethically aligned, socially equitable, and politically accountable. It moves beyond abstract ethical debate to offer a mathematically formalized, algorithmically precise roadmap for embedding fairness and accountability directly into the architecture of urban AI systems.

The chapter therefore begins by tackling the structural problem of algorithmic bias in urban AI systems, particularly in predictive policing, public service allocation, and transport optimization. Through a mathematically rigorous framework that includes human-centered AI, fairness-constrained optimization, and fairness-aware reinforcement learning, it advances the central argument that ethical principles acquire practical force only when they are translated into operational, auditable, and institutionally reviewable architectures. Fairness must be more than an aspiration; it must become a constraint. Accountability must be more than a declaration of responsibility; it must become a traceable relation between decisions, agents, standards, and consequences. Transparency must be more than a normative demand; it must become an explainability structure capable of supporting scrutiny, review, and contestation. These frameworks are not technical appendices to ethical reflection. They are the methodological core through which ethical urban AI moves from abstract discourse to computable civic governance, aligning mathematical formalization with the institutional conditions required for responsible deployment (Novelli et al. 2024; Bartsch et al. 2025; Ibitoye et al. 2025; Alrasbi and Ariaratnam 2026).

The first framework formalizes the computational essence of fairness and accountability through the Urban AI Bias Mitigation Framework (UABMF), a layered architecture that employs Generative Adversarial Networks (GANs) for

D. Sargiotis, *AI with Integrity*, https://doi.org/10.1007/978-3-032-31133-7_1

adversarial debiasing and a fairness-constrained optimization mechanism that dynamically enforces statistical equity thresholds during model training. This dual mechanism ensures that AI-driven decisions, such as predictive policing, energy distribution, or service allocation, are continuously recalibrated to prevent discrimination and promote inclusiveness across demographic and socioeconomic spectra.

Building on this foundation, the chapter introduces the Fairness-Aware AI Optimization for Urban Development (FAI-AOUD) framework, which integrates a fairness-constrained reward function within Proximal Policy Optimization (PPO). This advanced reinforcement learning structure enables AI systems to self-regulate their fairness in real time, dynamically penalizing biased outcomes and rewarding equitable ones. Through this formulation, fairness ceases to be an external constraint and becomes an intrinsic learning objective of the AI itself.

The discussion then advances toward the notion of accountability, exploring how AI's decision processes can be made transparent and interpretable to human stakeholders. The Reinforcement Learning-based Explainability for Urban Solutions (RL-XAI-US) framework merges Explainable Artificial Intelligence (XAI) with Proximal Policy Optimization (PPO) to ensure that every AI-generated decision is accompanied by a human-understandable justification. By introducing a policy-justification function, the framework assigns interpretability scores to each decision, transforming black-box predictions into reasoned, auditable outputs that enhance public trust and regulatory compliance.

Accountability, however, is incomplete without redress and oversight. Hence, the chapter develops the Multi-Agent Reinforcement Learning for AI Accountability in Urban Governance (MARL-AI-AG) framework. Here, Multi-Agent Reinforcement Learning (MARL) is integrated with a dispute resolution function, enabling citizens to challenge AI-driven decisions. The system embeds a governance oversight model that operates independently of the primary decision-making AI, creating a dual-layer accountability mechanism that combines adaptive correction with democratic transparency.

Finally, the chapter expands the ethical perimeter by examining the political-economic context of AI deployment through the Game-Theoretic Reinforcement Learning for Ethical AI in Urban Solutions (GTRL-EAI-US) framework. This model employs Game-Theoretic Reinforcement Learning (GTRL) and the Nash equilibrium principle to balance corporate incentives with public welfare objectives. An ethical reinforcement learning penalty function ensures that profit-driven decisions are counterbalanced by socially beneficial constraints, allowing AI systems to harmonize economic feasibility with ethical responsibility.

Through these interlocking frameworks, the chapter presents a comprehensive synthesis that spans mathematical formalization, algorithmic design, governance regulation, and ethical reasoning. Each framework contributes to a unified objective, ensuring that the intelligence embedded in urban infrastructures acts as a moral as well as technical agent capable of learning fairness, explaining itself, and being held accountable.

This chapter does more than discuss fairness and accountability; it constructs the algorithmic grammar of ethical urban intelligence. It demonstrates how fairness metrics can be encoded as optimization constraints, how interpretability can be mathematically quantified, how appeals can be algorithmically resolved, and how corporate power can be regulated through equilibrium theory. For the international reader whether policymaker, researcher, or city innovator, this work offers not merely theory, but an actionable blueprint for governing the algorithmic city of the twenty-first century, one where AI serves equity, not entropy; transparency, not opacity; and human dignity, not data dominance.

1.1 Formalizing Fairness and Accountability in Urban AI Systems

The concept of fairness and accountability in AI-driven urban solutions revolves around ensuring that AI technologies deployed in urban development are equitable, transparent, and just. The ethical dimensions of these technologies are critical, especially when AI systems make decisions that affect vast urban populations and infrastructures. Addressing fairness and accountability in AI applications requires a robust understanding of how biases, transparency, and mechanisms of control are embedded into these systems.

In urban environments, AI systems are increasingly used to optimize public services such as transportation, energy management, and public safety. These systems often rely on large datasets collected from urban populations, which may include personal and sensitive data. A key ethical challenge arises when these datasets reflect social biases, which can inadvertently result in AI systems reinforcing existing inequalities (Attard-Frost et al. 2023). For instance, predictive policing systems, designed to identify areas where crimes are likely to occur, may disproportionately target low-income neighborhoods or communities of color if the training data contains historical biases. These issues underscore the need for fairness in AI-driven urban solutions, ensuring that AI does not perpetuate or exacerbate social disparities (Trotta et al. 2023).

Predictive policing, public service allocation, and transport-resource distribution are not neutral administrative domains; they operate on historical data that may already contain patterns of unequal exposure, under-service, institutional neglect, or over-policing. The citizens most affected are those living in low-income neighborhoods, demographically marginalized communities, peripheral urban districts, and areas already overrepresented in administrative or security datasets. The governance risk is that historical inequality may be converted into automated authority, making biased allocations appear objective, efficient, or statistically justified. In this context, adversarial debiasing and fairness-constrained optimization provide a formal mechanism for detecting and suppressing discriminatory signal propagation while preserving predictive usefulness. This modeling sequence therefore translates fairness from a general ethical demand into a measurable constraint within urban decision-making.

To achieve bias mitigation and fairness optimization in AI-driven urban systems, an adversarial debiasing framework can be formulated. This AI model leverages Generative Adversarial Networks (GANs) to reduce discriminatory patterns in AI predictions while preserving accuracy. The adversarial framework consists of two competing networks: a predictor $f_{\mathrm{w}}(\mathrm{X})$ that estimates urban service allocations and an adversarial debiaser $g_{\theta}(X)$ that penalizes biased patterns, dynamically adjusting the model to ensure fair outcomes. The AI-driven adversarial debiasing model is defined as:

$$\underset{\mathrm{w}}{min}\ \underset{\theta}{max} \sum_{i=1}^{N} \ell(f_{\mathrm{w}}(X_i), Y_i) - \lambda \sum_{i=1}^{N} \ell_{\mathrm{adv}}(g_{\theta}(X_i), S_i)$$

where $f_{\mathrm{w}}(\mathrm{X})$ is the predictor model that estimates outcomes such as crime risk assessments or transportation resource allocations, $g_{\theta}(X)$ is the adversarial debiaser that identifies and corrects biased patterns, $\ell(f_{\mathrm{w}}(X_i), Y_i)$ represents the standard loss function measuring the accuracy of the AI model's predictions, $\ell_{\mathrm{adv}}(g_{\theta}(X_i), S_i)$ is the adversarial loss function designed to mitigate bias, and λ is a regularization parameter balancing fairness constraints against predictive performance. This equation defines an adversarial optimization framework, where the first term minimizes the prediction error, while the second term maximizes the debiasing effectiveness. The adversarial model learns dynamically, adjusting predictions to align with equitable urban service distribution while maintaining accuracy.

To ensure a more refined fairness-aware policy, a fairness-constrained optimization function is introduced to regulate AI-driven decision-making. The optimization function integrates fairness constraints into the predictive framework, ensuring that AI allocations adhere to predefined equity metrics:

$$\underset{\mathrm{w}}{min} \sum_{i=1}^{N} \ell(f_{\mathrm{w}}(X_i), Y_i) \text{subject to } D_{\mathrm{fair}}(f_{\mathrm{w}}(X), S) \leq \tau$$

where $D_{\mathrm{fair}}(f_{\mathrm{w}}(X), S)$ is the fairness disparity function measuring statistical bias across sensitive groups, and τ is a predefined fairness threshold ensuring that bias remains below acceptable limits.

This function enforces fairness during the optimization process, ensuring the AI-driven allocations are aligned with ethical principles, which are crucial for urban governance and resource management.

This fairness-constrained optimization ensures that AI-driven urban solutions actively enforce fairness criteria, preventing discriminatory service allocations. By dynamically adjusting prediction outcomes while adhering to fairness thresholds, AI systems can maintain ethical integrity without compromising decision accuracy. The combination of adversarial debiasing and fairness-constrained optimization in AI-driven urban solutions ensures that automated decision-making does not propagate historical biases but instead aligns with ethical urban governance principles.

These AI models dynamically regulate fairness by continuously learning from bias correction feedback while optimizing urban services based on real-world data.

Framework: Urban AI Bias Mitigation Framework (UABMF)

Layered AI Architecture for Bias Mitigation in Urban Services

Layer	Main functionality	AI module(s) and components	Operational scope (Input → Processing → Output)
Layer 1	Data collection and preprocessing: Collect and preprocess urban service data, including historical data, demographics, and sensor inputs	Data harmonization engine and feature normalization pipeline	Input → Raw sensor and demographic data → Output → Preprocessed data ready for AI models
Layer 2	Adversarial Debiasing: Minimize biased predictions while preserving accuracy in urban service allocations	Adversarial Debiasing (GANs)	Input → Preprocessed data → Output → Reduced bias in predictions
Layer 3	Fairness-constrained optimization: Incorporate fairness constraints into the predictive model to ensure equitable allocations	Fairness-constrained optimization	Input → Debiased predictions → Output → Fair urban service allocation
Layer 4	Monitoring and evaluation: Continuously monitor the model's performance and fairness in real time	Model performance auditor and fairness monitoring dashboard	Input → Performance data → Output → Performance reports and fairness evaluations

To promote fairness, AI systems in urban development must be carefully designed to detect and mitigate biases during their development and deployment phases. This includes employing algorithmic fairness techniques that reduce bias in both training data and decision-making processes. For example, bias detection algorithms can be used to ensure that AI models make decisions that do not disproportionately affect any specific demographic group. However, achieving true fairness goes beyond technical fixes; it also requires a comprehensive understanding of the social context in which these AI systems operate (Barredo Arrieta et al. 2020).

In this setting, the central urban problem is not limited to the presence of bias in historical datasets; it also concerns the continuous reproduction of inequity during real-time optimization. AI systems that allocate transport capacity, prioritize infrastructure maintenance, distribute energy resources, or recommend public service interventions may gradually reinforce unequal outcomes if their reward structures privilege efficiency without explicitly accounting for fairness. Citizens in underserved districts, low-connectivity areas, energy-vulnerable households, and communities with limited institutional visibility may therefore remain disadvantaged even when the system appears technically optimized. The governance risk is that adaptive AI may learn to reproduce inequity under the language of performance, treating unequal access as an efficient equilibrium rather than as a corrective priority. A fairness-aware reinforcement mechanism is therefore required to make equity part of the learning objective itself, not merely a retrospective audit after decisions have already been made.

To ensure that AI systems in urban development are rigorously fair and accountable, the fairness-aware AI optimization framework is introduced. This model integrates adversarial debiasing with fairness-constrained reinforcement learning to dynamically detect and mitigate biases during both training and deployment phases. Unlike traditional static fairness correction methods, this adaptive AI-driven approach continuously refines AI decision-making by incorporating fairness constraints into both learning and optimization processes. To achieve bias detection and correction, a fairness-constrained reward function is introduced within the AI optimization process. This function ensures that AI decisions remain unbiased by penalizing discriminatory outcomes while reinforcing equitable allocations. The fairness-aware reward function is defined as:

$$R(s_t, a_t) = r_t - \lambda D_{\text{fair}}(s_t, a_t)$$

where $R(s_t, a_t)$ represents the adjusted fairness-aware reward function, r_t is the original reward signal derived from the AI optimization process, $D_{\text{fair}}(s_t, a_t)$ is the fairness disparity function measuring bias in AI decision-making, and λ is a fairness penalty coefficient that controls the trade-off between optimizing performance and ensuring equitable outcomes. This equation ensures that AI models do not merely optimize for efficiency but also prioritize fairness by dynamically reducing disparities during both model training and real-time decision-making. The fairness-aware reward function penalizes discriminatory biases while reinforcing ethical AI-driven urban development.

To implement bias correction dynamically, a Proximal Policy Optimization (PPO) fairness adjustment is integrated within the AI-driven decision process. The fairness-optimized PPO policy update is given by:

$$\max_{\theta} \mathbb{E}[min(r_t(\theta)A_t, \text{clip}(r_t(\theta), 1-\epsilon, 1+\epsilon)A_t) - \lambda D_{\text{fair}}(s_t, a_t)]$$

where $\mathbb{E}$ is the expectation operator, $r_t(\theta)$ is the policy ratio measuring policy improvement, A_t is the advantage function representing expected benefit from action a_t, clip $(\cdot)$ is the PPO clipping function that stabilizes training, ϵ is the clipping threshold preventing excessive updates, and $D_{\text{fair}}(s_t, a_t)$ is the bias disparity function ensuring fairness-aware optimization.

This PPO fairness-adjusted policy update ensures that AI models dynamically adjust to fairness constraints while optimizing decision-making. Unlike static bias correction methods, this reinforcement learning-based approach continuously adapts fairness strategies based on real-world AI interactions, ensuring long-term, unbiased AI-driven urban development.

By integrating fairness-constrained reward functions with PPO fairness optimization, this AI-driven model guarantees that AI systems in urban development detect and mitigate biases dynamically. The reinforcement learning framework actively learns from fairness penalties, ensuring that bias is systematically reduced while maintaining optimal AI-driven urban operations. This method provides an adaptive, fairness-aware AI decision-making process, ensuring ethical and equitable AI deployment in urban governance.

Framework: Fairness-Aware AI Optimization for Urban Development (FAI-AOUD)

Layered AI Architecture for Bias Mitigation and Fairness Optimization in Urban Systems

Layer	Main functionality	AI module(s) and components	Operational scope (Input → Processing → Output)
Layer 1	Data collection and preprocessing: Gather and preprocess urban service data, including demographic information, sensor data, and historical records	Data integration gateway and feature engineering pipeline	Input → Raw demographic data, sensor data → Output → Preprocessed data ready for AI models
Layer 2	Fairness-aware reward function: Incorporate fairness constraints into the reward function to penalize biased decision-making while optimizing for performance	Fairness-constrained reward function	Input → Reward signal, bias disparity → Output → Adjusted fairness-aware reward
Layer 3	PPO fairness adjustment: Dynamically optimize policies with fairness constraints using Proximal Policy Optimization (PPO) to ensure fairness in decision-making	Proximal Policy Optimization (PPO)	Input → Current policy, advantage function → Output → Fairness-adjusted policy update
Layer 4	Monitoring and evaluation: Continuously monitor AI decisions and fairness metrics to assess the performance and fairness of AI systems in real-world scenarios	Decision Analytics Monitor and Fairness Compliance Dashboard	Input → Performance data, fairness metrics → Output → Evaluation reports on fairness and decision quality

Accountability in AI-driven urban solutions involves establishing clear lines of responsibility for the decisions made by AI systems. Since these systems can impact public services and urban planning on a large scale, it is crucial that there are structures in place to ensure transparency and oversight (Attard-Frost et al. 2023). One approach to enhancing accountability is the incorporation of explainable AI (XAI), which makes the decision-making processes of AI systems transparent and understandable to humans (Barredo Arrieta et al. 2020). By providing explanations for AI-driven decisions, such as why a particular urban area was prioritized for infrastructure development or why certain services were allocated more resources, XAI can help stakeholders ranging from city officials to the general public understand and trust these systems.

In urban governance, the problem is not only whether an AI decision is accurate, but whether its rationale can be reconstructed, justified, and challenged when it affects public priorities. Decisions concerning infrastructure investment, emergency response, traffic control, service allocation, or maintenance scheduling may redistribute benefits and burdens across neighborhoods without citizens or public authorities being able to trace the logic behind them. The affected stakeholders include residents of underserved districts, commuters, public administrators, service providers, and communities whose access to urban resources depends on algorithmic prioritization. The governance risk is opacity without responsibility; an AI decision may be operationally efficient while remaining institutionally unanswerable. A reinforcement learning-based explainability mechanism is therefore required to attach interpretable justifications to adaptive policy choices, making optimization legible to audit, oversight, and public review.

Against this governance risk, a reinforcement learning-based explainability framework is introduced to strengthen transparency and accountability in AI-driven urban solutions. This model integrates Explainable AI (XAI) with Proximal Policy Optimization (PPO) to provide clear, interpretable justifications for AI-driven decisions while optimizing urban planning and resource allocation. Unlike black-box AI systems, this framework ensures that every decision is accompanied by an interpretable explanation, increasing public trust and regulatory oversight.

To achieve explainable accountability, a policy-justification function is introduced, ensuring that AI-generated decisions include rationalized outputs that are interpretable by humans. The explainability-enhanced AI-driven policy function is formulated as:

$$J_{\mathrm{XAI}}(\pi) = \sum_{t=1}^{T} \mathbb{E}[\log \pi(a_t|s_t)A_t + \alpha I(s_t, a_t)]$$

where $J_{\mathrm{XAI}}(\pi)$ represents the explainability-enhanced policy function, $\pi(a_t| s_t)$ is the policy determining the probability of selecting action at a given state s_t, A_t is the advantage function quantifying the expected benefit of action a_t, α is the explainability weight regulating the importance of interpretable justifications, and $I(s_t, a_t)$ is the explanation function quantifying the interpretability score of the AI-driven decision. This function ensures that AI-driven decisions not only maximize efficiency but also optimize interpretability, allowing stakeholders to understand why specific actions were taken. The explanation function $I(s_t, a_t)$ assigns a transparency score to each decision, ensuring that AI outputs remain understandable and justifiable.

To enhance interpretability further, an explainability-optimized PPO policy update is introduced, integrating explainable decision-making into the reinforcement learning process. The XAI-optimized PPO update is defined as:

$$\underset{\theta}{max}\, \mathbb{E}[min(r_t(\theta)A_t, \mathrm{clip}(r_t(\theta), 1-\epsilon, 1+\epsilon)A_t) + \alpha(s_t, a_t)]$$

where $r_t(\theta)$ is the policy ratio evaluating policy improvement, A_t is the advantage function quantifying expected benefit, clip $(\cdot)$ is the PPO clipping function preventing large updates, ϵ is the clipping threshold, and $I(s_t, a_t)$ is the interpretability function ensuring decision transparency.

This explainability-optimized reinforcement learning framework ensures that AI-driven decisions are not only optimized for urban planning efficiency but also provide interpretable, stakeholder-friendly explanations. By incorporating XAI into reinforcement learning, this model enhances trust, transparency, and accountability in AI-driven urban solutions, ensuring that decision-making processes are both efficient and understandable for policymakers and the public.

Framework: Reinforcement Learning-based Explainability for Urban Solutions (RL-XAI-US)

Layered AI Architecture for Explainable AI and Accountability in Urban Planning

Layer	Main functionality	AI module(s) and components	Operational Scope (Input → Processing → Output)
Layer 1	Data collection and preprocessing: Gather and preprocess urban service data, such as historical records, demographic data, and urban sensor data	Data ingestion engine and semantic preprocessing pipeline	Input → Raw urban data, demographic information → Output → Preprocessed data ready for AI models
Layer 2	Policy-justification function: Integrate interpretability into AI-driven policy decisions by adding a justification layer that ensures explanations are provided for each decision	Policy-justification function	Input → Policy decisions, advantage function → Output → Rationalized and interpretable decisions
Layer 3	XAI-optimized PPO update: Incorporate explainable AI into Proximal Policy Optimization to ensure transparency and interpretability of decision-making	Proximal Policy Optimization (PPO), Explainable AI (XAI)	Input → Current policy, advantage function → Output → Fairness-optimized and explainable policy update
Layer 4	Monitoring and evaluation: Continuously monitor AI decisions and the transparency of the explanations provided, ensuring compliance with accountability standards	Explainability Audit Module and Transparency Evaluation Dashboard	Input → Decision data, transparency metrics → Output → Evaluation reports on decision explanations and transparency

Moreover, AI accountability in urban settings must also include mechanisms for redress. Citizens affected by AI-driven decisions, whether in policing, housing, or resource allocation, should have the ability to challenge and appeal those decisions. This calls for clear governance frameworks that outline who is responsible for AI system errors or biases, and how those responsible can be held accountable (Trotta et al. 2023). Without these governance structures, the deployment of AI in urban development risks becoming a technocratic process that undermines democratic participation and transparency. In urban governance, the central problem is that AI decisions may materially affect citizens while leaving them without a meaningful procedural pathway for appeal. A household may be deprioritized for infrastructure repair, a neighborhood may be classified as high-risk, a housing applicant may receive an adverse allocation decision, or a district may receive reduced service coverage without a clear mechanism for contesting the underlying model output. The affected stakeholders are not only individual citizens, but also communities, municipal agencies, service providers, and public authorities that must justify and correct AI-mediated decisions. The governance risk is procedural asymmetry: the system can decide, rank, allocate, or deny, while the affected public lacks a structured route to challenge the decision. A dispute resolution architecture is therefore required to transform accountability from passive explanation into active redress, review, and recalibration.

To ensure AI accountability in urban governance, a Multi-Agent Reinforcement Learning (MARL) framework with dispute resolution mechanisms is introduced. This model integrates governance-driven AI oversight with mechanisms for redress, ensuring that citizens can challenge AI-driven decisions effectively. Unlike traditional AI models that operate as black-box decision-makers, this framework incorporates appeal processes within AI-driven urban governance, ensuring transparent, accountable, and contestable AI decisions.

To formalize accountability in AI decision-making, a dispute resolution function is introduced, which dynamically assesses whether AI-generated decisions align with fairness and legal governance principles. The AI-driven appeal assessment model is formulated as:

$$V_{\text{account}}(s_t, a_t) = \mathbb{E}[R(s_t, a_t) - \beta C(s_t, a_t)]$$

where $V_{\text{account}}(s_t, a_t$ represents the accountability-adjusted value function, $R(s_t, a_t)$ is the reward function quantifying the urban service decision outcome, $C(s_t, a_t)$ is the citizen dispute cost function, and β is the penalty coefficient regulating the AI model's responsiveness to citizen appeals. This equation ensures that AI-driven urban systems are not solely optimized for efficiency but also incorporate accountability principles. The dispute cost function introduces an appeal-sensitive adjustment, ensuring that contested AI decisions are reassessed within an adaptive governance framework.

To enhance fairness in decision reviews, a multi-agent governance optimization function is introduced. This function ensures that AI decisions are assessed not only by a primary AI model but also by an independent review model, ensuring governance oversight. The governance-enhanced AI decision model is given by:

$$\min_{\theta} \sum_{i=1}^{N} \ell(f_{w}(X_i), Y_i) + \gamma \sum_{i=1}^{N} \ell_{gov}(g_{\theta}(X_i), Y_i)$$

where $f_w(X)$ represents the primary AI decision-making model, $g_\theta(X)$ is the governance oversight model that independently evaluates decisions, $\ell(f_w(X_i), Y_i)$ is the decision loss function assessing the primary AI system's correctness, $\ell_{gov}(g_\theta(X_i), Y_i)$ is the governance review loss function penalizing biased or incorrect decisions, and γ is the governance penalty coefficient adjusting the AI model's accountability responsiveness.

This multi-agent governance-driven AI framework ensures that AI-driven urban solutions integrate mechanisms for redress, allowing citizens to appeal and challenge AI decisions. The dispute resolution function dynamically adjusts AI decision-making based on legal and fairness criteria, while the governance oversight model independently evaluates AI decisions, ensuring democratic accountability, transparency, and fair urban AI deployment. By incorporating citizen appeals within AI-driven urban systems, this model prevents technocratic overreach and aligns AI governance with public accountability and ethical oversight.

Framework: Multi-Agent Reinforcement Learning for AI Accountability in Urban Governance (MARL-AI-AG)

Layered AI Architecture for Governance and Redress in AI-Driven Urban Systems

Layer	Main functionality	AI module(s) and components	Operational scope (Input → Processing → Output)
Layer 1	Data collection and preprocessing: Collect and preprocess data relevant to urban services, including historical records, citizen feedback, and urban sensor data	Secure data lake connector and citizen feedback preprocessing pipeline	Input → Raw citizen data, sensor data, service history → Output → Preprocessed data for governance and AI models
Layer 2	Dispute resolution function: Incorporate a dynamic function to assess and resolve disputes raised by citizens regarding AI decisions	Dispute resolution function	Input → Dispute data, AI decisions → Output → Adjusted decisions based on citizen appeals
Layer 3	Governance-enhanced AI decision-making: Incorporate governance oversight into the AI decision-making process to independently assess fairness and accuracy	Multi-Agent Reinforcement Learning (MARL), Governance Oversight Model	Input → Primary AI decision data, governance review data → Output → Fairness-optimized and overseen decisions

(continued)

Layer	Main functionality	AI module(s) and components	Operational scope (Input → Processing → Output)
Layer 4	Monitoring and evaluation: Continuously track AI decision outcomes, monitor governance compliance, and assess the effectiveness of redress mechanisms	Governance compliance monitor and accountability reporting dashboard	Input → Decision outcomes, governance metrics → Output → Evaluation reports on decision accountability and fairness

The political and economic contexts in which AI systems are developed and deployed also play a significant role in shaping their fairness and accountability. Corporate interests or competitive pressures may lead to shortcuts in ethical oversight, where AI systems are implemented without sufficient consideration of their long-term social impacts (Attard-Frost et al. 2023). Ethical AI frameworks must therefore extend beyond the technical aspects of algorithmic decision-making to consider the broader political economy of AI systems. This includes examining how business decisions, such as the prioritization of profit over public welfare, can influence the design and deployment of AI in urban contexts.

In urban governance, this problem becomes acute when AI systems are embedded in public-private infrastructures, procurement regimes, platform-based services, or smart-city partnerships where optimization objectives may be shaped by commercial incentives. Mobility platforms may prioritize profitable routes, energy systems may privilege cost efficiency over vulnerable households, surveillance technologies may expand under security or market pressures, and infrastructure analytics may direct investment toward areas of higher economic return rather than greater social need. The affected stakeholders include citizens, municipalities, regulated utilities, private vendors, marginalized communities, and future urban populations exposed to long-term infrastructural consequences. The governance risk is incentive capture: AI systems may appear technically rational while encoding economic priorities that narrow public accountability. A game-theoretic reinforcement learning framework is therefore required to model, constrain, and rebalance the interaction between profit-seeking behavior and public welfare obligations.

To ensure that AI-driven urban solutions remain ethically sound despite political and economic pressures, a Game-Theoretic Reinforcement Learning (GTRL) framework is introduced. This model integrates multi-agent decision-making with incentive-aligned governance, ensuring that AI systems prioritize social welfare alongside economic efficiency. Unlike conventional AI frameworks that optimize purely performance metrics, this approach explicitly incorporates ethical constraints and economic trade-offs, preventing AI systems from prioritizing corporate interests at the expense of public welfare.

To quantify the interplay between corporate incentives and ethical AI deployment, a multi-agent Nash equilibrium utility function is introduced. This function ensures that AI deployment decisions account for both profit-driven and public-interest stakeholders, balancing economic feasibility with long-term societal benefits. The AI-driven ethically constrained Nash equilibrium function is given by:

$$U^*(\pi_1, \pi_2) = \max_{\pi_1, \pi_2} \mathbb{E}\left[R_{\text{social}}(\pi_1) - \delta R_{\text{profit}}(\pi_2)\right]$$

where $U^*(\pi_1, \pi_2)$ represents the equilibrium utility function balancing social and economic incentives, π_1 is the policy optimizing for public welfare, π_2 is the policy optimizing for corporate profit, $R_{\text{social}}(\pi_1)$ is the reward function quantifying the societal benefits of AI-driven decisions, $R_{\text{profit}}(\pi_2)$ is the profit-based reward function reflecting corporate-driven AI objectives, and δ is the regulatory penalty parameter discouraging excessive profit-seeking behavior at the expense of fairness. This equation ensures that AI systems are not solely optimized for profitability but also account for ethical governance constraints, preventing corporations from deploying AI in ways that prioritize financial gain over public well-being. The equilibrium-driven incentive structure dynamically adjusts AI decision-making to maintain fairness and long-term urban sustainability.

To further enforce accountability in AI-driven corporate decisions, an ethical reinforcement learning penalty function is introduced. This function penalizes AI deployments that excessively prioritize economic returns at the cost of social welfare. The regulated AI economic penalty function is given by:

$$\mathcal{L}_{\text{ethics}} = \sum_{i=1}^{N} \ell(f_{\text{w}}(X_i), Y_i) + \lambda \sum_{i=1}^{N} P_{\text{reg}}(X_i, Y_i)$$

where $\mathcal{L}_{\text{ethics}}$ is the ethical AI loss function, $\ell(f_{\text{w}}(X_i), Y_i)$ represents the AI model's decision loss function, $P_{\text{reg}}(X_i, Y_i)$ is the regulatory penalty function discouraging profit-driven biases, and λ is the penalty coefficient regulating the impact of governance constraints.

This ethically constrained AI-driven optimization function ensures that AI models dynamically adjust their decision-making based on long-term social impacts, reducing the risk of corporate-influenced ethical shortcuts. By integrating game-theoretic reinforcement learning with incentive regulation, this AI-driven framework guarantees that urban AI deployments remain fair, transparent, and accountable, safeguarding against profit-driven biases while preserving economic feasibility. The equilibrium-based governance model provides a systematic mechanism to align AI incentives with public welfare, ensuring that AI deployment in urban contexts adheres to ethical and socially responsible standards.

Framework: Game-Theoretic Reinforcement Learning for Ethical AI in Urban Solutions (GTRL-EAI-US)

Layered AI Architecture for Ethical AI Deployment and Governance in Urban Contexts

Layer	Main functionality	AI module(s) and components	Operational scope (Input → Processing → Output)
Layer 1	Data collection and preprocessing: Gather and preprocess urban service	Socioeconomic data integration engine and	Input → Raw data on socioeconomic factors, corporate

(continued)

Layer	Main functionality	AI module(s) and components	Operational scope (Input → Processing → Output)
	data, including socioeconomic data, corporate incentives, and regulatory frameworks	regulatory mapping pipeline	policies → Output → Preprocessed data ready for AI and governance models
Layer 2	Ethically constrained Nash equilibrium: Balance social welfare and economic incentives within AI decision-making through the Nash equilibrium framework	Game-Theoretic Reinforcement Learning (GTRL), Nash equilibrium	Input → Policies optimizing social welfare and corporate profit → Output → Equilibrium utility balancing social and economic outcomes
Layer 3	Ethical reinforcement learning penalty: Penalize AI systems for prioritizing profit over social welfare using an ethical penalty function	Reinforcement learning, ethical penalty function	Input → AI decision outcomes, regulatory penalties → Output → Adjusted decisions balancing profit and social welfare
Layer 4	Monitoring and evaluation: Track AI performance, social impacts, and governance compliance to ensure fairness and accountability in urban AI systems	Ethical compliance monitor and governance impact assessment dashboard	Input → Performance metrics, social impact data → Output → Reports on ethical compliance and fairness in AI decision-making

Achieving fairness and accountability in AI-driven urban solutions requires a multidimensional approach that integrates technical, social, and ethical considerations. AI systems must be designed with fairness at their core, ensuring that they do not replicate or amplify existing inequalities in urban settings. At the same time, strong accountability mechanisms must be established to provide transparency, enable oversight, and ensure that AI systems are held to ethical standards. Only through this comprehensive approach can AI technologies truly contribute to the equitable and sustainable development of urban environments.

Figure 1.1 illustrates a structure centered on "Fairness and Accountability in AI-Driven Urban Solutions." The main concept branches into three primary nodes: Fairness in AI, accountability mechanisms, and political and economic context. Each primary node further divides into key areas. "Fairness in AI" covers bias mitigation, equity in decision-making, and contextual awareness. "Accountability Mechanisms" addresses transparency, oversight, governance, and redress. "Political and Economic Context" highlights ethical oversight risks, broader political economy, and the balance between profit and public welfare. Each sub-node explores critical factors for ensuring ethical, fair, and transparent AI deployment in urban environments.

Limitations and Governance Implications

The frameworks developed in this chapter strengthen the governance capacity of AI-driven urban systems by translating bias auditing, fairness constraints,

Fig. 1.1 Fairness and accountability in AI-driven urban solutions

explainability, contestability, and public-interest alignment into operational model architectures. The Urban AI Bias Mitigation Framework (UABMF), Fairness-Aware AI Optimization for Urban Development (FAI-AOUD), Reinforcement Learning-based Explainability for Urban Solutions (RL-XAI-US), Multi-Agent Reinforcement Learning for AI Accountability in Urban Governance (MARL-AI-AG), and Game-Theoretic Reinforcement Learning for Ethical AI in Urban Solutions (GTRL-EAI-US) offer complementary mechanisms for addressing algorithmic bias, inequitable prioritization, weak accountability, and incentive capture in urban governance and public service allocation. Yet these models should not be interpreted as substitutes for institutional responsibility, legal review, human judgment, democratic oversight, or public deliberation. Fairness metrics may identify disparities, explainability functions may clarify model reasoning, dispute resolution modules may support procedural redress, and game-theoretic penalties may constrain profit-driven distortions; nevertheless, the legitimacy of AI-mediated urban decisions ultimately depends on accountable public institutions, enforceable rights, transparent administrative procedures, and meaningful opportunities for affected communities to contest, revise, or reject algorithmic outcomes. The central governance implication is therefore that technical architectures can support ethical urban intelligence only when they remain embedded within civic accountability, legal proportionality, and human dignity.

1.2 Conclusion

In final analysis, this chapter establishes a rigorous scientific and ethical foundation for the next generation of AI-driven urban governance systems that not only compute and optimize but also explain, justify, and assume moral responsibility for their decisions. By moving beyond descriptive ethics toward mathematically encoded fairness and accountability, the chapter transforms normative principles into executable architectures, laying out a structured and reproducible path for aligning artificial intelligence with the values of justice, transparency, and human dignity within urban infrastructures.

The analytical trajectory of this work unfolds across five integrative frameworks, each addressing a critical dimension of ethical AI deployment in the city. The Urban AI Bias Mitigation Framework (UABMF) introduces a generative adversarial architecture capable of adversarial debiasing and fairness-constrained optimization, transforming bias mitigation from a peripheral audit process into an embedded learning mechanism. This marks the transition from reactive fairness to autonomous ethical adaptation, where the AI learns equity through continuous feedback loops grounded in statistical parity and socio-demographic awareness.

Advancing this foundation, the fairness-aware AI optimization for urban development (FAI-AOUD) elevates fairness from a constraint to an optimization objective by embedding fairness-sensitive reward functions within proximal policy optimization algorithms. In doing so, fairness becomes self-regulatory—a property the system internalizes as a condition for maximizing its cumulative utility. This is a crucial epistemic leap: fairness is no longer an exogenous regulation but an endogenous principle guiding reinforcement learning under evolving socio-technical contexts.

The chapter then extends the analytical horizon from fairness to interpretability and accountability through the Reinforcement Learning-based Explainability for Urban Solutions (RL-XAI-US) framework. Here, Explainable Artificial Intelligence (XAI) principles are formalized into a policy-justification function, ensuring that every decision carries a semantic rationale accessible to human oversight. Through this mechanism, the traditionally opaque reinforcement learning process becomes a dialogical system, one capable of reasoning about its own actions, thereby reintroducing epistemic transparency into the digital governance of cities. The interpretability function formalized within this framework ensures that AI systems remain subject to scrutiny, creating a computational basis for civic trust.

Building upon this interpretive foundation, the Multi-Agent Reinforcement Learning for AI Accountability in Urban Governance (MARL-AI-AG) framework operationalizes accountability and redress. By embedding dispute resolution and governance oversight directly into multi-agent reinforcement learning, it constructs an institutionalized form of algorithmic due process. AI decisions can now be appealed, reviewed, and recalibrated within the model itself, preserving democratic agency in digital urban governance. Through multi-agent governance optimization, AI systems evolve from unilateral decision engines to participatory governance instruments, aligning technical performance with procedural justice.

Finally, the Game-Theoretic Reinforcement Learning for Ethical AI in Urban Solutions (GTRL-EAI-US) framework introduces the crucial political-economic dimension of ethical AI. By modeling the interaction between corporate incentives and public welfare through ethically constrained Nash equilibria, this framework establishes a balance between profitability and social accountability. Its inclusion of regulatory penalty functions and ethical reinforcement learning penalties ensures that algorithmic behavior remains socially sustainable, economically rational, and normatively bounded. This structure not only protects against profit-driven bias but also proposes a scalable model for aligning AI-driven decision ecosystems with long-term urban ethics and sustainability.

These five frameworks compose an integrated governance continuum from bias detection and fairness regulation to interpretability, accountability, and ethical equilibrium. Together, they define the algorithmic morphology of fairness and accountability as both computational and civic phenomena. The chapter thus reveals that urban AI cannot merely function as an optimization instrument; it must operate as an ethical infrastructure adaptive, auditable, contestable, and socially responsive.

From an academic standpoint, the contribution of this chapter lies in its ability to transform abstract ethical imperatives into reproducible algorithmic structures. Each model is formulated with mathematically explicit equations and modular architectures that can be implemented as Software-as-a-Service (SaaS) components, ensuring applicability in industrial, governmental, and research domains alike. From a philosophical perspective, the chapter reinstates the concept of moral agency within artificial systems, framing AI as a participant in, rather than a product of, civic governance. From a global standpoint, it articulates a transdisciplinary paradigm for intelligent cities—one where computational intelligence and ethical consciousness coevolve toward human-centric resilience.

In synthesis, the chapter demonstrates that the future of AI in urban governance depends not merely on greater intelligence but on greater conscience embedded within the code itself. It shows that fairness can be learned, accountability can be computed, and transparency can be optimized without diminishing efficiency or innovation. Such architectures herald the emergence of the ethical algorithmic city; a socio-technical organism governed by intelligence that not only calculates outcomes but also understands consequences, serving humanity not as a system of control, but as a framework of care.

References

Alrasbi, O., & Ariaratnam, S. T. (2026). The organizational transformation of artificial intelligence in smart cities: An urban artificial intelligence governance maturity model. *Urban Science, 10*(1), Article 63. https://doi.org/10.3390/urbansci10010063

Attard-Frost, B., De los Rios, A., & Walters, D. R. (2023). The ethics of AI business practices: A review of 47 AI ethics guidelines. *AI and Ethics*, 3, 389–406. https://doi.org/10.1007/s43681-022-00156-6

Barredo Arrieta, A., Díaz-Rodríguez, N., Del Ser, J., Bennetot, A., Tabik, S., Barbado, A., Garcia, S., Gil-Lopez, S., Molina, D., & Herrera, F. (2020). Explainable artificial intelligence (XAI): Concepts, taxonomies, opportunities, and challenges toward responsible AI. *Information Fusion*, 58, 82–115. https://doi.org/10.1016/j.inffus.2019.12.012

Bartsch, S. C., Nguyen, L. H., Schmidt, J.-H., Du, G., Adam, M., Benlian, A., & Sunyaev, A. (2025). The present and future of accountability for AI systems: A bibliometric analysis. *Information Systems Frontiers, 27*, 2463–2484. https://doi.org/10.1007/s10796-025-10636-9

Ibitoye, A. O., Nkwo, M. S., & Orji, R. (2025). Rethinking responsible AI from ethical pillars to sociotechnical practice. *AI and Ethics, 5*, 6207–6223. https://doi.org/10.1007/s43681-025-00809-2

Novelli, C., Taddeo, M., & Floridi, L. (2024). Accountability in artificial intelligence: What it is and how it works. *AI & Society, 39*, 1871–1882. https://doi.org/10.1007/s00146-023-01635-y

Trotta, A., Ziosi, M., & Lomonaco, V. (2023). The future of ethics in AI: Challenges and opportunities. *AI & Society*, 38, 439–441. https://doi.org/10.1007/s00146-023-01644-x

Chapter 2
Transparency and Explainability in AI-Driven Urban Governance: Frameworks for Ethical, Accountable, and Participatory Intelligence

In the emerging era of algorithmic urbanism, transparency and explainability have become the ethical and operational cornerstones of Artificial Intelligence (AI) deployment in cities. This chapter, Transparency and Explainability in AI-Driven Urban Governance: Frameworks for Ethical, Accountable, and Participatory Intelligence, redefines how urban decision systems can be made both intelligent and interpretable moving decisively beyond opaque automation toward civic comprehension and trust. By integrating Explainable Artificial Intelligence (XAI) and Deep Reinforcement Learning (DRL), it establishes a new scientific foundation where AI models not only learn optimal strategies for urban management but also communicate the reasoning behind every decision.

The discussion begins by formalizing transparency through the Explainability-Driven Deep Reinforcement Learning (X-DRL) framework, which mathematically embeds interpretability within value and policy functions to ensure that every AI-generated action in urban planning is both efficient and explainable. The analysis then expands with the Interpretable Deep Reinforcement Learning (I-DRL) and Self-Interpretable Deep Reinforcement Learning (SI-DRL) architectures, each designed to convert black-box models into self-explaining systems capable of articulating cause, feature influence, and policy rationale. These frameworks allow policymakers and citizens to trace the full logic chain of AI decisions, transforming governance from mere algorithmic execution into transparent reasoning.

Accountability is then institutionalized through the Explainability-Guided Deep Reinforcement Learning (XG-DRL) model, where dynamic explainability constraints, bias-aware reward adjustments, and fairness metrics enable AI systems to self-audit, detect discriminatory outcomes, and autonomously recalibrate their decision logic. The chapter advances this foundation with the Participatory Explainable Artificial Intelligence (P-XAI) framework, which democratizes AI by embedding multi-stakeholder feedback, federated learning validation, and engagement-weighted accountability mechanisms. Through this participatory layer, urban AI evolves into a civic partner transparent, justifiable, and aligned with collective values rather than a technocratic enigma.

D. Sargiotis, *AI with Integrity*, https://doi.org/10.1007/978-3-032-31133-7_2

For post hoc transparency in complex models, the SHapley Additive exPlanations (SHAP) method is introduced as a mathematically grounded approach derived from cooperative game theory. SHAP decomposes each AI prediction into additive feature contributions, offering interpretable, fair, and visually communicable explanations for non-expert audiences. Finally, the Fairness-Constrained Explainability Model (FCEM) establishes a rigorous balance between model accuracy, interpretability, and privacy protection ensuring that explainability never compromises data confidentiality or fairness.

This chapter provides the most comprehensive, equation-driven roadmap for ethical transparency in AI-powered urban systems. It bridges the computational precision of reinforcement learning with the moral imperatives of public governance, offering a blueprint for cities seeking to implement intelligent yet accountable infrastructures. For international researchers, engineers, and policymakers, it represents an indispensable guide to designing explainable, bias-resistant, and participatory AI architectures capable of earning public trust while advancing sustainable urban transformation.

2.1 Explainability-Driven Deep Reinforcement Learning for Transparent Urban AI Systems

Transparency and explainability in AI applications are critical components of ethical AI deployment, especially in urban development. These concepts focus on ensuring that AI systems are understandable to humans and provide clear, accessible explanations for their decision-making processes. The importance of transparency and explainability cannot be overstated, as they build trust between AI systems and their users, foster accountability, and allow for informed oversight.

Transparency in AI refers to the ability of the system to clearly communicate how and why it made specific decisions. In urban development, this becomes particularly important because AI systems are used to make decisions that affect a large population, such as optimizing public services, infrastructure planning, and resource allocation. When AI decisions are transparent, stakeholders, including city planners, policymakers, and citizens, can better understand the rationale behind the system's choices, leading to increased trust and acceptance (Barredo Arrieta et al. 2020).

In urban governance, the central problem is that many AI-driven decisions are not isolated predictions but adaptive sequences of actions that reshape public space over time. A traffic-control system may repeatedly redirect flows away from congested areas, an infrastructure planning model may prioritize certain districts for maintenance, and a resource allocation system may continuously update service priorities according to changing demand. The affected stakeholders include commuters, residents, planners, emergency services, infrastructure operators, and communities whose access to mobility, safety, and public services depends on algorithmic choices. The governance risk is that optimization may become opaque across time;

even if a single decision appears reasonable, the cumulative policy trajectory may remain difficult to explain, audit, or contest. An explainability-driven reinforcement learning framework is therefore required to connect adaptive policy optimization with interpretable justifications at the level of both individual decisions and evolving urban strategies.

To ensure transparency and explainability in AI-driven urban development, an Explainability-Driven Deep Reinforcement Learning (X-DRL) framework is introduced. This model integrates Explainable AI (XAI) with Deep Reinforcement Learning (DRL) to enhance the interpretability of AI-driven decisions in urban planning. Unlike conventional AI models, which often function as black-box systems, this framework ensures that each AI decision is accompanied by an intelligible explanation, making urban AI governance more transparent and accountable.

To formally define transparent AI decision-making, an explainability-optimized value function is introduced. This function ensures that AI-driven urban solutions not only optimize performance but also provide human-interpretable justifications for each decision. The XAI-enhanced reward function is given by:

$$V_{\mathrm{XAI}}(s_t, a_t) = \mathbb{E}[R(s_t, a_t) + \alpha I(s_t, a_t)]$$

where $V_{\mathrm{XAI}}(s_t, a_t)$ is the explainability-enhanced value function, $R(s_t, a_t)$ is the standard reward function reflecting AI optimization criteria, $I(s_t, a_t)$ is the interpretability function measuring how well the AI decision can be explained, and αα is the explainability weight controlling the balance between performance and interpretability. This equation ensures that AI-driven urban planning decisions are not only optimized for efficiency but also include an explicit explainability component, allowing policymakers and citizens to understand why particular actions were taken.

To further enhance transparency in AI decision-making, a policy explainability loss function is introduced. This function penalizes AI models that fail to provide clear justifications for their actions, ensuring that urban AI deployments remain interpretable. The explainability-aware policy optimization function is given by:

$$\mathcal{L}_{\mathrm{XAI}} = \sum_{i=1}^{N} \ell(f_{\mathrm{w}}(X_i), Y_i) + \beta \sum_{i=1}^{N} I_{\mathrm{loss}}(X_i, Y_i)$$

where v is the explainability loss function, $\ell(f_{\mathrm{w}}(X_i), Y_i)$ represents the policy loss function assessing AI decision accuracy, $I_{\mathrm{loss}}(X_i, Y_i)$ is the explainability penalty function that quantifies how difficult it is to interpret a decision, and ββ is the penalty coefficient controlling the impact of the explainability constraint.

This equation ensures that AI models prioritize decision transparency, discouraging opaque, and uninterpretable AI outputs in urban governance. By integrating explainability-optimized value functions with explainability-aware policy optimization, this X-DRL framework enhances AI transparency, ensuring that urban development AI models remain interpretable, accountable, and aligned with public trust. The dual-layer approach of reinforcement learning and explainable AI provides human-interpretable justifications for AI decisions, allowing urban stakeholders to understand, trust, and regulate AI-driven urban solutions effectively.

Framework: Explainability-Driven Deep Reinforcement Learning for Urban Transparency (X-DRL)

Layered AI Architecture for Transparency and Explainability in Urban AI Systems

Layer	Main functionality	AI module(s) and components	Operational scope (Input → Processing → Output)
Layer 1	Data collection and preprocessing: Gather and preprocess urban service data, such as historical records, urban planning data, and citizen feedback	Data preprocessing engine, feature extraction module, data quality validator	Input → Raw urban planning data, citizen feedback → Output → Preprocessed data for AI models
Layer 2	Explainability-optimized value function: Integrate interpretability into AI decision-making by ensuring that each decision has a human-understandable justification	Explainable AI (XAI), Deep Reinforcement Learning (DRL)	Input → AI-driven decisions, performance criteria → Output → Optimized value with transparency and interpretability
Layer 3	Explainability-aware policy optimization: Ensure that AI policies prioritize decision transparency by penalizing opaque and difficult-to-interpret outputs	Deep Reinforcement Learning (DRL), Explainable AI (XAI)	Input → Policy decisions, reward signals → Output → Fair and transparent AI decision-making
Layer 4	Monitoring and evaluation: Continuously track AI decision outcomes and transparency metrics to assess the effectiveness of explainability and governance	Transparency monitoring dashboard, explainability evaluation module	Input → Decision outcomes, transparency reports → Output → Evaluation on decision transparency and interpretability

One of the key challenges in achieving transparency in AI is that many modern AI models, such as deep learning systems, function as "black boxes." Their decision-making processes are complex and opaque, even to the developers themselves. This opacity can be problematic, especially in high-stakes environments like urban planning, where the public expects clear, justifiable decisions (Trotta et al. 2023). Without transparency, AI systems can lose the trust of their users, and their deployment may raise ethical concerns about fairness and accountability.

In urban planning, the interpretability problem begins before the final decision is made; it begins with how the AI system represents the city itself. Variables such as congestion, land-use pressure, infrastructure vulnerability, service demand, demographic exposure, and environmental stress may be encoded into complex state representations that are difficult for planners, policymakers, or citizens to inspect. The affected stakeholders include residents whose neighborhoods are classified through these representations, planners who rely on model outputs, and public authorities responsible for justifying planning priorities. The governance risk is representational opacity; if the state of the city is encoded in a form that cannot be meaningfully examined, then the resulting policy cannot be fully understood or contested. An interpretable reinforcement learning framework is therefore required to make the internal representation of urban conditions legible before optimization proceeds.

To address the challenge of transparency in AI-driven urban planning, an Interpretable Deep Reinforcement Learning (I-DRL) framework is introduced. This model integrates Explainable AI (XAI) techniques with Deep Reinforcement Learning (DRL) to ensure that AI-driven decisions remain understandable and justifiable to urban stakeholders. Unlike conventional deep learning models, which often function as opaque "black boxes," this framework prioritizes interpretability while maintaining AI decision-making efficiency, ensuring that urban planning decisions are both effective and transparent.

To formally define interpretable AI decision-making, a transparency-enhanced state representation function is introduced. This function ensures that AI models represent urban planning decisions in a way that is explainable to humans. The state interpretability function is given by:

$$S_{\mathrm{XAI}}(t) = \sum_{i=1}^{N} \phi_i X_i$$

where $S_{\mathrm{XAI}}(t)$ represents the explainability-enhanced state representation, ϕ_i is the interpretability coefficient for feature X_i, and X_i is the urban planning feature being analyzed. This equation ensures that AI models structure their decision-making processes in an interpretable manner, allowing policymakers and city planners to understand which features influenced a particular decision.

To further enhance transparency in AI decision policies, a self-explaining policy optimization function is introduced. This function ensures that AI-generated decisions can be understood and justified by human stakeholders. The explainability-driven policy optimization function is defined as:

$$\pi^*(a_t|s_t) = \arg \max_{\pi} \mathbb{E}[R(s_t, a_t) + \gamma I(s_t, a_t)]$$

where $\pi^*(a_t| s_t)$ represents the optimal explainability-driven policy, $R(s_t, a_t)$ is the standard reward function reflecting AI optimization criteria, $I(s_t, a_t)$ is the explainability function quantifying how interpretable a given AI decision is, and v is the explainability trade-off coefficient balancing decision accuracy with transparency. This equation ensures that AI models do not prioritize efficiency at the expense of transparency but instead integrate self-explaining mechanisms within the decision-making process.

To further eliminate black-box AI characteristics, an interpretable reward adjustment function is introduced, ensuring that AI decision outputs include human-readable justifications. The explainability-augmented reward function is given by:

$$R_{\mathrm{XAI}}(s_t, a_t) = R(s_t, a_t) + \beta I(s_t, a_t)$$

where $R_{\mathrm{XAI}}(s_t, a_t)$ is the transparency-enhanced reward function, $R(s_t, a_t)$ is the original AI reward function, $I(s_t, a_t)$ is the explainability function ensuring interpretability, and β is the explainability weighting factor. This reward-augmented AI optimization function ensures that every AI-driven decision in urban planning includes an explicit human-understandable rationale, reducing ethical concerns related to fairness and accountability.

By using interpretable state representations, explainability-driven policy optimization, and explainability-augmented reward functions, this I-DRL framework guarantees that AI systems used in urban planning remain transparent, interpretable, and accountable. The self-explaining AI model dynamically adjusts decision transparency based on governance needs, ensuring that urban stakeholders including policymakers, city planners, and citizens can understand, evaluate, and trust AI-driven urban development decisions.

Framework: Interpretable Deep Reinforcement Learning for Transparent Urban Solutions (I-DRL)

Layered AI Architecture for Transparency and Interpretability in AI-Driven Urban Planning

Layer	Main functionality	AI module(s) and components	Operational scope (Input → Processing → Output)
Layer 1	Data collection and preprocessing: Collect and preprocess urban planning data, including socioeconomic factors, environmental data, and citizen feedback	Data integration engine, feature processing module, data validation service	Input → Raw data on socio-economic factors, environmental conditions → Output → Preprocessed data for AI models
Layer 2	Transparency-enhanced state representation: Represent urban planning decisions in an interpretable manner to ensure clarity for human stakeholders	Explainable AI (XAI), Deep Reinforcement Learning (DRL)	Input → Urban planning features → Output → Human-understandable state representation
Layer 3	Self-explaining policy optimization: Ensure that AI decisions are understandable by human stakeholders by incorporating explainability directly into the policy optimization	Deep Reinforcement Learning (DRL), Explainable AI (XAI)	Input → Policies, reward signals → Output → Transparent AI decision policies

(continued)

Layer	Main functionality	AI module(s) and components	Operational scope (Input → Processing → Output)
Layer 4	Monitoring and evaluation: Continuously monitor AI decisions, assess their transparency and interpretability, and ensure they align with governance and ethical standards	Interpretability monitoring system, transparency assessment dashboard	Input → AI decision outcomes, transparency evaluations → Output → Reports on decision transparency and interpretability

Explainability, a closely related concept, refers to the ability of an AI system to provide clear explanations of its outputs. Explainable AI (XAI) techniques aim to make AI models more interpretable by providing insights into how the model arrived at a specific decision. In the context of urban development, explainability is essential for ensuring that decisions related to public services, such as traffic management or housing allocations, can be understood by both experts and non-experts alike (Barredo Arrieta et al. 2020).

In urban development, the problem is not only that AI outputs may be difficult to understand, but that the factors influencing those outputs may remain hidden. A traffic management decision may be driven by congestion intensity, road-network topology, emergency-service priority, or predicted travel demand; a housing allocation decision may be shaped by eligibility criteria, spatial availability, vulnerability indicators, or historical service patterns. The affected stakeholders include residents, applicants, commuters, planners, and public officials who must understand why one option was prioritized over another. The governance risk is attributional opacity; without a clear account of which features influenced the decision, explanations remain superficial and accountability becomes difficult to operationalize. A decision attribution mechanism is therefore required to identify, quantify, and communicate the contribution of relevant urban features to AI-generated decisions.

To ensure explainability in AI-driven urban development, a Self-Interpretable Deep Reinforcement Learning (SI-DRL) framework is introduced. This model integrates Explainable AI (XAI) techniques with Deep Reinforcement Learning (DRL) to provide clear, human-understandable justifications for AI-driven decisions in public services such as traffic management and housing allocation. Unlike traditional black-box AI models, this framework ensures that AI decision processes remain interpretable to both experts and non-experts, fostering transparency and trust in urban governance.

To formally define explainable AI decision-making, a decision attribution function is introduced, ensuring that AI models clearly indicate which factors influenced a given decision. The explainability-aware decision attribution function is given by:

$$A_{\mathrm{XAI}}(s_t, a_t) = \sum_{i=1}^{N} \omega_i X_i$$

where $A_{\mathrm{XAI}}(s_t, a_t)$ is the explainability-driven decision attribution function, ω_i represents the importance weight of feature X_i in AI decision-making, and X_i denotes the urban planning feature influencing the AI-generated decision. This function ensures that AI models explicitly quantify the contribution of each input feature, allowing both policymakers and the general public to understand why a particular AI decision was made.

To further enhance interpretability in AI-driven policy selection, a self-explaining policy function is introduced, ensuring that AI policies include explanations alongside decision outputs. The explainable AI policy optimization function is defined as:

$$\pi^*_{\mathrm{XAI}}(a_t|s_t) = \arg \max_{\pi} \mathbb{E}[R(s_t, a_t) + \gamma I(s_t, a_t)]$$

where $\pi^*_{\mathrm{XAI}}(a_t|s_t)$ is the explainability-driven optimal policy, $R(s_t, a_t)$ is the original reinforcement learning reward function, $I(s_t, a_t)$ is the interpretability function ensuring that AI-generated decisions are understandable, and γ is the explainability weight controlling the balance between performance and transparency. This equation ensures that AI models in urban development do not merely optimize decision efficiency but also prioritize human interpretability, allowing non-expert stakeholders to understand the reasoning behind AI decisions.

To ensure that explainability is embedded directly within the AI training process, an explainability-constrained loss function is introduced, ensuring that AI models are penalized if their decisions are difficult to interpret. The explainability-augmented training loss function is formulated as:

$$\mathcal{L}_{\mathrm{XAI}} = \sum_{i=1}^{N} \ell(f_{\mathrm{w}}(X_i), Y_i) + \beta \sum_{i=1}^{N} I_{\mathrm{loss}}(X_i, Y_i)$$

where $\mathcal{L}_{\mathrm{XAI}}$ is the explainability-enhanced loss function, $\ell(f_{\mathrm{w}}(X_i), Y_i)$ represents the standard AI decision loss function, $I_{\mathrm{loss}}(X_i, Y_i)$ is the interpretability loss function penalizing non-transparent decisions, and ββ is the explainability penalty coefficient regulating the transparency constraint. This explainability-aware loss function ensures that AI models explicitly prioritize human interpretability during training, reducing reliance on complex, opaque decision structures.

By taking advantage of explainability-aware decision attribution, self-explaining policy functions, and explainability-constrained loss functions, this SI-DRL framework guarantees that AI-driven urban solutions remain fully interpretable to both experts and the public. The self-explaining AI model dynamically adjusts decision transparency, ensuring that AI-driven public service applications such as traffic management and housing allocation are understandable, explainable, and trusted by all urban stakeholders.

Framework: Self-Interpretable Deep Reinforcement Learning for Transparent Urban Solutions (SI-DRL)

Layered AI Architecture for Explainability and Interpretability in Urban AI Systems

Layer	Main functionality	AI module (s) and components	Operational scope (Input → Processing → Output)
Layer 1	Data collection and preprocessing: Gather and preprocess data for urban services, including features related to traffic, housing, and socioeconomic factors	Data preprocessing engine	Input → Raw urban data, traffic data, housing information → Output → Preprocessed data for AI models
Layer 2	Decision attribution function: Ensure AI models clearly indicate which factors influenced specific decisions in urban planning	Explainable AI (XAI), Deep Reinforcement Learning (DRL)	Input → Urban planning features, AI decision outputs → Output → Attributed decisions with human-understandable justifications
Layer 3	Self-explaining policy optimization: Integrate interpretability directly into AI decision-making policies to ensure decisions are transparent	Deep Reinforcement Learning (DRL), Explainable AI (XAI)	Input → Policies, reward signals → Output → Transparent and self-explaining policies
Layer 4	Monitoring and evaluation: Continuously track AI decisions and assess the level of transparency and interpretability of these decisions	Transparency evaluation dashboard	Input → Decision data, transparency metrics → Output → Evaluation on decision transparency and interpretability

Explainability plays a critical role in fostering accountability. When AI systems are explainable, it becomes easier to identify and address errors or biases in the system. For example, if an AI-driven traffic management system disproportionately directs traffic away from certain neighborhoods, explainability can help identify whether this decision was based on biased data or flawed algorithms. By making AI systems more interpretable, explainability empowers stakeholders to hold AI systems accountable for their actions (Trotta et al. 2023).

In urban governance, this problem becomes especially serious when biased or erroneous decisions are not isolated events but repeated policy adjustments produced by adaptive systems. A traffic management model may repeatedly displace congestion toward the same districts, a service-allocation system may systematically deprioritize communities with weaker data representation, or an infrastructure model may reinforce unequal maintenance cycles by learning from historically uneven investment patterns. The affected stakeholders include residents exposed to recurring service burdens, mobility users, municipal agencies, infrastructure operators, and communities whose disadvantage may be reproduced through automated policy updates. The governance risk is corrective blindness; the system may continue optimizing while failing to identify where bias enters, which features drive the error, and how the policy should be adjusted. An explainability-guided reinforcement learning framework is therefore required not only to explain decisions, but to attribute errors, adjust rewards, and guide policy correction in real time.

To ensure accountability through explainability in AI-driven urban solutions, an Explainability-Guided Deep Reinforcement Learning (XG-DRL) framework is introduced. This model integrates dynamic explainability constraints, bias-aware reward adjustment, self-interpreting policy optimization, and error attribution through fairness metrics to ensure that AI-driven decisions in urban governance are transparent, traceable, and accountable. Unlike conventional AI models that often function as opaque systems, this framework enables real-time error detection and bias identification, ensuring that AI-driven systems remain responsible and interpretable.

To formally define explainability-driven accountability, an error attribution function is introduced, ensuring that AI systems can systematically identify and correct biases in decision-making. The explainability-based error attribution function is given by:

$$E_{\text{XAI}}(s_t, a_t) = \sum_{i=1}^{N} \rho_i X_i$$

where $E_{\text{XAI}}(s_t, a_t)$ represents the explainability-driven error attribution function, ρ_i is the bias detection weight for feature X_i, and X_i is the input feature contributing to AI decision-making. This function ensures that AI models explicitly trace the source of biases or errors, making it easier for policymakers and system auditors to correct unintended discriminatory outcomes.

To further enhance accountability in AI decision-making, a bias-aware reward adjustment function is introduced, ensuring that AI models dynamically adjust their decisions based on fairness constraints. The fairness-constrained reward function is given by:

$$R_{\text{XAI}}(s_t, a_t) = R(s_t, a_t) - \lambda B(s_t, a_t)$$

where $R_{\text{XAI}}(s_t, a_t)$ is the bias-adjusted reward function, $R(s_t, a_t)$ is the original reward function measuring AI decision efficiency, $B(s_t, a_t)$ is the bias penalty function that quantifies detected biases in AI decisions, and λ is the bias penalty coefficient regulating the trade-off between performance and fairness. This function ensures that AI decisions are actively corrected when biases are detected, reinforcing ethical AI deployment in urban settings.

To ensure that AI-driven urban solutions remain transparent and interpretable, a self-interpreting policy optimization function is introduced. This function ensures that AI models justify their actions within the decision-making process. The explainability-driven policy optimization function is formulated as:

$$\pi^*_{\text{XAI}}(a_t|s_t) = \arg\max_{\pi} \mathbb{E}[R(s_t, a_t) + \gamma I(s_t, a_t)]$$

where $\pi^*_{\text{XAI}}(a_t|s_t)$ is the self-interpreting optimal policy, $R(s_t, a_t)$ is the reward function reflecting AI decision efficiency, $I(s_t, a_t)$ is the explainability function ensuring that AI-generated decisions remain understandable, and γ is the explainability weight controlling the balance between policy optimization and interpretability. This policy optimization function ensures that AI models provide explanations alongside decisions, allowing stakeholders to evaluate and challenge AI-driven urban planning choices.

Using explainability-based error attribution, bias-aware reward adjustment, and self-interpreting policy optimization, this XG-DRL framework guarantees that AI-driven urban solutions remain transparent, accountable, and bias-aware. The self-explaining AI model continuously audits its own decisions, allowing urban stakeholders, including policymakers, city planners, and citizens to detect, evaluate, and correct AI-driven urban development strategies effectively.

Framework: Explainability-Guided Deep Reinforcement Learning for Accountable Urban Solutions (XG-DRL)

Layered AI Architecture for Explainability, Accountability, and Bias Correction in Urban AI Systems

Layer	Main functionality	AI module (s) and components	Operational scope (Input → Processing → Output)
Layer 1	Data collection and preprocessing: Gather and preprocess data relevant to urban services, such as traffic data, socioeconomic features, and housing information.	Data processing engine	Input → Raw data (traffic, housing, socio-economic) → Output → Preprocessed data for AI models
Layer 2	Explainability-based error attribution: Trace the sources of biases or errors in AI decision-making, allowing for easy identification and correction of discriminatory outcomes	Explainable AI (XAI), Deep Reinforcement Learning (DRL)	Input → AI decision data, features → Output → Attributed errors and bias sources
Layer 3	Bias-aware reward adjustment: Dynamically adjust AI decisions based on fairness constraints and penalize detected biases to maintain ethical AI behavior	Deep Reinforcement Learning (DRL), Explainable AI (XAI)	Input → Reward function, detected biases → Output → Bias-adjusted decisions and rewards

(continued)

Layer	Main functionality	AI module (s) and components	Operational scope (Input → Processing → Output)
Layer 4	Self-interpreting policy optimization: Ensure that AI decisions include explanations that make the decision-making process transparent and understandable	Deep Reinforcement Learning (DRL), Explainable AI (XAI)	Input → Policies, reward signals → Output → Transparent AI policies with explanations
Layer 5	Monitoring and evaluation: Track AI decisions, monitor bias correction, and assess transparency and accountability in real time	Accountability monitoring dashboard	Input → Decision outcomes, evaluation metrics → Output → Performance and transparency reports

Furthermore, transparency and explainability contribute to the democratization of AI. When AI systems are transparent and their decisions are explainable, they allow a wider range of stakeholders to participate in discussions about their use. In urban development, this means that city officials, residents, and advocacy groups can better understand AI-driven decisions and engage in meaningful dialogue about their implications. This level of engagement is crucial for ensuring that AI systems are aligned with the values and needs of the communities they serve (Attard-Frost et al. 2023).

In this context, the urban problem is not merely technical opacity, but democratic distance. AI systems may shape zoning priorities, mobility interventions, housing eligibility, environmental monitoring, or public service distribution without affected communities having a meaningful role in interpreting or questioning the decision logic. The governance risk is participatory exclusion, where citizens are governed by algorithmic policies they cannot understand, evaluate, or influence. A participatory explainability framework is therefore required to transform explanation from a one-way technical disclosure into a multi-stakeholder process of interpretation, feedback, and civic accountability.

To promote transparency and explainability in AI-driven urban solutions, we propose the Participatory Explainable AI (P-XAI) framework. This model integrates stakeholder-inclusive AI decision transparency, Multi-level explainability, federated learning for decentralized decision justification, and engagement-driven AI accountability metrics to foster public participation in AI decision-making. Unlike conventional AI systems that operate without public insight, this framework ensures that city officials, residents, and advocacy groups can meaningfully engage with AI-driven urban policies, promoting democratic participation and ethical governance.

To define stakeholder-inclusive AI transparency, a community-informed decision attribution function is introduced. This function ensures that AI-generated urban planning decisions are explainable at multiple levels, making them accessible to both experts and non-experts. The participatory AI decision attribution function is given by:

$$A_{\mathrm{P-XAI}}(s_t, a_t) = \sum_{i=1}^{N} \theta_i X_i$$

where $A_{\mathrm{P-XAI}}(s_t, a_t)$ represents the participatory explainability function, θ_i is the transparency weight assigned to feature X_i based on stakeholder importance, and X_i denotes the urban planning feature influencing AI decisions. This function ensures that AI-driven urban solutions incorporate multi-level transparency, allowing stakeholders from different backgrounds to interpret and assess AI-generated outcomes.

To further enhance participatory engagement in AI decision-making, a federated learning transparency constraint is introduced. This function ensures that AI-driven decisions are validated by multiple stakeholder groups before deployment, preventing centralized control over decision-making. The federated explainability constraint function is given by:

$$\sum_{j=1}^{M} \mathbb{E}\left[I_j(s_t, a_t)\right] \geq \tau$$

where $\mathbb{E}\left[I_j(s_t, a_t)\right]$ represents the explainability function from stakeholder group j evaluating AI decisions, M is the total number of stakeholder groups participating in validation, and τ is the minimum transparency threshold required for AI decision approval. This function ensures that AI-driven urban solutions are reviewed by multiple community perspectives before being deployed, preventing biased or non-democratic decision-making.

To ensure that public engagement is actively embedded in AI governance, an engagement-driven AI accountability function is introduced. This function ensures that stakeholder participation is weighted into AI decision evaluations, reinforcing democratic oversight. The engagement-weighted accountability function is formulated as:

$$V_{\mathrm{P-XAI}}(s_t, a_t) = \mathbb{E}\left[R(s_t, a_t) + \gamma \sum_{k=1}^{K} \eta_k P_k(s_t, a_t)\right]$$

where $V_{\mathrm{P-XAI}}$ represents the stakeholder-weighted accountability function, $R(s_t, a_t)$ is the original AI reward function, $P_k(s_t, a_t)$ is the public engagement metric measuring how well AI decisions align with community preferences, η_k is the engagement weight for each participating group κ, and γ is the weighting factor adjusting AI decision alignment with public feedback. This function ensures that AI-driven urban planning remains accountable to public stakeholders, reinforcing democratic oversight and participatory governance.

Through the use of community-informed decision attribution, federated learning transparency constraints, and engagement-driven ai accountability functions, this P-XAI framework guarantees that AI-driven urban solutions remain democratic, explainable, and aligned with the needs of communities. The participatory AI model continuously adapts to public feedback, ensuring that urban planning AI systems remain inclusive, transparent, and trusted by all stakeholders.

Framework: Participatory Explainable AI for Democratic Urban Solutions (P-XAI)

Layered AI Architecture for Transparency, Accountability, and Public Engagement in Urban AI Systems

Layer	Main functionality	AI module (s) and components	Operational scope (Input → Processing → Output)
Layer 1	Data collection and preprocessing: Gather and preprocess data, including urban features, socioeconomic factors, and community input	Community data processing engine	Input → Raw data on urban features, community feedback → Output → Preprocessed data for AI models
Layer 2	Participatory AI DECISION ATTRIBUTION: PROVIDE Multi-level transparency in AI-driven urban decisions, ensuring that stakeholders can interpret and assess outcomes	Explainable AI (XAI), Deep Reinforcement Learning (DRL)	Input → Urban planning features, stakeholder importance → Output → Attributed AI decisions for stakeholder interpretation
Layer 3	Federated explainability constraint: Ensure that AI-driven decisions are validated by multiple stakeholder groups to maintain transparency and avoid centralized control	Federated Learning, Explainable AI (XAI)	Input → Stakeholder feedback on decisions → Output → Transparency-validated decisions
Layer 4	Engagement-driven AI accountability: Ensure that public engagement is embedded in AI decision-making, reinforcing democratic oversight and alignment with community needs	Deep Reinforcement Learning (DRL), Explainable AI (XAI)	Input → Community feedback, AI decision outcomes → Output → AI decisions adjusted based on public engagement
Layer 5	Monitoring and evaluation: Track AI decision transparency, assess community feedback, and ensure continuous adaptation based on stakeholder engagement	Stakeholder feedback monitoring dashboard	Input → Decision outcomes, transparency metrics → Output → Performance reports and transparency evaluations

In practice, achieving transparency and explainability in AI involves a variety of strategies. These can include the use of interpretable models, such as decision trees or linear models, which are inherently transparent. In cases where more complex models like neural networks are necessary, post hoc explainability methods, such as feature importance analysis or local surrogate models, can be used to provide insights into the model's decision-making process (Barredo Arrieta et al. 2020). Additionally, visualizations and natural language explanations can help non-expert users understand AI decisions. To provide transparency in high-capacity AI models such as neural networks used in smart-city healthcare, mobility, or infrastructure systems, the SHapley Additive exPlanations (SHAP) model offers a robust, mathematically grounded method for post hoc interpretability. Based on cooperative game theory, SHAP assigns a contribution value to each input feature, representing its impact on the model's output prediction. This enables users, planners, and non-expert stakeholders to understand not only what the model predicts, but which features contributed to that prediction.

In urban governance, this post hoc attribution capacity is critical because high-capacity AI systems often combine heterogeneous variables, including mobility flows, demographic indicators, infrastructure condition, health-service demand, environmental exposure, and historical service records. Without feature-level attribution, a prediction may appear technically reliable while remaining difficult to audit, explain, or challenge. The affected stakeholders include planners, citizens, public-health authorities, infrastructure managers, mobility operators, and communities whose access to services may depend on model outputs. The governance risk is interpretive opacity. Stakeholders may see the prediction without understanding which urban variables shaped it. SHAP therefore provides a structured mechanism for translating complex model outputs into reviewable feature contributions, supporting transparency, auditability, and public reasoning.

The central equation of SHAP defines the explanation model as an additive feature attribution function:

$$g(z') = \phi_0 + \sum_{i=1}^{M} \phi_i z'_i$$

where $g(z')$ is the interpretable surrogate model (typically linear), $z' \in \{0, 1\}^M$ is a binary vector indicating the presence or absence of each original input feature, ϕ_0 is the model's base value (i.e., expected output across all instances), and ϕ_i is the Shapley value quantifying the contribution of feature i to the prediction. This function expresses the original model's output as a sum of marginal feature contributions, where each ϕ_i corresponds to a weighted average of that feature's impact across all possible feature combinations. It ensures that the explanation is faithful to the underlying model, regardless of its complexity.

To compute ϕ_i, the SHAP framework uses the Shapley value formula derived from cooperative game theory:

$$\phi_i = \sum_{S \subseteq F \setminus \{i\}} \frac{|S|!(F| - |S| - 1)!}{|F|!} [f(S \cup \{i\}) - f(S)]$$

where F is the full set of features, S is a subset of F excluding feature i, and $f(S)$ is the model's output when only the features in S are known. The factorial terms compute the combinatorial weight of each subset, ensuring fair and exhaustive attribution. This equation evaluates the average marginal contribution of feature i across all possible orderings of feature inclusion, ensuring consistency and local accuracy. For complex models, SHAP values are estimated using approximation algorithms such as Kernel SHAP or Deep SHAP, which allow efficient computation even for deep learning networks.

To enhance non-expert interpretability, the SHAP output can be paired with visualization tools that render model logic as ranked feature bars, interaction plots, or decision dependence graphs highlighting which inputs most strongly influenced a prediction and how.

Together, the SHAP model provides a mathematically consistent and visually accessible explanation layer for opaque AI systems. The additive feature attribution function ensures comprehensibility, the Shapley formula provides fairness grounded in game theory, and the visual-narrative pairing allows integration into civic and health applications where trust and transparency are paramount. This architecture ensures that smart-city AI systems remain not only powerful but publicly accountable, aligning with ethical, legal, and practical imperatives for explainable urban intelligence.

Framework: SHAP-based AI Explainability in Smart-City Applications (SHAPE-SCA)

Layered AI Architecture for Explainable AI in Urban Intelligence

Layer	Main functionality	AI module (s) and components	Operational scope (Input → Processing → Output)
Layer 1	Data collection: Gathering input data for AI models from smart-city sensors	Sensor data processing engine	Input → Real-time sensor data (environmental, healthcare, mobility) → Output → Preprocessed input for AI models
Layer 2	Pre-AI interpretation: Using interpretable models to introduce transparency in AI decisions	Interpretable modeling module	Input → Historical sensor data → Output → Transparent predictions through interpretable models (decision trees)
Layer 3	Core AI: SHAP model used for post hoc interpretability of AI predictions	SHAP (Shapley Additive Explanations)	Input → Neural network prediction → Output → SHAP values explaining model prediction
Layer 4	Post-AI Explainability: Visualization of SHAP values for better user understanding of decisions	Explainability visualization interface	Input → SHAP output → Output → Visualized SHAP values (ranked features, interaction plots)

(continued)

Layer	Main functionality	AI module (s) and components	Operational scope (Input → Processing → Output)
Layer 5	Stakeholder engagement: Providing transparency and ensuring model accountability for non-experts	User engagement and accountability dashboard	Input → User interaction data, stakeholder feedback → Output → Refined transparency reports, user-explained decision-making

Despite the advances in XAI, there are still significant challenges to achieving full transparency and explainability in AI systems. For instance, there is often a trade-off between model complexity and interpretability—more complex models, which may offer higher accuracy, are typically harder to interpret (Barredo Arrieta et al. 2020).

Furthermore, transparency must be balanced with concerns about privacy and security, particularly when AI systems handle sensitive urban data. Overly transparent systems could expose vulnerabilities or sensitive information, creating risks for individuals and communities (Attard-Frost et al. 2023).

To address this critical issue, we propose a Fairness-Constrained Explainability Model (FCEM) that integrates explainability constraints while safeguarding privacy and security. In urban AI systems, the central governance challenge is not simply to make decisions transparent, but to ensure that transparency does not expose sensitive demographic, behavioral, infrastructural, or security-relevant information. The FCEM is therefore designed to balance three requirements: maintaining predictive accuracy, preserving interpretability, and preventing the disclosure of sensitive urban data. By embedding fairness and explainability constraints within the model architecture, the framework supports ethical urban governance in contexts where public accountability must coexist with privacy protection and system security.

The mathematical formulation of FCEM can be represented as follows:

$$\min_{\theta} \left(\mathcal{L}_{\text{fair}}(\theta) + \lambda \cdot \mathcal{L}_{\text{explainable}}(\theta)\right)$$

$$\text{where } \theta \in \mathbb{R}^d, \mathcal{L}_{\text{fair}}(\theta) = \sum_{i=1}^{n} \mid y_i - \hat{y}_i \mid, \mathcal{L}_{\text{explainable}}(\theta) = \sum_{j=1}^{m} \mid \phi_j - \hat{\phi}_j \mid$$

where θ represents the parameters of the AI model, $\mathcal{L}_{\text{fair}}(\theta)$ is the fairness loss function, $\mathcal{L}_{\text{explainable}}(\theta)$ is the explainability loss function, y_i and $\hat{y}_i$ represent the true and predicted outputs for instance i, ϕ_j and $\hat{\phi}_j$ are the true and predicted explainable features of the model, λ is the regularization parameter controlling the trade-off between fairness and explainability, n and m are the number of instances and features, respectively.

The proposed Fairness-Constrained Explainability Model (FCEM) integrates key AI-driven mechanisms to ensure that urban AI systems operate transparently, ethically, and securely. At its core, the model strikes a critical balance between model interpretability and the need to preserve privacy, especially in the context of sensitive urban data. By incorporating fairness constraints, the FCEM prevents potential biases in decision-making, ensuring equitable outcomes for all citizens. At the same time, the explainability constraints embedded within the system allow decision-making processes to be clearly understood and communicated to the public, fostering transparency and accountability in urban governance. The model also prioritizes privacy by preventing the exposure of sensitive information, ensuring that transparency does not come at the cost of individual rights. This framework demonstrates a proactive approach to AI ethics in urban systems, where AI decisions are not only optimized for performance but also aligned with the principles of fairness, explainability, and privacy protection. The integration of these components creates an urban AI ecosystem that is both trustworthy and accountable, offering a sustainable model for the ethical deployment of AI in urban environments.

Framework: Fairness-Constrained Explainability Model (FCEM)

Layered AI Architecture for Urban AI Systems

Layer	Main functionality	AI module (s) and components	Operational scope
Layer 1	Data collection: Gather urban data (e.g., traffic, resource usage)	Urban data processing engine	Input → Raw urban data (traffic, energy usage, etc.) → Output → Preprocessed data for AI models
Layer 2	Learning: Fairness and Explainability through FCEM	Fairness-Constrained Explainability Model (FCEM)	Input → Raw urban data, fairness constraints, explainability constraints → Output → Trained model outputs with fairness and explainability integration
Layer 3	Optimization: Fairness and explainability balancing	Fairness-Constrained Explainability Model (FCEM)	Input → Training model outputs → Output → Optimized fairness-explainability trade-off in decision-making
Layer 4	Transparency and privacy protection: Explainable AI decision-making	Privacy-aware transparency module	Input → Model outputs, privacy constraints → Output → Transparent, interpretable decisions that maintain privacy
Layer 5	Model evaluation: Assess model accuracy, fairness, and explainability	Ethical performance evaluation system	Input → Model outputs, evaluation criteria → Output → Performance evaluation and ethical compliance checks
Layer 6	Governance and deployment: Final deployment and post-implementation monitoring	Governance monitoring dashboard	Input → AI model decision outcomes → Output → Deployment of AI system and ongoing monitoring for ethical compliance

Figure 2.1 presents a structured view of "Transparency and Explainability in AI Applications," organized around three main nodes: Transparency in AI, Explainability in AI, and Broader Implications. "Transparency in AI" focuses on clear communication of decisions, stakeholder trust, and challenges such as black-box model complexity and ethical concerns. "Explainability in AI" highlights

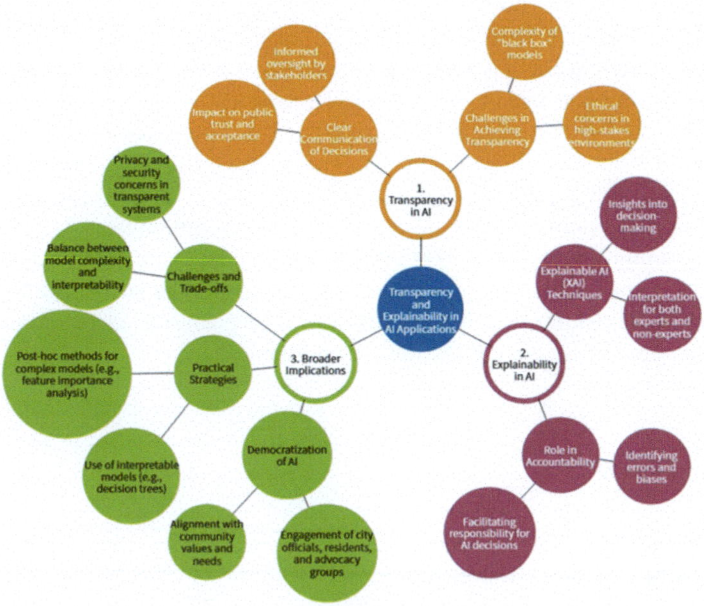

Fig. 2.1 Transparency and explainability in AI applications

Explainable AI (XAI) techniques, the role in accountability, and error/bias identification. "Broader Implications" covers democratization of AI, alignment with community values, practical strategies like interpretable models, and challenges balancing complexity and interpretability with privacy concerns.

Transparency and explainability are vital for ensuring that AI systems used in urban development are ethical, trustworthy, and accountable. These principles not only foster public trust and engagement but also enable better oversight and control over AI systems that make decisions affecting urban populations. As AI continues to play a larger role in shaping cities, achieving transparency and explainability will be crucial for aligning AI-driven solutions with the broader goals of fairness, accountability, and sustainability in urban development.

Limitations and Governance Implications

The explainability-oriented frameworks advanced in this chapter address a central vulnerability of AI-driven urban decision-making: the tendency of complex optimization systems to produce technically efficient yet publicly unintelligible outcomes. Through Explainability-Driven Deep Reinforcement Learning (X-DRL), Interpretable Deep Reinforcement Learning (I-DRL), Self-Interpretable Deep Reinforcement Learning (SI-DRL), Explainability-Guided Deep Reinforcement Learning (XG-DRL), Participatory Explainable AI (P-XAI), SHAP-based AI Explainability in Smart-City Applications (SHAPE-SCA), and the Fairness-Constrained Explainability Model (FCEM), the chapter develops mechanisms for interpretability, participatory validation, post hoc attribution, and fairness-aware explanation. These mechanisms can make urban AI systems more legible to citizens, planners, policymakers, auditors, and community representatives; however, explanation alone cannot guarantee democratic legitimacy. A technically coherent explanation may still be incomplete, inaccessible, strategically selective, or socially unpersuasive if it is not situated within institutional processes that allow questioning, correction, and collective evaluation. The governance implication is therefore that explainability must be treated not merely as a model property, but as a civic relation between system designers, public authorities, and affected communities. Transparent AI becomes meaningful only when explanations are usable, contestable, and connected to real procedural rights, including the capacity to challenge decisions, demand justification, and require institutional revision where automated recommendations conflict with equity, accountability, or public trust.

2.2 Conclusion

This chapter reveals that transparency and explainability are not auxiliary ethical gestures but the epistemic foundation of intelligent urban governance. Through the sequence of frameworks presented from Explainability-Driven Deep Reinforcement Learning (X-DRL) and Interpretable Deep Reinforcement Learning (I-DRL) to Self-Interpretable Deep Reinforcement Learning (SI-DRL), Explainability-Guided Deep Reinforcement Learning (XG-DRL), Participatory Explainable Artificial Intelligence (P-XAI), SHapley Additive exPlanations in Smart-City Applications (SHAPE-SCA), and the Fairness-Constrained Explainability Model (FCEM), the chapter establishes a coherent and mathematically grounded architecture for transparent, accountable, and participatory AI ecosystems. Each framework contributes a distinct layer in a progressive ethical continuum: beginning with technical interpretability, evolving through accountability mechanisms, and culminating in democratic engagement and fairness protection.

The analysis demonstrates that explainability can be formalized as a computational constraint within reinforcement learning, where interpretability becomes an intrinsic property of the optimization process rather than an afterthought. The X-DRL and I-DRL architectures pioneer this integration by embedding interpretability within the value and policy functions themselves, ensuring that AI models deliver not only optimal but also human-understandable decisions. The SI-DRL model deepens this paradigm by rendering AI self-reflective, transforming every policy decision into a narrative that humans can trace and audit.

The XG-DRL framework elevates transparency to institutional accountability, integrating bias-aware reward structures, error attribution functions, and fairness metrics that allow AI systems to self-audit and auto-correct. This transition from interpretability to accountability represents a critical ethical evolution turning AI systems into self-governing instruments capable of justifying, adjusting, and defending their behavior within complex socio-technical environments.

Building upon this, the Participatory Explainable Artificial Intelligence (P-XAI) framework introduces a paradigm of civic co-intelligence, where human stakeholders—citizens, policymakers, engineers, and advocacy groups co-validate algorithmic decisions through federated transparency constraints and engagement-weighted accountability. Here, the city becomes not merely a site of AI application but an arena of deliberative algorithmic governance, where public participation and computational learning are fused within a continuous feedback loop of explanation, critique, and refinement.

To provide a universal post hoc interpretability standard, the chapter integrates the SHapley Additive exPlanations (SHAP) model derived from cooperative game theory. This ensures that every feature in a complex neural network carries a measurable and fair attribution to the final outcome offering clarity even in high-dimensional decision spaces. SHAP operationalizes the mathematical fairness principle that each variable's contribution must be traceable, thus translating theoretical equity into computational transparency.

Finally, the Fairness-Constrained Explainability Model (FCEM) provides the normative closure of the chapter: a unified optimization framework where fairness, interpretability, and privacy coexist within a single mathematically bounded objective. FCEM ensures that explainability does not endanger privacy and that transparency does not devolve into exposure. It creates equilibrium among ethical, legal, and performance-driven imperatives realizing a self-regulating AI environment that meets industrial, civic, and academic standards simultaneously.

The chapter demonstrates that the pursuit of transparent and explainable AI is not a secondary concern but the organizing logic of the intelligent city itself. It repositions explainability from the periphery of AI ethics to the core of algorithmic design, providing a reproducible, equation-driven methodology for integrating interpretability across all layers of decision-making. The frameworks collectively show how AI can be both technically powerful and morally interpretable, how machine learning can serve human reasoning without eclipsing it, and how ethical oversight can be operationalized rather than merely proclaimed.

From an integrative perspective, this work stands as a foundational reference for international researchers, engineers, and policymakers seeking to align artificial intelligence with the democratic, transparent, and sustainable imperatives of twenty-first century urbanism. It invites the reader to envision a future where every algorithmic action in a city—whether in energy optimization, mobility management, or housing allocation—is not only efficient but also intelligible, accountable, and fair. In this synthesis, transparency and explainability cease to be constraints on AI; they become its defining virtues, transforming the urban system into a space where intelligence, ethics, and humanity converge in measurable harmony.

References

Attard-Frost, B., De los Rios, A., & Walters, D. R. (2023). The ethics of AI business practices: A review of 47 AI ethics guidelines. *AI and Ethics*, 3, 389–406. doi:https://doi.org/10.1007/s43681-022-00156-6

Barredo Arrieta, A., Díaz-Rodríguez, N., Del Ser, J., Bennetot, A., Tabik, S., Barbado, A., Garcia, S., Gil-Lopez, S., Molina, D., & Herrera, F. (2020). Explainable artificial intelligence (XAI): Concepts, taxonomies, opportunities, and challenges toward responsible AI. *Information Fusion*, 58, 82–115. doi:https://doi.org/10.1016/j.inffus.2019.12.012

Trotta, A., Ziosi, M., & Lomonaco, V. (2023). The future of ethics in AI: Challenges and opportunities. *AI & Society*, 38, 439–441. doi:https://doi.org/10.1007/s00146-023-01644-x

Chapter 3
Ethically Governed AI Frameworks for Fair, Transparent, and Sustainable Urban Resource Management

The chapter "Ethically Governed AI Frameworks for Fair, Transparent, and Sustainable Urban Resource Management" opens a new frontier in the ethical operationalization of Artificial Intelligence (AI) for the management of urban infrastructures, offering both a moral compass and a computational roadmap for cities entering the era of algorithmic governance. Rather than treating ethics as a peripheral concern, this chapter embeds ethical imperatives: fairness, sustainability, transparency, accountability, and privacy directly into the mathematical and algorithmic fabric of AI models that manage critical urban resources such as water, energy, waste, and public space.

The discussion begins by redefining fairness in AI through the Fairness-Constrained Multi-Agent Reinforcement Learning (FCRL-MARL) framework, which formalizes equitable resource distribution as an optimization problem constrained by ethical thresholds. By integrating fairness-constrained optimization, multi-agent decision coordination, and privacy-preserving federated learning, it ensures that intelligent urban systems distribute resources justly while safeguarding citizen data. Each AI agent in this architecture is not merely a decision-maker but an ethical participant trained to balance performance with responsibility.

Sustainability, the second pillar of this chapter, is modeled through the Sustainability-Constrained Reinforcement Learning (SC-RL) framework. Here, multi-objective optimization ensures that AI systems achieve energy, water, and waste efficiency without compromising human comfort or environmental welfare. Complemented by comfort-aware AI adjustments and green federated learning, the model introduces a self-regulating mechanism where urban AI systems learn to conserve energy during training and operation, embodying a "low-carbon intelligence" ethos essential for future climate-resilient cities.

The third ethical axis—transparency—is addressed through the Explainable Reinforcement Learning for Urban Resource Management (XRL-URM) framework. By unifying Explainable AI Policy Optimization, Shapley Additive Explanations (SHAP), Local Interpretable Model-Agnostic Explanations (LIME), and Federated Learning for Distributed Transparency, the chapter operationalizes explainability as

D. Sargiotis, *AI with Integrity*, https://doi.org/10.1007/978-3-032-31133-7_3

a measurable variable rather than an aspirational concept. The framework enables city officials and citizens alike to trace, interpret, and audit AI decisions in real time, ensuring that no algorithm acts as a "black box." Transparency here becomes an algorithmic design principle—a prerequisite for democratic legitimacy in data-driven governance.

Accountability is then formalized through the Accountability-Constrained Reinforcement Learning (AC-RL) framework, where ethical oversight and corrective feedback loops are encoded within the AI's decision function. By defining AI responsibility functions, fairness-constrained redress mechanisms, and explainable accountability metrics, the framework ensures that errors, inefficiencies, or biases trigger immediate adaptive responses. This establishes a governance architecture where AI systems are not only self-optimizing but also self-correcting and capable of explaining and justifying their actions within the bounds of legal and ethical norms.

Finally, the chapter culminates with the Fairness-Constrained Federated Learning (FC-FL) framework, which ensures ethical data handling across decentralized urban infrastructures. By introducing fairness regularization into federated optimization, this model allows multiple city agencies, districts, or utilities to collaborate without exchanging sensitive data. It enforces demographic parity across local nodes, minimizing algorithmic discrimination and preserving community data sovereignty. Here, privacy and fairness are mathematically encoded as constraints that govern both local learning and global aggregation—a decisive move toward privacy-preserving democracy in smart-city ecosystems.

Collectively, these five frameworks form a comprehensive architecture for ethically governed AI in urban resource systems. They transform normative principles into operational equations, bridging moral philosophy with computational logic. Each framework is layered, modular, and interoperable, enabling cities to customize ethical AI deployments according to their resource domains, governance maturity, and regulatory environment.

From an integrative perspective, this chapter demonstrates that ethical AI is not an external audit function but a self-regulating intelligence architecture one that unites fairness, sustainability, transparency, accountability, and privacy into a coherent urban ethics of computation. It offers global policymakers, engineers, and researchers a reproducible scientific foundation to design AI systems that are not only efficient but just; not only powerful but explainable; and not only innovative but deeply human-centered.

Thus, the chapter stands as both a philosophical manifesto and a technical blueprint for a new generation of trustworthy, regulation-ready, and ethically autonomous AI systems—the cornerstone of tomorrow's intelligent, equitable, and sustainable cities.

3.1 Fairness-Constrained Multi-Agent Reinforcement Learning for Ethical Urban Resource Management

Ethical considerations in the deployment of AI for urban resource management are pivotal, given that AI systems are increasingly relied upon to optimize resources such as water, energy, waste, and public spaces. These systems promise more efficient and sustainable cities, but their deployment also raises significant ethical concerns. The key ethical dimensions of AI in urban resource management include issues of equity, environmental sustainability, transparency, accountability, and the protection of privacy.

One of the most pressing ethical challenges in AI-driven urban resource management is ensuring fairness in the allocation of resources. AI systems often rely on large datasets to optimize the distribution of resources like water or energy. However, these datasets can be biased, reflecting historical inequities in resource distribution (Attard-Frost et al. 2023; Sargiotis 2024). This can result in the reinforcement of existing inequalities, where underserved communities receive fewer resources or are excluded from the benefits of AI optimizations. To address this, it is essential to integrate fairness as a core principle in the design of AI systems, ensuring that all urban residents have equal access to essential resources regardless of their socio-economic status or geographic location. Against this challenge, a Fairness-Constrained Multi-Agent Reinforcement Learning (FCRL-MARL) framework is introduced. This model integrates equitable resource allocation via fairness-constrained optimization, multi-agent decision coordination for sustainable AI governance, privacy-preserving federated learning, and explainability-driven AI decision transparency to ensure that AI-driven urban resource distribution remains fair, sustainable, and accountable while protecting citizen privacy. Unlike conventional resource optimization models that may reinforce existing inequalities, this framework actively mitigates biases, ensures environmental responsibility, and upholds ethical governance.

To formally define fair resource allocation in AI-driven urban management, a fairness-constrained optimization function is introduced. This function ensures that AI systems do not disproportionately allocate resources based on historical inequities, instead enforcing fairness constraints dynamically. The fairness-constrained resource allocation function is given by:

$$\min_{\mathrm{w}} \sum_{i=1}^{N} \ell(f_{\mathrm{w}}(X_i), Y_i) \text{subject to } D_{\mathrm{fair}}(f_{\mathrm{w}}(X), S) \leq \tau$$

where $D_{\mathrm{fair}}(f_{\mathrm{w}}(X), S)$ is the fairness disparity function measuring inequities in AI-driven resource allocation, and τ is the predefined fairness threshold ensuring equitable urban resource distribution. This function ensures that AI-driven urban resource management actively enforces fairness, preventing systemic biases from affecting essential resource allocations such as water, energy, and public infrastructure.

To enhance sustainability and ethical coordination in AI resource distribution, a multi-agent resource coordination function is introduced. This function ensures that multiple AI decision-making agents collaborate dynamically to optimize urban resource allocation while maintaining environmental balance. The multi-agent sustainability optimization function is defined as:

$$V_{\mathrm{MARL}}(s_t, a_t) = \sum_{k=1}^{K} \mathbb{E}[R_k(s_t, a_t) - \lambda E_k(s_t, a_t)]$$

where $V_{\mathrm{MARL}}(s_t, a_t)$ represents the multi-agent sustainability-adjusted value function, $R_k(s_t, a_t)$ is the reward function for AI agent k optimizing resource efficiency, $E_k(s_t, a_t)$ is the environmental impact penalty function regulating AI-driven resource consumption, and λ is the penalty coefficient balancing efficiency with sustainability. This function ensures that AI-driven urban resource management integrates sustainability constraints, preventing over-utilization and ensuring responsible environmental governance.

To protect citizen privacy in AI-driven urban systems, a privacy-preserving federated learning function is introduced. This function ensures that AI models optimize urban resource allocation without requiring centralized access to sensitive personal data, preserving individual privacy. The privacy-constrained AI optimization function is formulated as:

$$\min_{\mathbf{w}} \sum_{i=1}^{N} \ell(f_{\mathbf{w}}(X_i), Y_i) + \beta \sum_{i=1}^{N} P_{\mathrm{privacy}}(X_i, Y_i)$$

where $P_{\mathrm{privacy}}(X_i, Y_i)$ is the privacy-preserving function regulating AI access to sensitive data, and β is the privacy weight ensuring AI decision-making respects data protection laws. This function ensures that AI-driven urban solutions remain privacy-compliant, preventing unauthorized data usage while optimizing resource management.

Via the combination of fairness-constrained resource allocation, multi-agent sustainability optimization, and privacy-preserving AI optimization, this FCRL-MARL framework ensures that AI-driven urban resource management remains ethically responsible, transparent, and aligned with long-term sustainability goals. The multi-agent fairness-aware AI model dynamically adjusts resource allocation strategies, ensuring that urban residents have equitable access to essential resources without bias or environmental harm.

Framework: Fairness-Constrained Multi-Agent Reinforcement Learning for Urban Resource Allocation (FCRL-MARL)

Layered AI Architecture for Fair, Sustainable, and Privacy-Preserving Urban Resource Management

Layer	Main functionality	AI module(s) and components	Operational scope (Input → Processing → Output)
Layer 1	Data collection and fairness constraints: Gathering data while ensuring fairness in urban resource allocation	Ethical data acquisition and fairness-aware preprocessing engine	Input → Resource usage data, historical inequities → Output → Fairness-constrained data for AI-driven resource distribution
Layer 2	Fairness-constrained optimization: Applying fairness constraints to resource allocation to prevent biases	Fairness-constrained optimization	Input → Raw allocation data, fairness parameters → Output → Optimized, equitable resource distribution
Layer 3	Multi-agent coordination: Ensuring multiple AI agents collaborate to optimize urban resource distribution sustainably	Multi-Agent Reinforcement Learning (MARL)	Input → Resource data, agent coordination policies → Output → Dynamic coordination of multi-agent decisions for resource distribution
Layer 4	Privacy-preserving federated learning: Optimizing resource allocation while maintaining privacy through federated learning	Privacy-preserving federated learning	Input → Local data on resource usage → Output → Optimized global model respecting privacy constraints
Layer 5	Post-AI explainability: Providing transparency in decision-making to stakeholders and the public	Explainable decision analytics and stakeholder transparency dashboard	Input → AI decision outputs → Output → Transparent decision-making, visualizations, and stakeholder reports

Environmental sustainability is another critical ethical concern. AI systems are often deployed to enhance the sustainability of cities by reducing energy consumption, optimizing waste management, or improving water use efficiency. While these systems offer great potential for environmental benefits, their implementation must balance efficiency with human and environmental impacts. For example, AI systems that prioritize energy efficiency in buildings must also consider the well-being and comfort of the occupants (Trotta et al. 2023). If the pursuit of efficiency leads to energy-saving measures that compromise human health or comfort, the ethical implications of these systems become questionable. AI must be used to promote sustainability in a way that also respects human welfare. Against this ethical trade-off, a Sustainability-Constrained Reinforcement Learning (SC-RL) framework is introduced to align environmental optimization with human welfare. This model integrates multi-objective optimization for energy, water, and waste efficiency, comfort-aware AI adjustments for human-centric building management, green federated learning for low-energy AI training, and ethical trade-off modeling to ensure that AI-driven sustainability initiatives optimize resource use without compromising occupant comfort and well-being. Unlike purely efficiency-driven AI models, this framework ensures sustainable AI deployment that is both environmentally responsible and human-centered. To formally define sustainability-conscious AI decision-making, a multi-objective optimization function is introduced. This function ensures that AI models balance energy efficiency, water conservation, and waste reduction while maintaining human welfare constraints. The sustainability-constrained optimization function is given by:

$$V_{\mathrm{SC-RL}}(s_t, a_t) = \sum_{i=1}^{N} \mathbb{E}[R_i(s_t, a_t) - \lambda C_i(s_t, a_t)]$$

where $V_{\mathrm{SC-RL}}(s_t, a_t)$ represents the sustainability-adjusted value function, $R_i(s_t, a_t)$ is the reward function for optimizing a specific resource (e.g., energy, water, waste), $C_i(s_t, a_t)$ is the human comfort cost function ensuring that AI decisions do not negatively impact occupant well-being, and λ is the trade-off coefficient regulating the balance between efficiency and comfort. This function ensures that AI-driven sustainability models optimize energy and resource use while dynamically adjusting policies to protect human comfort and welfare.

To further enhance human-centric sustainability in AI-driven urban systems, a comfort-aware adjustment function is introduced. This function ensures that AI models dynamically adapt resource-saving measures to prevent discomfort for building occupants. The comfort-optimized AI function is defined as:

$$C_{\mathrm{adjust}}(s_t, a_t) = \alpha H(s_t, a_t) + \beta E(s_t, a_t)$$

where $C_{\mathrm{adjust}}(s_t, a_t)$ is the comfort-aware AI adjustment function, $H(s_t, a_t)$ represents human-centric parameters (e.g., indoor temperature, air quality, and lighting levels), $E(s_t, a_t)$ is the energy efficiency optimization parameter, and α,βα,β are weighting coefficients ensuring a balance between comfort and efficiency. This function ensures that AI-driven building management does not compromise occupant comfort for the sake of energy savings, reinforcing ethical AI sustainability practices.

To ensure that AI sustainability models remain energy-efficient, a green federated learning function is introduced. This function ensures that AI models are trained in a decentralized manner, reducing computational energy consumption. The federated low-energy AI training function is formulated as:

$$\mathcal{L}_{\mathrm{GFL}} = \sum_{i=1}^{N} \ell(f_{\mathbf{w}}(X_i), Y_i) + \gamma \sum_{i=1}^{N} P_{\mathrm{energy}}(X_i, Y_i)$$

where $\mathcal{L}_{\mathrm{GFL}}$ is the energy-efficient AI training loss function, $f_{\mathbf{w}}(X_i)$, $Y_i)$ represents the standard AI model loss function, $P_{\mathrm{energy}}(X_i, Y_i)$ is the energy consumption penalty function minimizing AI computational overhead, and γ is the penalty coefficient regulating AI training energy use. This function ensures that AI models supporting sustainability initiatives do not generate excessive computational energy consumption, reinforcing a low-carbon AI training approach.

By employing multi-objective optimization for sustainable AI, comfort-aware AI adjustments, and green federated learning, this SC-RL framework ensures that AI-driven sustainability models remain human-centered, environmentally responsible, and ethically balanced. The adaptive AI system dynamically regulates resource-saving strategies, ensuring that AI-driven urban sustainability remains aligned with human welfare and long-term ecological responsibility.

Framework: Sustainability-Constrained Reinforcement Learning for Sustainable Urban Resource Management (SC-RL)

Layered AI Architecture for Sustainable, Human-Centered, and Energy-Efficient Urban Resource Management

Layer	Main functionality	AI module(s) and components	Operational scope (Input → Processing → Output)
Layer 1	Data collection and sustainability constraints: Gathering data while ensuring sustainability and human comfort	Sustainable data integration and human-centric contextualization engine	Input → Resource usage data, human welfare parameters → Output → Sustainability-constrained data for AI-driven resource management
Layer 2	Multi-objective optimization: Balancing energy, water, and waste efficiency with human comfort	Multi-objective optimization	Input → Resource consumption data, comfort parameters → Output → Optimized energy, water, and waste distribution while considering human welfare
Layer 3	Comfort-aware AI adjustments: Ensuring AI decisions do not compromise occupant comfort for efficiency	Comfort-aware AI adjustments	Input → Human-centric parameters (temperature, air quality), efficiency data → Output → AI-adjusted comfort parameters integrated into resource management
Layer 4	Green federated learning: Decentralized AI training for low-energy consumption while maintaining model accuracy	Green Federated Learning (GFL)	Input → Decentralized local data for resource management → Output → Trained AI models minimizing energy consumption
Layer 5	Post-AI transparency and ethical trade-offs: Providing transparency in AI decisions, ensuring ethical resource allocation	Ethical decision auditing and sustainability visualization dashboard	Input → AI-generated resource allocation decisions → Output → Transparent ethical decision-making and trade-off analysis reports

Transparency and explainability are fundamental to the ethical use of AI in urban resource management. The decisions made by AI systems must be understandable to city officials, policymakers, and citizens (Barredo Arrieta et al. 2020). This is particularly important when AI systems are used to manage public resources, as the public has a right to know how decisions affecting their access to resources are made. AI systems that operate as "black boxes" without offering explanations for their decisions risk undermining public trust and accountability. Therefore, ensuring that AI systems in urban resource management are transparent and their decision-making processes are explainable is a key ethical imperative. Against this risk of black-box resource governance, an Explainable Reinforcement Learning for Urban Resource Management (XRL-URM) framework is introduced. This model integrates policy-level explainability, feature attribution using SHAP and LIME, federated learning for distributed AI transparency, and trustworthy AI governance to ensure that AI-driven decisions remain clear, interpretable, and publicly accountable. Unlike black-box AI systems, this framework ensures stakeholder engagement and regulatory oversight, reinforcing public trust in AI-driven urban planning.

To formally define explainability in AI decision-making for urban resource management, a policy explainability function is introduced. This function ensures that AI-generated policies are traceable and interpretable, allowing decision-makers to understand how AI systems allocate resources. The explainable AI policy function is given by:

$$\pi^*_{\mathrm{XRL}}(a_t|s_t) = \arg\max_{\pi} \mathbb{E}[R(s_t, a_t) + \gamma I(s_t, a_t)]$$

where $\pi^*_{\mathrm{XRL}}(a_t|s_t)$ represents the explainability-driven optimal policy, $R(s_t, a_t)$ is the reward function reflecting AI optimization objectives, $I(s_t, a_t)$ is the interpretability function ensuring decision transparency, and γ is the explainability weight controlling the balance between efficiency and interpretability. This function ensures that AI-driven urban planning policies remain transparent, preventing opaque decision-making processes that could undermine public trust.

To enhance feature-level explainability in AI decision-making, a feature attribution function is introduced. This function ensures that AI models explicitly quantify the importance of each input variable in decision-making using SHAP (Shapley Additive Explanations) and LIME (Local Interpretable Model-Agnostic Explanations). The SHAP-based feature attribution function is formulated as:

$$A_{\mathrm{XRL}}(s_t, a_t) = \sum_{i=1}^{N} \phi_i X_i$$

where $A_{\mathrm{XRL}}(s_t, a_t)$ represents the explainability-enhanced feature attribution function, ϕ_i is the importance weight assigned to feature X_i, and X_i denotes the urban planning feature influencing AI decisions. This function ensures that AI systems provide granular transparency by revealing which factors influence urban resource allocation, ensuring decision accountability.

To ensure distributed AI transparency and prevent centralized opacity, a federated learning explainability function is introduced. This function ensures that AI-driven urban management decisions are validated across decentralized data sources, preventing biased decision-making. The federated explainability constraint function is given by:

$$\sum_{j=1}^{M} \mathbb{E}\left[I_j(s_t, a_t)\right] \geq \tau$$

where $\mathbb{E}\left[I_j(s_t, a_t)\right]$ represents the explainability function from stakeholder group j evaluating AI decisions, M is the total number of decentralized validation nodes, and τ is the minimum explainability threshold required for AI decision approval. This function ensures that AI-driven urban resource decisions are validated across multiple governance entities, reinforcing decentralized transparency and fairness.

By embedding explainable AI policy optimization, SHAP-based feature attribution, and federated learning for AI transparency, this XRL-URM framework ensures that AI-driven urban resource management remains transparent, accountable, and publicly interpretable. The multi-level explainability approach dynamically adjusts AI decision transparency, ensuring that AI-driven urban policies are understood, evaluated, and trusted by both policymakers and the public.

Framework: Explainable Reinforcement Learning for Urban Resource Management (XRL-URM)

Layered AI Architecture for Transparent, Interpretable, and Accountable Urban Resource Management

Layer	Main functionality	AI module(s) and components	Operational scope (Input → Processing → Output)
Layer 1	Data collection and explainability constraints: Gathering data and ensuring decision transparency in AI	Transparent data ingestion and interpretability-ready preprocessing engine	Input → Urban resource usage data → Output → Transparent and traceable data for AI-driven decisions
Layer 2	Policy explainability: Ensuring AI-generated policies are traceable and interpretable for decision-makers	Explainable AI policy optimization	Input → Urban planning objectives, optimization rewards → Output → Explainable optimal AI policy for resource allocation
Layer 3	Feature attribution: Explaining the importance of each feature in AI decision-making using SHAP and LIME	SHAP, LIME (Local Interpretable Model-Agnostic Explanations)	Input → Features affecting urban resource allocation → Output → Feature attribution weights, decision transparency
Layer 4	Federated learning for distributed transparency: Ensuring decentralized validation of AI-driven urban decisions	Federated learning	Input → Decentralized resource management data → Output → Federated validation of AI decisions to ensure transparency
Layer 5	Post-AI transparency and stakeholder validation: Ensuring AI decisions are validated by multiple governance entities	Stakeholder validation Interface and explainability reporting dashboard	Input → AI-generated decisions, stakeholder feedback → Output → Validated AI decisions, transparency reports

Accountability is closely linked to transparency and is another critical ethical consideration. AI systems must be designed with clear lines of accountability so that when they make errors or cause harm, there is a clear mechanism for addressing those issues (Attard-Frost et al. 2023). In the context of urban resource management, this could involve mechanisms for redress when AI-driven systems lead to unfair distribution of resources, inefficient use of energy, or environmental harm. Without accountability, the deployment of AI across these critical domains risks eroding public trust and exacerbating existing challenges in urban governance. Against this accountability risk, an Accountability-Constrained Reinforcement Learning (AC-RL) framework is introduced This model integrates multi-agent AI oversight for ethical decision monitoring, fairness-constrained redress mechanisms for AI error resolution, explainable accountability metrics, and regulatory compliance optimization to ensure that AI-driven urban systems remain transparent, justifiable, and responsive to errors or biases. Unlike conventional AI models, which may lack clear accountability structures, this framework enforces systematic oversight, stakeholder-driven auditing, and redress mechanisms to maintain public trust.

To formally define accountability in AI decision-making, an AI responsibility function is introduced. This function ensures that AI-driven urban resource allocation decisions can be traced and justified, allowing governance bodies to address system errors, biases, and negative consequences. The accountability-aware decision function is given by:

$$V_{\mathrm{AC-RL}}(s_t, a_t) = \mathbb{E}[R(s_t, a_t) - \lambda P_{\mathrm{error}}(s_t, a_t)]$$

where $V_{\mathrm{AC-RL}}(s_t, a_t)$ represents the accountability-adjusted value function, $R(s_t, a_t)$ is the reward function quantifying the effectiveness of AI-driven resource allocation, $P_{\mathrm{error}}(s_t, a_t)$ is the error penalty function ensuring that AI models are held responsible for biases or inefficiencies, and λ is the penalty coefficient regulating accountability trade-offs. This function ensures that AI-driven urban systems incorporate responsibility constraints, dynamically adjusting resource distribution decisions to prevent unintended harm.

To provide systematic redress mechanisms for AI-driven errors, a fairness-constrained redress function is introduced. This function ensures that when AI-driven decisions result in unfair resource allocation or inefficiencies, there is an adaptive correction mechanism. The AI redress mechanism function is given by:

$$\min_{\mathbf{w}} \sum_{i=1}^{N} \ell(f_{\mathbf{w}}(X_i), Y_i) + \beta \sum_{i=1}^{N} R_{\mathrm{redress}}(X_i, Y_i)$$

where $R_{\mathrm{redress}}(X_i, Y_i)$ is the fairness redress function ensuring that AI systems actively correct unjust decisions, and β is the penalty coefficient enforcing fairness adjustments. This function ensures that AI-driven urban governance does not merely detect errors but actively corrects them, reinforcing equitable and ethical AI resource management.

To enhance public accountability and AI transparency, an explainability-driven accountability function is introduced. This function ensures that AI systems justify their decisions with clear, interpretable reasoning, preventing black-box decision-making. The explainable accountability function is formulated as:

$$A_{\text{AC-RL}}(s_t, a_t) = \sum_{i=1}^{N} \phi_i X_i$$

where $A_{\text{AC-RL}}(s_t, a_t)$ represents the explainability-driven accountability function, ϕ_i is the accountability weight assigned to feature X_i, and X_i denotes the urban planning feature affecting AI decisions. This function ensures that AI-driven urban resource management is transparent and auditable, reinforcing public oversight and regulatory compliance.

Using accountability-aware AI decision functions, fairness-constrained redress mechanisms, and explainability-driven accountability metrics, this AC-RL framework ensures that AI-driven urban governance remains justifiable, corrective, and publicly accountable. The adaptive AI system continuously monitors and corrects errors, ensuring that AI-driven urban resource management aligns with ethical governance principles and maintains public trust.

Framework: Accountability-Constrained Reinforcement Learning for Urban Resource Management (AC-RL)

Layered AI Architecture for Ethical, Accountable, and Transparent Urban Resource Management

Layer	Main functionality	AI module(s) and components	Operational scope (Input → Processing → Output)
Layer 1	Data collection and accountability constraints: Gathering data while ensuring accountability for AI decisions	Accountability-aware data governance and provenance tracking engine	Input → Resource usage data, historical biases → Output → Accountability-constrained data for fair resource distribution
Layer 2	AI responsibility: Ensuring AI-driven resource allocation decisions are traceable and justifiable	AI responsibility function	Input → Resource allocation decisions, AI errors → Output → Responsibility-adjusted decisions with penalty for biases
Layer 3	Fairness-constrained redress: Correcting unfair or inefficient AI decisions through adaptive redress mechanisms	Fairness-constrained redress function	Input → Unfair decisions, resource allocation data → Output → Corrected decisions, fairness enforcement
Layer 4	Explainability-driven	Explainable accountability function	Input → AI decisions, decision features → Output → Transparent,

(continued)

Layer	Main functionality	AI module(s) and components	Operational scope (Input → Processing → Output)
	accountability: Justifying AI decisions with interpretable and transparent reasoning		accountable decisions with clear reasoning
Layer 5	Post-AI accountability: Ensuring that AI systems are responsive to errors and subject to oversight	Oversight monitoring interface and adaptive compliance dashboard	Input → AI decisions, stakeholder feedback → Output → Monitored and corrected AI decisions, public accountability

The ethical use of data in AI-driven urban resource management also raises significant concerns. AI systems require vast amounts of data to function effectively, and in urban settings, this often includes data collected from residents. Ensuring that this data is collected and used ethically, with respect to privacy rights and data protection laws, is crucial (Barredo Arrieta et al. 2020). Furthermore, cities must ensure that AI systems do not disproportionately surveil or extract data from marginalized communities, which could exacerbate existing inequalities. To uphold ethical standards in AI-driven urban resource management, Fairness-Constrained Federated Learning (FC-FL) provides a robust framework that allows multiple urban data sources such as neighborhoods, health centers, and mobility hubs to collaboratively train AI models without sharing raw data. This decentralized approach ensures compliance with data protection laws while minimizing risks of over-surveillance or bias against marginalized communities.

At the heart of the FC-FL framework lies the federated optimization objective augmented with a fairness constraint:

$$\underset{\theta}{min} \sum_{k=1}^{K} w_k \cdot \mathcal{L}_k(\theta) \text{subject to } \mathcal{F}(\theta) \leq \delta$$

where θ are the shared global model parameters, K is the number of participating urban entities (e.g., districts or agencies), $\mathcal{L}_k(\theta)$ is the local loss function for node k, w_k is the weight reflecting population or data volume at node k, $\mathcal{F}(\theta)$ is a fairness constraint measuring demographic parity or equal opportunity, and δ is the maximum allowable fairness deviation. This formulation ensures that while the model is optimized for overall performance across the city, it does not privilege one demographic or geographic group over another. The fairness constraint $\mathcal{F}(\theta)$ is typically defined as the absolute difference in error rates or resource allocations across sensitive attributes such as race, income level, or zip code.

To preserve individual privacy, each node performs local updates and communicates only model gradients or parameter updates. The shared global model $\theta^{(t)}$ is updated using the federated averaging algorithm:

$$\theta^{(t+1)} = \sum_{k=1}^{K} \frac{n_k}{n} \cdot \theta_k^{(t)}$$

where $\theta_k^{(t)}$ is the locally trained model at node k, n_k is the number of data samples at node k, and $n = \sum_k n_k$ is the total number of samples across all nodes. This federated mechanism ensures that sensitive data especially from underserved or privacy-sensitive communities remains localized and under control of the data owner, reducing exposure risks and reinforcing data sovereignty.

To guard against indirect group inference or fairness violations during training, the FC-FL model incorporates a regularized fairness penalty into each local objective:

$$\mathcal{L}_k^{\text{fair}}(\theta) = \mathcal{L}_k(\theta) + \lambda \cdot \text{Var}_{g \in G}\left(\mu_{k,g}\right)$$

where $\mu_{k,\ g}$ is the average prediction or allocation for group g in node k, G is the set of sensitive demographic groups, and λ is the fairness regularization parameter. This function penalizes disparate treatment across groups and encourages the model to produce equitable outcomes regardless of race, income, or age ensuring ethical service delivery in domains like energy allocation, healthcare access, and transportation policy.

Together, this FC-FL architecture offers privacy-compliant, and ethically sound foundation for AI in smart cities. The federated learning scheme protects personal data, the fairness constraints address demographic equity, and the local training preserves community autonomy. Through this model, cities can harness the predictive power of AI while upholding justice, privacy, and trust—core values for inclusive and democratic urban innovation.

Framework: Fairness-Constrained Federated Learning for Ethical Urban Resource Management (FC-FL)

Layered AI Architecture for Privacy-Preserving, Fair, and Transparent Urban Resource Management

Layer	Main functionality	AI module (s) andcomponents	Operational scope (Input → Processing → Output)
Layer 1	Data collection and federated learning constraints: Gathering data while ensuring privacy and fairness	Federated data ingestion and privacy-aware preprocessing engine	Input → Urban data (neighborhood, health centers, mobility hubs) → Output → Decentralized model updates while protecting individual privacy
Layer 2	Federated optimization with fairness constraints: Optimizing model performance while ensuring demographic parity	Federated learning	Input → Local model parameters, fairness constraints → Output → Global model updates with fairness-enforced optimization

(continued)

Layer	Main functionality	AI module (s) andcomponents	Operational scope (Input → Processing → Output)
Layer 3	Privacy-preserving federated learning: Ensuring that personal data is never shared, only model parameters are exchanged	Federated averaging	Input → Local data models → Output → Federated model updates while preserving individual data privacy
Layer 4	Fairness regularization: Penalizing disparate treatment across sensitive demographic groups during model training	Fairness regularization	Input → Local predictions for different demographic groups → Output → Adjusted model to ensure equity in outcomes
Layer 5	Post-AI monitoring: Ensuring AI decisions remain fair, transparent, and privacy-compliant throughout deployment	Fairness compliance monitoring and federated audit dashboard	Input → AI-driven decisions and resource allocations → Output → Monitoring reports and compliance audits for fairness and privacy

Figure 3.1 presents a simplified structure of ethical considerations in AI for urban resource management, focusing on five main areas: Equity in Resources, Environmental Sustainability, Transparency, Accountability, and Data Ethics. “Equity in Resources” emphasizes fair access and avoiding bias. “Environmental Sustainability” highlights balancing efficiency with environmental impact and human needs. “Transparency” centers on clear decisions and trust-building through openness. “Accountability” involves responsibility, correcting errors, and establishing mechanisms to address issues. “Data Ethics” focuses on privacy protection and avoiding excessive surveillance, ensuring ethical data use in urban AI applications.

The ethical considerations surrounding AI for urban resource management are multifaceted, requiring careful attention to fairness, sustainability, transparency, accountability, and privacy. As AI becomes increasingly integrated into the fabric of urban governance, addressing these ethical concerns will be critical to ensuring that AI technologies contribute to the equitable, sustainable, and ethical management of urban resources.

Limitations and Governance Implications

The resource management frameworks proposed in this chapter respond to one of the most demanding governance tensions in AI-driven cities: the need to allocate energy, water, waste management capacity, and infrastructural resources in ways that are efficient, sustainable, and socially fair. Fairness-Constrained Multi-Agent Reinforcement Learning for Urban Resource Allocation (FCRL-MARL), Sustainability-Constrained Reinforcement Learning for Sustainable Urban Resource Management (SC-RL), Explainable Reinforcement Learning for Urban Resource Management (XRL-URM), Accountability-Constrained Reinforcement Learning for Urban Resource Management (AC-RL), and Fairness-Constrained Federated Learning for Ethical Urban Resource Management (FC-FL) provide a structured basis for balancing distributive equity, environmental responsibility, transparent allocation, accountable optimization, and privacy-preserving coordination. Their value, however, remains conditional on the institutional setting in which they are deployed. A model may

Fig. 3.1 Ethical considerations in AI for urban resource management

identify an efficient allocation strategy, reduce resource waste, or improve sustainability metrics, yet still fail normatively if vulnerable households, peripheral districts, or essential-service users experience disproportionate burdens. For this reason, technical optimization must be accompanied by public service guarantees, regulatory supervision, social-impact assessment, and mechanisms through which municipalities, utilities, infrastructure operators, and affected communities can review allocation outcomes. The governance implication is that sustainability-aware urban AI cannot be assessed only through efficiency gains or aggregate environmental performance; it must also be judged by its capacity to preserve social protection, procedural transparency, and equitable access to essential urban resources.

3.2 Conclusion

This chapter constitutes a rigorous and multidimensional synthesis of how Artificial Intelligence (AI) can be ethically governed across the entire life cycle of urban resource management. Through a sequential yet interlinked presentation of five mathematically grounded frameworks each representing a distinct ethical vector, it demonstrates that ethical AI governance is not a philosophical abstraction but a technical necessity for sustainable and democratic cities. What emerges is a coherent architecture of fairness-constrained, explainable, accountable, and privacy-preserving intelligence systems capable of aligning computational efficiency with social justice, environmental stewardship, and human dignity.

In analytical perspective, the chapter's first contribution lies in its reconceptualization of fairness as an optimization constraint within AI decision systems. The Fairness-Constrained Multi-Agent Reinforcement Learning (FCRL-MARL) model provides a formalized mechanism for equitable resource allocation, transforming fairness from a moral aspiration into a quantifiable, algorithmic property. By embedding multi-agent coordination and privacy-preserving federated learning, it proposes a scalable governance model wherein fairness is enforced dynamically, not post hoc, ensuring that distributive justice becomes a structural feature of smart-city automation rather than a reactive correction.

Viewed through the sustainability lens, the chapter's second framework the Sustainability-Constrained Reinforcement Learning (SC-RL) model introduces a profound rebalancing between energy efficiency and human comfort. It formalizes sustainability not as a collateral outcome but as a decision boundary within AI optimization, integrating comfort-aware functions, multi-objective trade-offs, and low-carbon federated training. This approach redefines technological progress as a harmony between ecological restraint and human well-being, advancing the notion of "ethical sustainability" as a measurable performance index within adaptive urban systems.

From a transparency standpoint, the Explainable Reinforcement Learning for Urban Resource Management (XRL-URM) framework institutionalizes interpretability as a design principle rather than a compliance afterthought. By operationalizing explainability through Shapley Additive Explanations (SHAP), Local Interpretable Model-Agnostic Explanations (LIME), and distributed transparency mechanisms, the chapter ensures that every algorithmic action can be traced, justified, and debated. This transforms AI from a technocratic instrument into a participatory governance tool—one that fortifies public trust through algorithmic clarity and civic accountability.

From an ethical governance perspective, the Accountability-Constrained Reinforcement Learning (AC-RL) model advances the discourse by embedding oversight, redress, and compliance functions directly into AI logic. Its formulation of accountability-aware decision and redress functions introduces the concept of "adaptive responsibility" a dynamic correction mechanism allowing AI systems to self-identify errors and initiate compensatory adjustments. Through this, accountability evolves from static auditing into a living process of continuous ethical calibration.

In the final analytical dimension, the Fairness-Constrained Federated Learning (FC-FL) framework secures the ethical integrity of data ecosystems that underpin AI-driven urban services. It demonstrates that privacy and equity can coexist through decentralized training and fairness regularization, ensuring that learning processes neither exploit data nor amplify societal inequities. This federated paradigm restores data sovereignty to communities and institutions, enabling collaborative intelligence without central surveillance—a foundational requirement for digital democracy and trust-based innovation in smart cities.

From a holistic vantage, the synthesis of these five frameworks yields a self-regulating ethical infrastructure for AI governance—an architecture that operationalizes moral philosophy through formal mathematics and computational logic. The chapter transcends the dichotomy between technical optimization and ethical compliance, showing that the two can converge into a unified paradigm of moral computation. By introducing fairness constraints, sustainability metrics, explainability functions, accountability penalties, and federated privacy safeguards, it establishes a reproducible system of governance that is at once scientifically rigorous, normatively grounded, and practically deployable.

In reflective depth, what this chapter demonstrates is that the future of AI in urban management depends not merely on algorithmic intelligence but on institutionalized ethics within those algorithms. The frameworks proposed herein provide a technical lingua franca for engineers, policymakers, and industry partners to co-design intelligent infrastructures that are ethically interpretable and operationally verifiable. Each model can be integrated as a Software-as-a-Service (SaaS) component within broader civic ecosystems interoperable, auditable, and scalable across energy, water, mobility, and public administration.

From a philosophical synthesis, the chapter stands as an affirmation that intelligence, when detached from justice, leads to imbalance; yet when constrained by ethical design, it becomes a generative force for social equity and planetary responsibility. The interplay of reinforcement learning, multi-agent coordination, and federated optimization presented here signals the emergence of an AI governance paradigm in which technological advancement is inseparable from moral intentionality.

In final analysis, this chapter contributes not only to the scientific corpus of AI ethics but also to the civic imagination of sustainable urban futures. It articulates a vision where ethical computation forms the backbone of governance where transparency ensures trust, fairness secures inclusion, accountability enforces integrity, and privacy preserves autonomy. In doing so, it offers the academic community, industry innovators, and global policymakers a transformative framework: a codified, implementable ethics of intelligence designed to guide cities toward resilience, justice, and human-centered progress in the algorithmic age.

References

Attard-Frost, B., De los Rios, A., & Walters, D. R. (2023). The ethics of AI business practices: A review of 47 AI ethics guidelines. *AI and Ethics*, 3, 389–406. https://doi.org/10.1007/s43681-022-00156-6

Barredo Arrieta, A., Díaz-Rodríguez, N., Del Ser, J., Bennetot, A., Tabik, S., Barbado, A., Garcia, S., Gil-Lopez, S., Molina, D., & Herrera, F. (2020). Explainable artificial intelligence (XAI): Concepts, taxonomies, opportunities, and challenges toward responsible AI. *Information Fusion*, 58, 82–115. https://doi.org/10.1016/j.inffus.2019.12.012

Sargiotis, D. (2024). Ethical AI in Information Technology: Navigating Bias, Privacy, Transparency, and Accountability. *Advances in Machine Learning & Artificial Intelligence*, 5(3), 01–14. https://doi.org/10.33140/AMLAI.05.03.03

Trotta, A., Ziosi, M., & Lomonaco, V. (2023). The future of ethics in AI: Challenges and opportunities. *AI & Society*, 38, 439–441. https://doi.org/10.1007/s00146-023-01644-x

Chapter 4
AI-Driven Privacy, Fairness, and Transparency Frameworks for Urban Data Governance

The chapter "AI-Driven Privacy, Fairness, and Transparency Frameworks for Urban Data Governance" opens a profound exploration into how Artificial Intelligence (AI) can transform city governance without compromising citizens' digital rights. It argues that future urban intelligence will not be measured merely by computational power or data abundance, but by how ethically, securely, and equitably AI systems manage the data lifeblood of cities. The narrative begins with a fundamental tension: the more interconnected and sensor-rich a city becomes, the more its infrastructure relies on the continuous harvesting of personal, behavioral, and environmental data. Within this tension lies the ethical mandate to develop AI systems that preserve privacy, ensure fairness, guarantee accountability, and maintain transparency while still enabling innovation and predictive capability.

To resolve this dilemma, the chapter introduces an AI-architectural ecosystem built upon mathematically formalized governance frameworks. The first framework, the Federated Learning-Based Privacy-Preserving AI Model (FL-PPAM), decentralizes learning across edge devices, ensuring that sensitive citizen data remain local. Through Federated Learning (FL), models are trained collaboratively without centralizing raw data, while Differential Privacy (DP) introduces Gaussian noise to prevent re-identification. This system is reinforced by blockchain validation, guaranteeing the immutability and verifiability of all data interactions, creating a cryptographically trusted foundation for smart-city governance.

The discussion then advances to the Consent-Aware Federated Learning framework (CA-FL), which operationalizes the ethical principle of "privacy by design." Here, user consent becomes a mathematical parameter within the federated loss function. Citizens can opt out, and their data are automatically excluded from training while the model retains its predictive integrity through dynamic re-weighting and data utility control. This mechanism transforms legal consent into a live computational process—an active contract embedded into the AI model itself.

Building upon these privacy layers, the chapter introduces the AI-driven anonymization framework for urban governance, where anonymization is achieved

D. Sargiotis, *AI with Integrity*, https://doi.org/10.1007/978-3-032-31133-7_4

through a synergy of differential privacy, Generative Adversarial Networks (GANs), k-Anonymity, and t-Closeness. Differential privacy ensures individual indistinguishability, GANs synthesize statistically accurate but artificial datasets, k-Anonymity groups similar records to prevent identification, and t-Closeness maintains uniformity in sensitive attribute distributions. The result is an irreversible anonymization pipeline that preserves analytic value while neutralizing identity risks, transforming raw data into privacy-safe digital capital for urban analytics.

To secure accountability within this vast data ecosystem, the AI-Driven Accountability Framework (AI-AG) is presented. It combines blockchain-based audit trails for tamper-proof data histories, anomaly detection models for recognizing unauthorized access, and Reinforcement Learning (RL) algorithms for adaptive policy enforcement and penalty optimization. This tri-layered model ensures that every data transaction is traceable, every misuse detectable, and every violation penalized with proportional precision embedding ethical compliance directly into the system's operational core.

Equity and justice emerge as the next thematic axis. The Fairness-Enhanced AI for Urban Governance (FEA-UG) introduces fairness constraints into the AI loss functions, reducing demographic disparities across predictions. Adversarial debiasing networks further detect and mitigate hidden biases, while Explainable Artificial Intelligence (XAI) using Shapley Values provides interpretability of decision factors, guaranteeing that each feature's contribution is fair and transparent. This ensures that urban AI does not reinforce structural inequalities but actively corrects them, supporting inclusive development across all social strata.

Finally, the chapter culminates in the Transparency-Enhanced AI Framework (TEA-UG), where transparency becomes a quantifiable, auditable, and reportable attribute of AI systems. Using information-theoretic transparency scores, derived from the mutual information between model outputs and their public explanations, the framework measures how comprehensible AI decisions are to citizens. Traceable decision logs document every transformation in the inference process, while automated transparency reports aggregate interpretability metrics and publish them for civic oversight. Transparency thus transitions from an aspirational ideal to an operational metric of trustworthiness.

In synthesis, the chapter constructs a holistic architecture of ethical AI governance spanning privacy protection, consent enforcement, data anonymization, accountability assurance, fairness optimization, and transparency quantification. Each framework is mathematically rigorous, legally aware, and operationally deployable within the context of smart-city infrastructures. Together, they redefine urban intelligence as a civic institution where algorithms become accountable citizens, data systems uphold human dignity, and technological progress aligns with democratic governance.

This work invites policymakers, AI researchers, and urban planners to view the city not as a data mine, but as a living ethical organism whose digital metabolism must respect privacy, equity, and public trust. It promises a future where Artificial Intelligence, empowered by Federated Learning (FL), Differential Privacy (DP), Generative Adversarial Networks (GANs), Reinforcement Learning (RL), and

Explainable Artificial Intelligence (XAI), evolves into a transparent and just instrument of governance positioning the smart city as both technological marvel and moral achievement.

4.1 AI-Driven Frameworks for Privacy, Fairness, Accountability, and Transparency in Urban Data Governance

Privacy and data governance are crucial ethical considerations in the deployment of AI systems in urban environments. Urban AI systems, which manage everything from traffic control to energy distribution, rely on vast amounts of data, often collected from citizens. This raises significant privacy concerns, especially when sensitive personal data is involved. The integration of AI into urban infrastructures requires a comprehensive approach to data governance that balances innovation with the protection of citizens' rights (Sargiotis 2024a, b).

One of the primary challenges in preserving privacy within urban AI systems is the sheer volume and variety of data collected. AI systems used in smart cities gather data from multiple sources, including sensors, surveillance cameras, mobile devices, and social media platforms (Barredo Arrieta et al. 2020). Such data may include highly sensitive information about individuals' movements, habits, and interactions. Ensuring that this information is collected and processed in ways that respect privacy requires robust governance frameworks with clear rules for data collection, storage, access, and use. To address these requirements, a federated learning-based privacy-preserving AI model is formulated, enabling decentralized urban data governance while maintaining predictive accuracy. Federated learning (FL) allows AI systems to train decentralized models without directly accessing raw data, enhancing privacy by processing data locally on edge devices rather than transmitting it to centralized repositories. The AI-driven model incorporating federated learning for privacy-preserving urban data governance is formulated as:

$$\mathrm{w}_{t+1} = \mathrm{w}_t - \eta \sum_{i=1}^{N} \frac{n_i}{n} \nabla F_i(\mathrm{w}_t)$$

where w_{t+1}, updated global model parameters, w_t, previous model parameters, η, learning rate controlling model update steps, N, number of edge devices participating in federated learning, n_i, number of data points at device i, n, total data points across all devices, and $\nabla F_i(\mathrm{w}_t)$, local gradient computed at device i.

This federated learning model ensures privacy by allowing local AI agents to train models on distributed data sources without centralizing sensitive information. The model aggregates gradients rather than raw data, minimizing exposure while retaining predictive capabilities. However, federated learning systems are susceptible to data heterogeneity and adversarial attacks. To counter these vulnerabilities, a differential privacy-enhanced model update mechanism is incorporated:

$$\mathrm{w}_{t+1} = \mathrm{w}_t - \eta \sum_{i=1}^{N} \frac{n_i}{n} \left(\nabla F_i(\mathrm{w}_t) + \mathcal{N}(0, \sigma^2) \right)$$

where $\mathcal{N}(0, \sigma^2)$, Gaussian noise ensuring differential privacy, σ^2, noise variance controlling privacy-preserving noise strength.

By introducing noise to the model updates, this mechanism preserves privacy while maintaining urban AI performance. It ensures that no individual data point significantly influences the global model, mitigating re-identification risks. However, differential privacy alone does not fully secure decentralized systems against adversarial threats. To reinforce security, a blockchain-based data integrity validation mechanism is integrated:

$$H_t = H\left(H_{t-1} \middle\| D_t\right)$$

where H_t, cryptographic hash of the current data block, H_{t-1}, hash of the previous block, D_t, encrypted urban AI data at time t, $H(\cdot)$, cryptographic hash function ensuring tamper resistance.

This blockchain-enhanced governance model ensures that all AI-driven urban decisions remain verifiable and immutable, preventing unauthorized modifications. The federated learning framework combined with differential privacy and blockchain mechanisms establishes a privacy-aware AI model for smart-city governance, ensuring that AI-driven decisions respect citizen data rights while optimizing urban services.

By harnessing federated learning, differential privacy, and blockchain-based security, the proposed AI framework safeguards urban AI systems against data exploitation, regulatory non-compliance, and cyber threats. It fosters ethical AI deployment in smart cities by ensuring that data remains decentralized, obfuscated, and cryptographically secured. This comprehensive approach advances AI-driven governance by balancing urban innovation with robust privacy protection.

Framework: Federated Learning-Based Privacy-Preserving AI Model for Urban Governance (FL-PPAM)

Layered AI Architecture for Privacy Protection, Decentralized Learning, and Secure Urban Data Governance

Layer	Main functionality	AI module (s) andcomponents	Operational scope (Input → Processing → Output)
Layer 1	Data collection and federated learning constraints: Collecting data without violating privacy, enabling decentralized learning	Edge data ingestion and local preprocessing engine	Input → Sensor, surveillance, mobile, and social media data → Output → Localized model updates preserving privacy

(continued)

Layer	Main functionality	AI module (s) andcomponents	Operational scope (Input → Processing → Output)
Layer 2	Federated optimization: Aggregating model gradients from decentralized devices to update the global model	Federated learning	Input → Local model parameters from edge devices → Output → Aggregated global model updates
Layer 3	Differential privacy enhancement: Introducing noise to preserve privacy and mitigate data re-identification risks	Differential privacy (Gaussian Noise)	Input → Local model gradients → Output → Privacy-preserving gradients with noise, global model updates
Layer 4	Blockchain security: Ensuring tamper-proof model updates and verifiable decision-making	Blockchain	Input → Local encrypted model data → Output → Secured, cryptographically validated model updates
Layer 5	Post-AI monitoring: Monitoring for privacy compliance and performance verification after deployment	Model governance and compliance monitoring system	Input → AI-driven urban decisions → Output → Privacy audits, transparency reports, and model verifications

To address privacy concerns, data governance frameworks for urban AI systems must ensure that data collection is lawful, transparent, and accountable. Citizens should be informed about what data is collected, how it will be used, and who will have access to it (Attard-Frost et al. 2023). Consent mechanisms are therefore a critical component of urban data governance, ensuring that individuals retain the right to refuse or withdraw participation in data collection activities. Moreover, cities should implement privacy-by-design principles, embedding privacy protections into AI systems from the outset rather than treating them as an afterthought. To operationalize lawful, transparent, and consent-respecting data governance in urban AI systems, a Consent-Aware Federated Learning (CA-FL) model is proposed. This framework integrates user consent mechanisms, differential privacy, and data utility control functions at the architectural level, ensuring that privacy protection is embedded from the outset as a core element of the privacy-by-design principle.

At the heart of the model lies a dynamically weighted federated loss function that incorporates a binary consent mask for each user's data point:

$$\mathcal{L}_{\mathrm{CA}}(\theta) = \sum_{k=1}^{K} \frac{1}{n_k} \sum_{i=1}^{n_k} c_{k,i} \cdot \ell\left(f_\theta(x_{k,i}), y_{k,i}\right)$$

where θ are the global model parameters, $x_{k,\ i}$ and $y_{k,\ i}$ are the input-output pairs for data point i in node k, $c_{k,\ i} \in \{0, 1\}$ is the consent mask indicating whether the individual has opted in (1) or opted out (0), $\ell\ (\cdot)$ is the loss function, and K is the number of decentralized urban nodes. This function ensures that only data from consenting individuals influences the model's training, respecting the legal right to opt out while preserving model integrity.

To satisfy legal transparency and minimize re-identification risks, the system incorporates a differential privacy mechanism into the model updates before aggregation:

$$\tilde{g}_k^{(t)} = g_k^{(t)} + \mathcal{N}(0, \sigma^2 I)$$

where $g_k^{(t)}$ is the raw gradient update from node k at time t, $\tilde{g}_k^{(t)}$ is the noise-perturbed gradient, and $\mathcal{N}(0, \sigma^2 I)$ is Gaussian noise calibrated to the privacy budget ε, ensuring differential privacy. This formulation ensures that no single citizen's behavior can be inferred from the aggregate model update, even in worst-case inference attacks, thus fulfilling GDPR-style legal protections for sensitive urban data.

Additionally, the framework incorporates a data utility control function to prevent performance degradation when high levels of consent withdrawal occur:

$$U(\theta) = \mathbb{E}_{x \sim \mathcal{D}_{\text{consented}}} \left[\mathbb{1}_{\ell(f_\theta(x), y) < \delta} \right]$$

where $\mathcal{D}_{\text{consented}}$ is the subset of the dataset with valid consent, 1 is the indicator function, and δ is an accuracy threshold. This function monitors the model's expected performance given only the data that has been ethically and legally authorized for use.

Together, this consent-aware federated model fulfills three core objectives of ethical data governance in smart cities: (1) it limits data use strictly to authorized, informed individuals; (2) it preserves privacy through formal guarantees; and (3) it monitors utility to ensure the system remains operational under real-world consent dynamics. By embedding these mechanisms directly into the architecture, urban AI systems evolve from reactive data custodians to proactive protectors of civic trust and digital rights.

Framework: Consent-Aware Federated Learning for Urban Data Governance (CA-FL)

Layered AI Architecture for Privacy Protection, Consent Compliance, and Ethical Data Governance in Smart Cities

Layer	Main functionality	AI module (s) and components	Operational scope (Input → Processing → Output)
Layer 1	Data collection and consent mechanisms: Collecting data with explicit user consent	Consent management and data access control engine	Input → Urban data (sensor, mobile, etc.), consent information → Output → Consent-aware data for training
Layer 2	Federated learning with consent-aware loss: Training global model with user consent data	Federated learning (with consent-aware loss)	Input → Local data from nodes, user consent masks → Output → Federated global model updates considering consent

(continued)

Layer	Main functionality	AI module (s) and components	Operational scope (Input → Processing → Output)
Layer 3	Differential privacy: Adding noise to gradients for privacy protection	Differential privacy (Gaussian Noise)	Input → Raw gradients from local nodes → Output → Noised gradients for privacy-preserving aggregation
Layer 4	Data utility control: Monitoring performance impact of consent withdrawal	Data utility control (indicator function)	Input → Data with valid consent, model loss → Output → Utility score for data quality and performance
Layer 5	Post-AI monitoring and compliance: Auditing model updates for privacy and legal compliance	Compliance auditing and regulatory reporting system	Input → AI-driven model decisions → Output → Privacy audits, transparency reports, and compliance checks

Data anonymization is another key strategy for protecting privacy in urban AI systems. By removing or masking personally identifiable information, cities can still leverage AI to optimize services without unnecessarily exposing individual citizens to privacy risks (Trotta et al. 2023). Yet anonymized data may still be vulnerable to re-identification, especially when combined or cross-referenced with auxiliary datasets. This risk is particularly relevant in urban environments, where mobility traces, sensor records, service-use patterns, and demographic indicators may reveal sensitive information even after direct identifiers have been removed. Robust privacy-preserving techniques are therefore required to reduce re-identification risk while preserving the analytical utility of urban datasets.

One of the primary methods for strengthening privacy protection is differential privacy, which limits the extent to which the inclusion or exclusion of any single data point can affect the output of an analysis. This is achieved by introducing calibrated privacy-preserving noise into AI models or statistical queries, thereby reducing the ability of adversaries to link anonymized records to specific individuals. The AI-driven differential privacy model for urban data anonymization is formulated as:

$$\tilde{X} = X + \mathcal{N}(0, \sigma^2)$$

where $\tilde{X}$, differentially private dataset, X, original dataset, $\mathcal{N}(0, \sigma^2)$, Gaussian noise with mean zero and variance σ^2 ensuring privacy-preserving obfuscation.

This model prevents re-identification risks by adding stochastic noise to sensitive data attributes, thereby making it statistically indistinguishable from the original dataset while preserving essential analytical properties. However, differential privacy alone does not fully address correlation risks when anonymized data is cross-referenced with external datasets. To enhance privacy, a generative adversarial network (GAN)-based synthetic data generation model is introduced, producing artificially generated urban datasets that maintain statistical properties without revealing real individual records. The GAN-driven anonymization model is formulated as:

$$\min_{G} \max_{D} \mathbb{E}_{X \sim P_{\text{real}}}[\log D(X)] + \mathbb{E}_{Z \sim P_Z}[\log(1 - D(G(Z)))]$$

where $G(Z)$, generator function synthesizing artificial data samples, $D(X)$, discriminator function distinguishing real data from synthetic data, P_{real}, probability distribution of real data, P_Z, prior distribution of latent variables, and $\mathbb{E}\ [\cdot]$, expectation operator ensuring adversarial training.

The GAN framework ensures that synthetic urban datasets retain statistical fidelity to real-world data while eliminating re-identification risks, thereby providing a robust anonymization mechanism for AI-driven urban governance. However, synthetic data generation does not inherently guarantee compliance with strict data protection regulations. To further reinforce anonymization security, a k-anonymity mechanism is applied to ensure that each data entry is indistinguishable from at least kk other records, preventing unique identification:

$$k = \min_{\forall i} \mid \{x_j \in X : x_j \sim x_i\} \mid$$

where k, anonymity threshold ensuring each individual record is indistinguishable from at least k other records, x_i, data entry, X, anonymized dataset, $\sim$, similarity function defining indistinguishable records.

By ensuring that each record is grouped with at least kk others, k-anonymity minimizes the risk of identity linkage even when adversaries possess external information. Nonetheless, k-anonymity alone does not prevent inference attacks where adversaries deduce sensitive attributes based on correlations. To mitigate this, a t-closeness constraint is implemented, ensuring that the distribution of sensitive attributes within each group remains within a predefined threshold:

$$D(P_{\text{group}}, P_{\text{global}}) \leq t$$

where $D\ (\cdot,\cdot)$, statistical distance function measuring the divergence between group and global distributions, P_{group}, probability distribution of sensitive attributes within a data group, P_{global}, overall dataset distribution, t, predefined privacy threshold ensuring attribute indistinguishability.

By enforcing t-closeness, this mechanism prevents attribute inference by maintaining the statistical uniformity of sensitive information across different data groups. The integration of differential privacy, GAN-based synthetic data, k-anonymity, and t-closeness constraints establishes a multi-layered AI-driven anonymization framework that ensures urban datasets remain irreversibly anonymized while retaining their analytical utility.

This AI-driven anonymization framework enables cities to leverage AI for urban optimization without compromising individual privacy. Differential privacy ensures dataset obfuscation, GANs generate realistic yet privacy-preserving synthetic data, k-anonymity protects against identity linkage, and t-closeness prevents inference attacks. By combining these advanced techniques, AI-driven urban governance achieves a balance between data utility and privacy protection, mitigating re-identification risks while ensuring compliance with regulatory standards.

Framework: AI-Driven Anonymization for Urban Governance (AIDAG)

Layered AI Architecture for Data Anonymization, Privacy Protection, and Secure Urban Data Governance

Layer	Main functionality	AI module (s) and components	Operational scope (Input → Processing → Output)
Layer 1	Data collection and anonymization: Collecting data and applying privacy-preserving techniques	Differential privacy	Input → Raw urban data (sensor, social media, etc.) → Output → Privacy-preserving data
Layer 2	GAN-based synthetic data generation: Generating synthetic data that mimics real data	Generative Adversarial Network (GAN)	Input → Real urban data distribution, latent variable prior → Output → Synthetic urban datasets
Layer 3	k-anonymity: Ensuring that data entries are indistinguishable from other records to protect identity	k-Anonymity	Input → Anonymized dataset → Output → Data groups that protect individual identity
Layer 4	t-Closeness: Ensuring statistical uniformity across data groups to prevent attribute inference	t-Closeness	Input → Data groups with sensitive attributes → Output → Protected distributions with consistent statistics
Layer 5	Post-AI auditing: Verifying compliance with privacy and data protection regulations	Privacy compliance and data protection audit platform	Input → AI-generated decisions and anonymized data → Output → Privacy audits and regulatory reports

The ethical use of data in AI-driven urban systems also requires accountability mechanisms to ensure that data is not misused or accessed without authorization. Strong governance structures should include oversight bodies tasked with monitoring data usage and ensuring compliance with privacy laws and regulations (Attard-Frost et al. 2023). These bodies must also be empowered to enforce penalties for violations and ensure that citizens have recourse if their data privacy is breached. A critical AI-driven approach to ensuring accountability is the blockchain-based audit trail system, which provides transparent, immutable records of all data interactions within urban AI systems. Blockchain technology enables secure, decentralized recording of every action taken on data, offering an auditable trail that ensures all interactions are tracked, and unauthorized access is detectable. The AI-driven blockchain-based accountability model is formulated as:

$$H_t = H\left(H_{t-1} \middle\| D_t \middle\| A_t\right)$$

where H_t, cryptographic hash of the current data block, H_{t-1}, hash of the previous block, D_t, data associated with a specific AI-driven transaction, A_t, access record or metadata associated with the data transaction, $H(\cdot)$, cryptographic hash function ensuring tamper resistance.

This blockchain model ensures that each data access event, whether it be the use of a citizen's data for traffic optimization, the monitoring of energy consumption, or resource allocation decisions, can be traced back through an unalterable, time-stamped record. This transparency minimizes the risk of unauthorized data access and enhances accountability, as any attempt to modify or falsify data logs would be immediately detectable by the system.

To further reinforce ethical data governance, the introduction of AI-based anomaly detection systems allows for the identification of suspicious activities or deviations from expected data usage patterns. These systems leverage machine learning techniques to analyze historical data access behaviors and flag anomalies that could indicate potential misuse or unauthorized data retrieval. The AI-driven anomaly detection model is formulated as:

$$P(x_t|\theta) = \frac{1}{\sigma\sqrt{2\pi}} \exp\left(-\frac{(x_t - \mu)^2}{2\sigma^2}\right)$$

where $P(x_t|\theta)$, probability density function of anomalous access behavior at time t, x_t, observed data access pattern at time t, μ, mean of historical data access behavior, σ, standard deviation representing typical deviations in data access, θ, model parameters optimizing anomaly detection.

The model allows AI to predict normal access behavior and flag any deviations as potential violations. This enables the system to identify patterns consistent with unauthorized access, such as unusual data access spikes or accesses outside of authorized time windows, and alert appropriate oversight bodies.

Moreover, reinforcement learning (RL)-based policy enforcement can be integrated to automatically adjust and enforce penalties when violations of data usage rules occur. RL algorithms can optimize decision-making to ensure that violators are penalized proportionally to the severity of their breaches, ensuring that deterrents are in place to prevent unauthorized access. The RL-based policy enforcement model is formulated as:

$$R(s_t, a_t) = r_t + \gamma \max_{a_{t+1}} Q(s_{t+1}, a_{t+1})$$

where $R(s_t, a_t)$, reward function of state s_t and action α_t, r_t, immediate reward for taking action a_t in state s_t, γ, discount factor controlling the importance of future rewards, $Q(s_{t+1}, a_{t+1})$, Q-value function predicting future rewards based on state s_{t+1} and action a_{t+1}.

The RL model enables the system to dynamically adjust penalties or rewards based on the context and nature of data usage violations, reinforcing the ethical boundaries of AI-driven urban systems and ensuring that governance mechanisms remain responsive to emerging risks. By incorporating reinforcement learning, oversight bodies are empowered to implement adaptive enforcement strategies that are responsive to the evolving landscape of data privacy concerns.

Together, blockchain, anomaly detection, and reinforcement learning form a comprehensive AI-driven accountability model for urban systems. The blockchain framework ensures transparency and tamper-proof records, while anomaly detection systems identify unauthorized data access behaviors, and reinforcement learning enforces penalties dynamically. This integrated framework establishes a robust system for ensuring that data in AI-driven urban environments is used ethically, responsibly, and in compliance with privacy laws, creating accountability structures that promote trust and fairness in smart cities.

Framework: AI-Driven Accountability for Urban Governance (AI-AG)

Layered AI Architecture for Data Accountability, Privacy Protection, and Ethical Oversight in Smart Cities

Layer	Main functionality	AI module (s) and components	Operational scope (Input → Processing → Output)
Layer 1	Data collection and governance framework: Collecting data with transparency and compliance to privacy laws	Transparent data governance and access control engine	Input → Data from sensors, social media, and citizens → Output → Data with privacy protections and consent mechanisms
Layer 2	Blockchain-based audit trail: Ensuring transparent and tamper-proof records of data interactions	Blockchain	Input → Data and access records → Output → Immutable, time-stamped blockchain audit trail
Layer 3	AI-driven anomaly detection: Identifying suspicious activities or deviations from normal data usage patterns	Anomaly detection (Gaussian distribution)	Input → Historical data access behavior → Output → Flagged anomalies for potential violations
Layer 4	Reinforcement learning for policy enforcement: Optimizing and enforcing penalties for data usage violations	Reinforcement learning (Q-learning)	Input → Current state of data usage, penalty rules → Output → Dynamic adjustment of penalties based on violation severity
Layer 5	Accountability audits and legal compliance monitoring: Ensuring legal compliance and maintaining public trust through audits	Legal compliance and accountability audit platform	Input → Data usage and access records → Output → Compliance reports, transparency audits, legal evaluations

Another critical aspect of data governance in urban AI systems is the equitable use of data. AI systems often rely on large datasets to make decisions about resource allocation, public safety, and urban planning. If these datasets are biased or incomplete, they can lead to unfair outcomes that disproportionately affect marginalized communities (Barredo Arrieta et al. 2020). Ethical AI frameworks must therefore require representative data practices and regular audits of model fairness and accuracy. To address these challenges, a fairness-enhanced AI model can be deployed to reduce bias in decision-making processes. One effective approach is the use of fairness constraints within the optimization functions of AI models. These constraints prevent AI-driven decisions from disproportionately harming specific groups, particularly marginalized populations. The fairness-augmented model is formulated as:

$$\min_{\mathrm{w}} L(\mathrm{w}) + \lambda \sum_{i=1}^{n} [\mathrm{fairness\_loss}(y_i, \widehat{y}_i)]$$

where $L(\mathrm{w})$, loss function of the AI model, w, model parameters, λ, regularization parameter balancing fairness and accuracy, n, number of data points, y_i, actual value for the i-th data point, $\widehat{y}_i$, predicted value for the $i-th$ data point, $\mathrm{fairness\_loss}(y_i, \widehat{y}_i)$, fairness loss function quantifying the disparity between predicted and actual outcomes with respect to demographic attributes.

The fairness loss function minimizes discrepancies in the model's predictions across different demographic groups, ensuring that AI decisions do not inadvertently favor one group over another. By incorporating fairness constraints into the optimization process, AI systems become more equitable in their outcomes, offering a safeguard against biases that could negatively impact marginalized communities.

However, fairness in AI systems extends beyond model design to include regular auditing and validation processes to assess the fairness and accuracy of predictions. One method for conducting these audits is through the use of adversarial debiasing techniques, which identify and mitigate bias by introducing adversarial networks that attempt to find and correct biases in the model's predictions. The adversarial debiasing model is formulated as:

$$\min_{G} \max_{D} \mathbb{E}_{x \sim P_{\mathrm{data}}}[\log D(G(x))] + \mathbb{E}_{x \sim P_{\mathrm{data}}}[\log(1 - D(x))]$$

where $G(x)$, generator function modifying the model's predictions to reduce bias, $D(x)$, discriminator function attempting to detect biased predictions, P_{data}, data distribution representing input features, $\mathbb{E}\,[\cdot]$, expectation operator optimizing the adversarial training process.

This adversarial debiasing method utilizes a generator to propose modifications to the model's output in such a way that the discriminator cannot distinguish between biased and unbiased results. The result is a model that is actively trained to minimize discrimination in its predictions, further reinforcing fairness in urban AI decision-making processes.

In addition to fairness constraints and adversarial debiasing, the integration of explainable AI (XAI) models is essential for ensuring the transparency and interpretability of AI systems in urban governance. XAI techniques allow stakeholders to understand the rationale behind AI-driven decisions, providing insights into how the model uses data to reach conclusions. This transparency is critical for verifying that the AI system is functioning equitably. One approach to explaining AI decisions is through Shapley values, which provide a measure of each feature's contribution to the final decision. The Shapley value-based explanation model is formulated as:

$$\phi_i(f) = \sum_{S \subseteq N \setminus \{i\}} \frac{|S|!(|N| - |S| - 1)!}{|N|!} [f(S \cup \{i\}) - f(S)]$$

where $\phi_i(f)$, Shapley value for feature i, $f(S)$, function output for a subset S of features, N, total set of features, $|S|$, number of elements in subset S, $|N|$, total number of features.

The Shapley value provides a fair attribute to how much each feature contributes to a particular decision, ensuring that all features, particularly those related to demographic factors, are treated impartially. By incorporating Shapley values into urban AI systems, stakeholders can verify that the model is making decisions based on fair and representative data, helping to ensure that no group is unfairly disadvantaged.

Finally, combining fairness constraints, adversarial debiasing, and explainable AI provides a comprehensive AI-driven fairness and equity model for urban governance. This model ensures that AI systems making critical decisions about resource allocation, urban planning, and public safety are built on unbiased, representative data and regularly audited for fairness. The integration of fairness-enhancing techniques with explainable AI promotes transparency and accountability, helping to ensure that urban AI systems serve all communities equitably, particularly marginalized and historically underserved groups.

This AI-driven fairness framework fosters equitable outcomes in AI-based urban systems by addressing data biases, improving model transparency, and ensuring regular fairness audits. By embedding fairness principles within the AI model's core structure, this approach promotes justice in the decision-making processes that shape urban environments, empowering policymakers to ensure that all citizens benefit equitably from AI-driven services.

Framework: Fairness-Enhanced AI for Urban Governance (FEA-UG)

Layered AI Architecture for Fair and Equitable Urban Data Governance and Decision-Making

Layer	Main functionality	AI module (s) and components	Operational scope (Input → Processing → Output)
Layer 1	Data collection and fairness constraints: Ensuring that data used in decision-making is representative and unbiased	Bias detection and demographic data balancing engine	Input → Urban data with demographic representation → Output → Fairly balanced data for training AI models

(continued)

Layer	Main functionality	AI module (s) and components	Operational scope (Input → Processing → Output)
Layer 2	Fairness-augmented loss function: Incorporating fairness constraints in the optimization of AI models	Fairness constraints	Input → Model predictions and actual values → Output → Fairness-enhanced model parameters
Layer 3	Adversarial debiasing: Mitigating bias by training the model to produce fair predictions	Adversarial networks (GANs)	Input → Biased predictions → Output → Debiased predictions using adversarial training
Layer 4	Shapley value-based explanation: Providing interpretable, fair attributions for AI-driven decisions	Shapley Values (XAI)	Input → AI model decisions → Output → Transparent feature attributions for decision explanation
Layer 5	Post-AI fairness audits: Auditing the AI model's performance and fairness in real-world applications	Fairness compliance and model audit platform	Input → AI-driven decisions, model outputs → Output → Fairness audit reports and accuracy validation

Moreover, transparency in data governance is essential for building public trust in urban AI systems. Citizens should have access to information about how their data is used and by whom. This can be achieved through open data initiatives, where anonymized datasets are made available for public scrutiny, or through transparency reports that document the operation of AI systems in cities (Trotta et al. 2023). Transparent governance mechanisms not only enhance accountability but also help demystify AI technologies, making them more accessible and understandable to the public. To translate transparent data governance into measurable and enforceable AI practice, an auditable AI architecture integrated with information-theoretic transparency metrics is proposed. This model provides mathematically structured accountability for how decisions are made and how data is used, facilitating public scrutiny through open data portals and automated transparency reporting systems.

At the core of this system is a traceable decision function, where each AI output is accompanied by an interpretable log of data dependencies and decision layers:

$$T(x) = \{f_1(x), f_2(f_1(x)), \ldots, f_n(\cdot)\}$$

where $T(x)$ is the trace of the decision path for input x, and each f_i represents a transformation or model layer contributing to the final output. This traceable structure forms the basis for auditable logs that are automatically generated and stored during model inference.

To evaluate transparency quantitatively, the model includes an information-theoretic transparency score based on mutual information between the AI decision output y and the explanation ee provided to the public:

$$\mathcal{T}(x) = \frac{I(y; e \mid x)}{H(y \mid x)}$$

where $\mathcal{T}(x)$ is the transparency score for input x, $I(y; e \mid x)$ is the mutual information between the model's decision y and its explanation e, conditioned on the input x, and $H(y \mid x)$ is the conditional entropy of the model's output given the input. This metric quantifies how much of the model's output can be understood through its explanation. A transparency score of 1 means the explanation perfectly reveals the reasoning behind the decision, while lower scores indicate opaque or partial disclosures.

To support civic oversight, the system generates automated transparency reports summarizing system activity across time. The report aggregates trace records and transparency scores over batches of urban AI decisions:

$$\mathcal{R}_{\text{public}} = \{\text{mean}(\mathcal{T}(x_i)), \quad \log(T(x_i))\}_{i=1}^{N}$$

where $\mathcal{R}_{\text{public}}$ is the public-facing report containing the average transparency score across N decisions and the set of logged decision traces. This enables citizens and researchers to audit system behavior, identify biases or drift, and hold public AI systems accountable.

Using these functions into the core of urban AI systems, transparency is no longer an abstract aspiration but a computable, reportable, and auditable system feature. Traceable decisions foster internal accountability, mutual information scores quantify public interpretability, and open reports ensure civic oversight. These mechanisms together transform transparency from policy language into operational infrastructure strengthening public trust in AI as a democratic tool for equitable city governance.

Framework: Transparency-Enhanced AI for Urban Governance (TEA-UG)

Layered AI Architecture for Transparent, Accountable, and Ethical Urban Data Governance

Layer	Main functionality	AI module (s) and components	Operational scope (Input → Processing → Output)
Layer 1	Data collection and transparency governance: Collecting and processing data with full transparency and public access	Transparent data processing and public access management engine	Input → Urban data from multiple sources → Output → Transparent data logs for public scrutiny
Layer 2	Traceable decision function: Generating an interpretable log of data dependencies and decision layers	Traceable decision logs	Input → AI model output → Output → Traceable decision paths with data dependencies

(continued)

Layer	Main functionality	AI module (s) and components	Operational scope (Input → Processing → Output)
Layer 3	Information-theoretic transparency score: Quantifying the transparency of AI decisions using mutual information	Information-theoretic transparency	Input → Model output and explanation → Output → Transparency score reflecting model interpretability
Layer 4	Automated transparency reports: Summarizing system activity and transparency metrics for public oversight	Automated reporting system	Input → Trace logs and transparency scores → Output → Public-facing transparency reports
Layer 5	Post-AI transparency auditing: Auditing system behavior to ensure ongoing transparency and accountability	Transparency compliance and continuous audit platform	Input → AI-driven decisions, transparency reports → Output → Audited results and compliance reports

Privacy and data governance are central to the ethical deployment of AI systems in urban environments. Ensuring that these systems respect citizens' privacy, operate transparently, and use data responsibly is critical for building sustainable and equitable smart cities. By embedding privacy protections and governance structures into AI systems from the outset, cities can harness the power of AI to improve urban living while safeguarding individual rights.

Figure 4.1 illustrates the key aspects of privacy and data governance in urban AI systems, structured around five main areas: privacy protections, data anonymization, accountability, equity in data use, and transparency. "Privacy Protections" include informing citizens about data use and allowing opt-out options. "Data Anonymization" focuses on removing personal details and preventing re-identification. "Accountability" involves oversight, compliance, and penalties for misuse. "Equity in Data Use" highlights the need for fair, unbiased data, and regular fairness checks. "Transparency" emphasizes open access to anonymized data and public information on AI operations, fostering trust in AI deployment within urban environments.

Incorporating AI into urban development comes with the promise of increased efficiency, sustainability, and innovation. However, ethical considerations surrounding fairness, transparency, accountability, and privacy are paramount to ensure that AI benefits all citizens equitably and does not exacerbate existing social inequalities. As cities continue to integrate AI into their development strategies, adhering to ethical AI frameworks will be critical for fostering sustainable, inclusive, and accountable urban growth.

Limitations and Governance Implications

The data governance frameworks developed in this chapter confront the infrastructural conditions under which urban AI becomes possible: the collection, processing,

Fig. 4.1 Privacy and data governance in urban AI systems

linkage, anonymization, and circulation of data across citizens, municipalities, service providers, data controllers, and regulatory bodies. Federated Learning-Based Privacy-Preserving AI Model for Urban Governance (FL-PPAM), Consent-Aware Federated Learning for Urban Data Governance (CA-FL), AI-Driven Anonymization for Urban Governance (AIDAG), AI-Driven Accountability for Urban Governance (AI-AG), Fairness-Enhanced AI for Urban Governance (FEA-UG), and Transparency-Enhanced AI for Urban Governance (TEA-UG) provide mechanisms for privacy preservation, consent management, anonymization, accountability, and transparency reporting. Nevertheless, privacy-preserving computation cannot by itself resolve the broader political and legal questions raised by urban datafication. Federated learning may reduce the movement of raw data, anonymization may lower re-identification risk, and transparency reports may improve institutional visibility; yet none of these mechanisms can substitute for lawful purpose limitation, informed consent, proportional data use, independent supervision, or enforceable rights against misuse. The governance implication is that urban data governance must be treated as a socio-legal architecture rather than a purely technical control layer. AI systems that depend on civic data must therefore remain subject to continuous legal review, data-minimization duties, public accountability, and institutional mechanisms through which citizens can understand, question, restrict, or contest the use of their data in automated urban decision-making.

4.2 Conclusion

This chapter stands as a comprehensive scholarly treatise on how Artificial Intelligence (AI) can be architecturally aligned with the ethical imperatives of privacy, fairness, accountability, and transparency in contemporary urban governance. Through rigorous mathematical formalization and systemic synthesis, it demonstrates that the evolution of smart cities depends not only on the sophistication of AI algorithms but, more profoundly, on their moral and legal embedding within the civic fabric. The chapter does not treat privacy or fairness as abstract moral appeals but as quantifiable, reproducible, and computationally enforceable design principles.

The discussion begins with the Federated Learning-Based Privacy-Preserving AI Model (FL-PPAM), a cornerstone framework where distributed intelligence becomes an ethical safeguard. By enabling model training across edge devices without transferring raw data, the system constructs a decentralized architecture of trust. Federated Learning (FL) ensures locality and autonomy of data, Differential Privacy (DP) mathematically guarantees individual indistinguishability through calibrated Gaussian noise, and blockchain validation provides immutable integrity checks. This triadic integration of FL, DP, and blockchain transforms privacy from a compliance obligation into a technical guarantee that sustains both predictive efficiency and civic confidence.

The narrative advances to the Consent-Aware Federated Learning framework (CA-FL), which redefines legal consent as a computational construct. The inclusion of consent masks and data utility control functions converts the citizen's right to opt in or out into a dynamic mathematical variable of the federated optimization process. Here, privacy-by-design becomes more than a policy; it becomes an algorithmic invariant. Each training iteration embodies democratic choice, ensuring that the city's AI learns only from data voluntarily entrusted to it, and dynamically adapts when that consent is withdrawn.

A further progression emerges in the AI-Driven Anonymization Framework for Urban Governance (AIDAG). Through the integration of Differential Privacy, Generative Adversarial Networks (GANs), k-Anonymity, and t-Closeness, the chapter formulates a layered anonymization pipeline capable of producing analytically robust yet irreversibly anonymized datasets. GANs generate synthetic data statistically equivalent to the real, while k-Anonymity and t-Closeness safeguard against linkage and inference attacks. Together, these mechanisms constitute a verifiable privacy architecture where the analytical utility of urban data is preserved without violating human identity boundaries—a prerequisite for responsible innovation in urban analytics and smart infrastructure design.

In the dimension of ethical responsibility, the AI-Driven Accountability Framework (AI-AG) establishes the technical infrastructure for civic oversight. By embedding Blockchain-based audit trails, Anomaly Detection systems, and Reinforcement Learning (RL) for adaptive penalty calibration, it creates a self-corrective environment where data misuse is neither invisible nor without consequence. Every digital interaction becomes traceable, every anomaly detectable, and every violation enforceable. This accountability system transforms AI from a silent administrator of data flows into an ethically responsive actor within a governed ecosystem.

Addressing equity, the Fairness-Enhanced AI for Urban Governance (FEA-UG) framework operationalizes justice through algorithmic constraints and adversarial correction. Fairness Constraints within the loss function mathematically bound predictive disparities, while adversarial debiasing networks act as ethical correctors, continuously learning to identify and mitigate hidden biases. Explainable Artificial Intelligence (XAI) with Shapley Values further ensures interpretability, enabling policymakers to understand and justify AI decisions in human-comprehensible terms. In doing so, the chapter advances fairness auditing from qualitative rhetoric to quantitative governance, ensuring that every AI-enabled decision becomes traceable to its ethical logic.

Finally, the Transparency-Enhanced AI Framework (TEA-UG) brings epistemic clarity to AI decision systems through formal transparency metrics. The introduction of information-theoretic transparency scores, traceable decision logs, and automated reporting systems transforms transparency from a normative expectation into a measurable operational property. Mutual information functions quantify how well model explanations align with outputs, while automated reports enable public auditability. Transparency thus emerges not as the opposite of complexity but as its disciplined revelation—an algorithmic openness that reinforces democracy within digital governance.

The chapter constructs a complete epistemic architecture for ethical AI-driven urban governance. It establishes a continuum where privacy preservation, consent, anonymization, accountability, fairness, and transparency are not parallel concerns but interdependent layers of one cohesive system. Each framework supports the next, forming an ethical supply chain of algorithms that collectively ensure legitimacy, explainability, and resilience in city-scale AI systems.

From a philosophical and governance perspective, this work redefines urban AI as a civic institution rather than a technological tool. It demonstrates that the sustainable city of the future must be governed by algorithms that can be held accountable, audited for fairness, and verified for transparency. Such a transformation signifies a paradigm shift from "smart cities" built on data extraction to wise cities governed through ethical intelligence.

In final analysis, this chapter stands as both a theoretical foundation and a blueprint for implementation. It offers governments, engineers, and researchers a reproducible pathway toward Software-as-a-Service (SaaS) deployable frameworks that reconcile computational performance with social responsibility. By aligning federated learning, differential privacy, generative adversarial networks, reinforcement learning, and explainable AI into a unified governance matrix, it lays the groundwork for a new social contract between citizens and machines—one where Artificial Intelligence evolves not merely as an instrument of efficiency, but as a guarantor of human dignity within the algorithmic city.

References

Attard-Frost, B., De los Rios, A., & Walters, D. R. (2023). The ethics of AI business practices: A review of 47 AI ethics guidelines. *AI and Ethics*, 3, 389–406. https://doi.org/10.1007/s43681-022-00156-6

Barredo Arrieta, A., Díaz-Rodríguez, N., Del Ser, J., Bennetot, A., Tabik, S., Barbado, A., Garcia, S., Gil-Lopez, S., Molina, D., & Herrera, F. (2020). Explainable artificial intelligence (XAI): Concepts, taxonomies, opportunities, and challenges toward responsible AI. *Information Fusion*, 58, 82–115. https://doi.org/10.1016/j.inffus.2019.12.012

Sargiotis, D. (2024a). Ethical AI in Information Technology: Navigating Bias, Privacy, Transparency, and Accountability. *Advances in Machine Learning & Artificial Intelligence*, 5(3), 01–14. https://doi.org/10.33140/AMLAI.05.03.03

Sargiotis, D. (2024b). Data Security and Privacy: Protecting Sensitive Information. In: Data Governance. Springer, Cham. https://doi.org/10.1007/978-3-031-67268-2_6

Trotta, A., Ziosi, M., & Lomonaco, V. (2023). The future of ethics in AI: Challenges and opportunities. *AI & Society*, 38, 439–441. https://doi.org/10.1007/s00146-023-01644-x

Chapter 5
Ethical Challenges of AI in Civil Engineering, Urban Development, and Governance

Artificial Intelligence (AI) is rapidly reshaping the way cities are planned, infrastructures are designed, and public services are governed. In domains as high-stakes as civil engineering, urban development, and policy decision-making, the promise of AI efficiency, scale, and predictive acuity arrives tethered to profound ethical demands. This chapter takes those demands seriously. It moves beyond slogans and into practice, specifying how fairness, transparency, explainability, accountability, and sociotechnical context must be engineered directly into AI systems that allocate resources, prioritize investments, and shape everyday urban life.

From the outset, we frame fairness and equity not as aspirational ideals but as enforceable design properties. Readers are introduced to concrete routes by which bias creeps in historical data legacies, sampling asymmetries, and model architecture choices and to corrective techniques that make equity auditable and durable. These include reweighing strategies for data balancing, adversarial debiasing to neutralize discriminatory signals during learning, and Fairness-Constrained Reinforcement Learning (FCRL), which embeds distributive-justice penalties into sequential decision policies. The discussion extends to Multi-Objective Optimization (MOO), explicitly trading off efficiency with equity, and to Human-in-the-Loop (HITL) governance that situates expert oversight where algorithmic decisions have outsized social impact.

Because legitimacy requires legibility, the chapter then turns to transparency and explainability as operational standards. We present a toolkit of Explainable Artificial Intelligence (XAI) methods: Shapley value attribution for global feature contribution, Local Interpretable Model-agnostic Explanations (LIME) for local surrogate reasoning, Layer-Wise Relevance Propagation (LRP) for deep-network introspection, and counterfactual analysis for policy simulation—each mapped to urban use-cases such as transport prioritization, zoning, and utility access. We also introduce confidence scoring, interpretable rule extraction, and audit-ready logging to ensure that models are not only high-performing but inspectable by engineers, regulators, and citizens.

D. Sargiotis, *AI with Integrity*, https://doi.org/10.1007/978-3-032-31133-7_5

Ethics, however, is not solved by algorithms alone. The chapter situates AI within sociotechnical realities of power, history, and participation, arguing that equitable outcomes demand participatory data governance, community-weighted objective functions, and accountability pathways that outlive any single deployment. Readers will find formal mechanisms for responsibility attribution across developers, deploying organizations, and public authorities; liability modeling that distinguishes technical, operational, and regulatory responsibility; and compliance auditing that produces public-facing transparency reports. In short, we connect who builds, who deploys, and who is affected, then show how to measure and govern those relationships.

To aid immediate implementation, the chapter consolidates these elements into layered blueprints such as Fairness-Enhanced AI for Urban Governance (FEA-UG), AI-Driven Governance for Urban Systems (AI-GUS), AI Transparency and Governance for Urban Systems (AI-TGUS), Fairness-Constrained Urban AI Governance (FCUAG), and allied architectures. Each blueprint specifies inputs, processing modules, outputs, and post-deployment audits, enabling readers to move from principles to production with clarity and speed.

What international readers will gain is threefold. First, a diagnostic lens for recognizing when AI systems entrench inequity in built environments. Second, a methodological arsenal from data curation through model training to oversight for enforcing fairness, transparency, and accountability without sacrificing technical performance. Third, a governance grammar that aligns institutional responsibility, legal compliance, and civic legitimacy with the realities of intelligent infrastructure. For leaders determined to harness AI's power while honoring public trust, this chapter is a purchase in rigor: a route to systems that are not only smart and efficient, but just, auditable, and worthy of the cities they serve.

5.1 Fairness and Equity

Fairness and equity are foundational ethical imperatives in the conception, development, and deployment of Artificial Intelligence (AI) systems, particularly in the highly consequential domains of civil engineering, urban development, and governance. Within these spheres, AI serves as both an enabler of efficiency and a catalyst for transformation, optimizing the allocation of resources, enhancing infrastructure resilience, and strengthening evidence-based policy design. Yet, the very mechanisms that empower AI to model, predict, and optimize can also reproduce and amplify deep-seated social and structural inequities. The data-driven nature of contemporary AI renders it exquisitely sensitive to the biases encoded in historical datasets, institutional practices, and algorithmic architectures. When such biases are left unexamined, the resulting systems can generate discriminatory outcomes, reinforce spatial and socioeconomic disparities, and disproportionately affect marginalized populations. This risk is especially acute in civil and urban systems, where

AI-informed decisions ranging from zoning optimization and infrastructure investment to risk prediction and public resource distribution directly shape the lived environment and access to opportunity.

In this context, fairness and equity must transcend their status as aspirational values and become mathematically expressible, auditable design constraints within AI models. Formal fairness-aware frameworks such as demographic parity, equalized odds, and counterfactual fairness must be embedded at the algorithmic level to ensure that predictive accuracy does not come at the expense of distributive justice. Equally, the notion of procedural equity demands transparency in model training, validation, and deployment pipelines, guaranteeing that diverse stakeholders can interpret, contest, and recalibrate AI-driven decisions that affect their communities. The ethical architecture of AI for civil engineering and urban governance must therefore integrate not only statistical debiasing methods and equitable data curation but also institutional accountability mechanisms that monitor the long-term societal impacts of AI interventions.

Viewed through this lens, fairness and equity are not static criteria but dynamic equilibria, requiring continuous monitoring as AI systems evolve and adapt. Algorithmic audits, fairness-constrained reinforcement learning, and participatory data governance provide pathways toward ensuring that AI-enabled infrastructures remain inclusive, just, and aligned with the public good. As Dimitrios Sargiotis (2024d) has emphasized, the next generation of civil and urban AI must be ethically anticipatory, not merely minimizing harm ex post but embedding normative intelligence ex ante, so that technological innovation coevolves with social responsibility. The pursuit of fairness and equity in AI, thus, is not an auxiliary objective but a constitutive condition for sustainable, human-centered urban futures.

5.1.1 Bias in Data and Algorithms

One of the primary challenges to achieving fairness in AI systems is the risk of bias in the data used to train machine learning models. Bias can originate from various sources, including historical inequalities, incomplete datasets, or flawed data collection methodologies (Barredo Arrieta et al. 2020; Sargiotis 2024a). For example, in urban development, if AI systems are trained on datasets that reflect patterns of inequality, such as housing policies that have historically favored certain demographics, the resulting decisions may reinforce these disparities rather than mitigate them. This can manifest in biased resource allocation, where certain neighborhoods receive preferential treatment in infrastructure development or service provision, while others are overlooked. To address these challenges, a bias correction AI model can be deployed to mitigate the impact of biased data on urban AI systems. One effective method is the use of Reweighed Learning, which adjusts the weight of underrepresented or disadvantaged groups in the training data to ensure that the model does not favor overrepresented populations. The reweighed learning model is formulated as:

$$\min_{\mathrm{w}} L(\mathrm{w}, D) + \lambda \sum_{i=1}^{n} w_i \cdot \mathrm{loss}(y_i, \hat{y}_i)$$

where $L(\mathrm{w}, D)$, loss function based on model parameters w and dataset D, w_i, weight assigned to the i-th data point to counteract bias, $\mathrm{loss}(y_i, \hat{y}_i)$, traditional loss function comparing predicted values $\hat{y}_i$ with actual outcomes y_i, λ, regularization parameter ensuring balance between model accuracy and fairness.

By adjusting the weights assigned to data points from marginalized groups, this model prevents the AI system from favoring historically advantaged groups, ensuring that the algorithm's predictions and resource allocations are more balanced and equitable across different demographic categories.

Furthermore, to address algorithmic bias which arises from biases inherent in the model architecture or training process, adversarial debiasing can be employed. Adversarial debiasing involves training an adversarial network to detect and reduce biases in the model's predictions. The adversarial debiasing model is formulated as:

$$\min_{G} \max_{D} \mathbb{E}_{x \sim P_{\mathrm{data}}}[\log D(G(x))] + \mathbb{E}_{x \sim P_{\mathrm{data}}}[\log(1 - D(x))]$$

where $G(x)$, generator function modifying the model's output to reduce bias, $D(x)$, discriminator function attempting to identify biased predictions, P_{data}, data distribution representing input features, $\mathbb{E}\,[\cdot]$, expectation operator optimizing adversarial training.

This adversarial network effectively minimizes the model's exposure to bias by generating modified outputs that reduce the likelihood of discrimination in the AI system's predictions. By using adversarial debiasing, AI systems can be trained to make decisions that are less influenced by discriminatory patterns in the data, ensuring fairness in predictions and decisions regarding urban development.

To ensure that these models operate equitably over time, regular fairness audits are essential. AI systems must undergo continuous evaluations to check for any emerging biases as new data is introduced or as decisions are made over time. One approach for conducting fairness audits is through Fairness-Constrained Reinforcement Learning (FCRL), which applies fairness constraints within the reinforcement learning process to ensure that AI-driven decisions maintain equity. The FCRL model is formulated as:

$$R(s_t, a_t) = r_t + \gamma \max_{a_{t+1}} Q(s_{t+1}, a_{t+1}) + \lambda \cdot \mathrm{fairness\_penalty}(s_t, a_t)$$

where $R(s_t, a_t)$, reward function in state s_t and action a_t, r_t, immediate reward for taking action a_t, γ, discount factor for future rewards, $Q(s_{t+1}, a_{t+1})$, Q-value function predicting future rewards for subsequent actions, λ, regularization parameter, fairness _ penalty(s_t, a_t), fairness penalty applied to decisions that disproportionately affect marginalized groups.

The inclusion of a fairness penalty within the reinforcement learning framework ensures that actions taken by the AI model do not exacerbate inequities. By incorporating fairness constraints, the model can be guided toward outcomes that promote more balanced and inclusive urban development, preventing AI from perpetuating existing biases or creating new forms of discrimination.

The combination of reweighed learning, adversarial debiasing, and fairness-constrained reinforcement learning forms a comprehensive AI-driven fairness model for urban systems. These techniques work together to ensure that AI systems in urban planning and governance make decisions that are both accurate and equitable, mitigating the risks of reinforcing historical disparities.

This integrated fairness framework allows urban AI systems to make resource allocation and infrastructure decisions that are not only data-driven but also just and inclusive. By applying these fairness-enhancing techniques, AI can be used to drive policies and decisions that address systemic inequalities, contributing to more equitable urban environments where all communities benefit equally from technological advancements.

Framework: Fairness-Enhanced AI for Urban Governance (FEA-UG)

Layered AI Architecture for Equitable Urban Data Governance and Decision-Making

Layer	Main functionality	AI module(s) and components	Operational scope (Input → Processing → Output)
Layer 1	Data collection and bias mitigation: Ensuring data is representative and adjusting for historical bias	Data quality validation and bias correction engine	Input → Urban data with potential biases → Output → Balanced dataset for AI model training
Layer 2	Reweighed learning: Adjusting data point weights to correct for biased representation in the model	Reweighed learning	Input → Biased data with weights → Output → Bias-corrected model predictions
Layer 3	Adversarial debiasing: Training a generator to modify predictions and a discriminator to detect bias	Adversarial Networks (GANs)	Input → Biased predictions → Output → Debiased model predictions
Layer 4	Fairness-constrained reinforcement learning: Applying fairness penalties within the RL process to ensure equitable outcomes	Reinforcement Learning (Q-learning) with fairness constraints	Input → Current state of urban environment, data usage policies → Output → Fair and equitable decisions
Layer 5	Post-AI fairness audits: Continuously evaluating AI decisions to ensure fairness and equity in real-world applications	Fairness monitoring and compliance audit system	Input → AI-driven decisions, model outputs → Output → Fairness audit reports and compliance validations

5.1.2 Equity in Resource Distribution

In civil engineering and urban development, AI systems play a crucial role in optimizing the distribution of resources such as transportation, energy, and public services (Sargiotis 2024d). These technologies enable cities to enhance efficiency, reduce operational costs, and improve service delivery. However, without explicit consideration of equity, AI-driven optimizations risk prioritizing efficiency at the expense of fairness, leading to unequal access to essential services (Floridi et al. 2018). AI models designed for traffic flow optimization or energy distribution, for instance, may disproportionately favor wealthier areas where infrastructure is already well developed, while failing to address the needs of underserved communities that require greater support. This imbalance can exacerbate existing disparities, reinforcing historical patterns of exclusion rather than mitigating them.

To counteract these risks, AI systems must explicitly integrate equity considerations into resource distribution. Achieving fairness in AI-driven urban systems requires incorporating fairness constraints into optimization models, ensuring that AI allocates resources equitably across all communities, particularly those that have been historically marginalized (Manzoor et al. 2021). Additionally, decision-makers must actively engage with diverse stakeholders, including residents from disadvantaged areas, urban planners, and policymakers, to ensure that AI systems align with the needs and priorities of all communities.

One effective approach for embedding fairness into AI-driven decision-making is the fairness-constrained optimization model, which modifies standard optimization functions to account explicitly for equity in resource allocation. This model prevents AI from allocating resources solely on the basis of efficiency metrics and instead incorporates social and economic equity into urban planning decisions. The fairness-constrained optimization model is formulated as:

$$\min_{\mathrm{w}} L(\mathrm{w}, D) + \lambda \cdot \text{fairness\_penalty}(D)$$

where $L(\mathrm{w}, D)$, loss function based on model parameters w and dataset D, λ, regularization parameter balancing the primary objective and fairness, fairness _ penalty(D), penalty term added to enforce equity in resource allocation decisions.

The inclusion of the fairness penalty term ensures that AI systems do not optimize solely for efficiency, but also for equitable outcomes. This forces the model to consider the needs of underserved communities, even if their infrastructure is less developed, and ensures that their needs are not overlooked in favor of wealthier or more developed areas. By incorporating fairness constraints into optimization algorithms, this approach seeks to reduce the potential for exacerbating inequalities in the distribution of resources.

Moreover, multi-objective optimization (MOO) frameworks can be employed to explicitly balance both efficiency and equity in resource allocation. Multi-objective optimization allows decision-makers to consider multiple, often conflicting, objectives simultaneously, such as maximizing efficiency while ensuring fairness. The multi-objective optimization model for equitable resource distribution is formulated as:

$$\min_{w} \left[\alpha L_{\text{efficiency}}(w, D) + \beta L_{\text{equity}}(w, D)\right]$$

where $L_{\text{efficiency}}(w, D)$, loss function related to optimizing efficiency in resource allocation, $L_{\text{equity}}(w, D)$, loss function ensuring equitable distribution of resources, α, weight for efficiency objective, β, weight for equity objective.

By using multi-objective optimization, the AI system can simultaneously optimize efficiency (e.g., minimizing travel time or energy consumption) and equity (e.g., ensuring that disadvantaged neighborhoods receive sufficient resources). This balance allows for the creation of more just urban environments, where AI optimizes public services not only for the most developed areas but also for historically underserved communities.

In addition to algorithmic fairness, stakeholder engagement plays a vital role in ensuring equitable AI-driven resource distribution. Decision-makers must actively involve marginalized communities in the development and implementation of AI systems, ensuring that their needs and priorities are considered. This process can be supported by participatory AI frameworks, which enable communities to provide input on the goals and outcomes of AI systems. The participatory AI framework for equitable resource distribution is formulated as:

$$\text{Participatory\_Objective} = \sum_{i=1}^{n} w_i \cdot f(\text{Community\_Input}_i)$$

where w_i, weight assigned to the input of the i-th community member, Community _ Input$_i$, feedback or preferences of the i-th community member, $f(\cdot)$, function translating community input into actionable objectives.

This framework ensures that AI systems are not solely driven by efficient metrics but are also aligned with the values and needs of diverse communities. Through participatory engagement, marginalized groups have a voice in shaping the decisions that impact their neighborhoods, which leads to more inclusive and equitable outcomes.

The integration of fairness constraints, multi-objective optimization, and participatory AI forms a comprehensive model for equitable resource distribution in urban AI systems. This approach ensures that AI systems optimize both efficiency and fairness, while actively engaging with affected communities to ensure that their needs are considered in decision-making processes.

By embedding these principles into the design of AI-driven urban systems, cities can ensure that their resources, whether for transportation, energy, or public services, are distributed in a way that benefits all residents, particularly those from disadvantaged communities. This approach not only promotes fairness in the allocation of resources but also contributes to the development of more resilient, inclusive, and sustainable urban environments.

Framework: Fairness-Enhanced AI for Equitable Urban Resource Distribution (FEA-ERD)

Layered AI Architecture for Fair and Equitable Resource Allocation in Urban Systems

Layer	Main functionality	AI module (s) and components	Operational scope (Input → Processing → Output)
Layer 1	Data Collection and Bias Mitigation: Ensuring that the data used in decision-making is representative and fair	Data Quality Validation and Bias Correction Engine	Input → Urban data with diverse representation → Output → Data prepared for fairness-enhanced training
Layer 2	Fairness-constrained optimization: Integrating fairness penalties to ensure equitable resource distribution	Fairness-constrained optimization	Input → Biased training data → Output → Fair model parameters accounting for equity considerations
Layer 3	Multi-objective optimization: Simultaneously optimizing efficiency and equity in resource distribution	Multi-Objective Optimization (MOO)	Input → Data with efficiency and equity goals → Output → Optimized solutions balancing efficiency and fairness
Layer 4	Participatory AI framework: Incorporating feedback from diverse stakeholders to shape equitable AI-driven decisions	Stakeholder feedback integration platform	Input → Community input on AI objectives → Output → Participatory objectives integrated into model
Layer 5	Post-AI fairness audits: Continuously evaluating AI decisions to ensure fairness and equity in real-world applications	Fairness monitoring and compliance audit system	Input → AI-driven decisions, fairness audits → Output → Fairness reports and compliance checks

5.1.3 Accountability and Governance for Fairness

Ensuring fairness and equity in AI-driven urban systems requires robust accountability and governance frameworks designed to identify and mitigate biases while holding decision-makers accountable for the consequences of AI-driven decisions. Without structured oversight, AI models risk perpetuating systemic inequalities rather than addressing them. Governance mechanisms must incorporate transparent decision-making processes, provide avenues for scrutiny, and enforce compliance with fairness and equity principles to ensure AI technologies serve all communities equitably (Mökander 2023; Sargiotis 2024e).

A key element of AI governance is transparency and explainability, which enable both policymakers and citizens to scrutinize AI-driven decisions and challenge unfair or discriminatory outcomes. Without clear insights into how AI models operate, biases may remain undetected, leading to inequitable distributions of urban resources, public services, and infrastructure. Explainable AI (XAI) techniques offer a crucial solution by making AI models interpretable, thereby fostering trust and accountability in urban governance. XAI techniques ensure that AI-generated decisions are comprehensible, allowing policymakers to assess whether fairness principles are being upheld and take corrective action when necessary (Barredo Arrieta et al. 2020).

Beyond explainability, governance frameworks must also include regular fairness audits to proactively detect biases in AI models. These audits ensure that AI systems comply with fairness and equity standards by evaluating their decisions for potential discriminatory patterns before they negatively impact vulnerable populations. Bias detection mechanisms, such as statistical fairness assessments and adversarial debiasing, play a key role in auditing AI models to ensure equitable decision-making.

A particularly useful method for strengthening transparency and fairness in AI-driven urban governance is the use of Shapley value-based feature attribution models, which quantify the contribution of different input variables to an AI system's decision-making process. This approach supports accountability by revealing which features most significantly influence model predictions, enabling decision-makers to identify, examine, and mitigate potential sources of bias. The Shapley value-based explainability model is formulated as:

$$\phi_i(f) = \sum_{S \subseteq N \backslash \{i\}} \frac{|S|!(|N| - |S| - 1)!}{|N|!} [f(S \cup \{i\}) - f(S)]$$

where $\phi_i(f)$, Shapley value measuring the contribution of feature i to the AI model's decision, $f(S)$, function output for a subset S of features, N, total set of features, $|S|$, number of elements in subset S, $|N|$, total number of features.

This model ensures that AI-driven urban governance systems can be held accountable by revealing the influence of different factors in decision-making processes. By quantifying the contribution of each input feature, stakeholders can assess whether AI models are making decisions fairly or if certain demographic or geographical factors are disproportionately influencing outcomes.

In addition to explainability, auditing AI systems is a critical governance practice for ensuring fairness and compliance with equity standards. Regular audits allow policymakers and regulators to proactively detect and correct biases in data and algorithms before they result in harmful consequences for marginalized communities. A widely used approach for fairness auditing is Fairness-Constrained Risk Assessment (FCRA), which applies statistical fairness constraints during the evaluation of AI models. The FCRA fairness auditing model is formulated as:

$$\max_{\mathbf{W}} \sum_{i=1}^{n} [\text{accuracy}(y_i, \hat{y}_i) - \lambda \cdot \text{bias\_penalty}(y_i, \hat{y}_i)]$$

where $\text{accuracy}(y_i, \hat{y}_i)$, accuracy score of AI predictions compared to actual outcomes, λ, regularization parameter controlling the trade-off between accuracy and fairness, $\text{bias\_penalty}(y_i, \hat{y}_i)$, penalty function quantifying disparities in predictions across different demographic groups.

This model ensures that AI models are not only evaluated for performance but also for fairness. By applying bias penalties, the auditing framework actively discourages models from making unfair predictions, ensuring compliance with fairness standards.

Beyond explainability and auditing, governance frameworks must also incorporate reinforcement learning-based adaptive governance mechanisms, which allow AI-driven regulatory bodies to dynamically adjust oversight strategies based on evolving fairness metrics. The adaptive governance optimization model is formulated as:

$$R(s_t, a_t) = r_t + \gamma \max_{a_{t+1}} Q(s_{t+1}, a_{t+1}) - \beta \cdot \text{governance\_penalty}(s_t, a_t)$$

where $R(s_t, a_t)$, reward function for governance oversight decisions, r_t, immediate reward for taking governance action a_t in state s_t, γ, discount factor prioritizing long-term fairness, $Q(s_{t+1}, a_{t+1})$, future expected reward from policy action a_{t+1}, β, regularization term balancing fairness and efficiency, governance _ penalty(s_t, a_t), penalty for failing to enforce fairness standards.

This model ensures that AI governance mechanisms dynamically adjust based on detected fairness violations, making oversight bodies more responsive to emerging risks. If an AI system is found to be making biased decisions, the governance framework can reinforce stricter monitoring and intervention measures, ensuring ongoing compliance with fairness objectives.

The integration of explainable AI, fairness audits, and reinforcement learning-based governance forms a comprehensive AI-driven accountability framework for urban systems. Explainability mechanisms ensure transparency in AI decisions, fairness audits detect and correct biases, and adaptive governance policies dynamically adjust oversight strategies to maintain compliance with ethical standards.

By embedding these mechanisms into AI-driven urban governance, cities can establish trustworthy, accountable, and equitable AI systems that optimize resources while ensuring fairness. This governance model empowers policymakers, regulators, and citizens to scrutinize AI-driven decisions, challenge unfair outcomes, and foster AI deployment practices that uphold ethical and social justice principles in urban development.

Framework: AI-Driven Governance for Urban Systems (AI-GUS)

Layered AI Architecture for Transparent, Fair, and Accountable Urban Data Governance

Layer	Main functionality	AI module (s) and components	Operational scope (Input → Processing → Output)
Layer 1	Data collection and transparency governance: Ensuring transparency in data usage and governance practices	Data provenance and transparency management system	Input → Urban data, governance policies → Output → Transparent data collection and usage protocols

(continued)

Layer	Main functionality	AI module (s) and components	Operational scope (Input → Processing → Output)
Layer 2	Shapley value-based explainability: Quantifying the contribution of features to AI-driven decisions	Shapley Values (XAI)	Input → AI decision outputs, features → Output → Feature attributions revealing decision rationale
Layer 3	Fairness-constrained risk assessment: Evaluating AI models for fairness while maximizing accuracy	Fairness-Constrained Risk Assessment (FCRA)	Input → AI model predictions, accuracy metrics → Output → Fairness audited AI model evaluation
Layer 4	Reinforcement learning for governance oversight: Dynamically adjusting governance strategies based on fairness metrics	Reinforcement Learning (Adaptive Governance)	Input → AI governance decisions, fairness metrics → Output → Adaptive governance policies to reinforce fairness
Layer 5	Post-AI transparency audits: Auditing AI systems for transparency and legal compliance	Governance transparency and compliance audit platform	Input → AI decisions, legal compliance metrics → Output → Audited transparency reports for public scrutiny

5.1.4 Sociotechnical Considerations

Achieving fairness and equity in AI-driven systems extends beyond technical refinements, necessitating a sociotechnical perspective that accounts for the broader social and political contexts in which these technologies operate (Sargiotis 2024a). AI systems are not developed in isolation; rather, they are embedded within existing power structures that can either reinforce or challenge systemic inequalities, depending on their design, governance, and implementation (Floridi et al. 2018). Consequently, ensuring fairness requires a holistic approach that integrates technical solutions with social and political considerations, aligning AI deployment with broader goals of social justice and equity.

In civil engineering and urban development, AI models are widely used to optimize infrastructure projects, streamline resource allocation, and enhance public services. However, if these models prioritize efficiency without addressing underlying socioeconomic disparities, they risk perpetuating systemic inequities rather than mitigating them. For example, an AI system designed to optimize transportation networks may inadvertently favor well-developed urban centers, neglecting underserved areas that lack historical investments in infrastructure. To counteract this risk, AI models must integrate social fairness constraints, ensuring that their recommendations align with inclusive urban development goals rather than reinforcing existing disparities.

A critical approach for embedding equity into AI driven civil engineering is the sociotechnical fairness-constrained optimization model, which modifies standard infrastructure planning algorithms to account for both efficiency and equity. The fairness-augmented optimization model is formulated as:

$$\min_{\mathrm{w}} \left[\alpha L_{\text{efficiency}}(\mathrm{w}, D) + \beta L_{\text{social\_equity}}(\mathrm{w}, D)\right]$$

where $L_{\text{efficiency}}(\mathrm{w}, D)$, loss function optimizing technical efficiency, $L_{\text{social \_ equity}}(\mathrm{w}, D)$, loss function ensuring equitable infrastructure development, α, weight assigned to efficiency objectives, β, weight assigned to social equity considerations.

This model ensures that AI-driven urban development does not focus exclusively on cost-effectiveness and performance metrics but also prioritizes socially inclusive and just infrastructure planning. By integrating equity as an optimization objective, AI models can be directed to address disparities in infrastructure investments while still maintaining technical feasibility.

Beyond algorithmic fairness, stakeholder engagement and participatory decision-making are essential for ensuring that AI-driven urban development aligns with community needs and social priorities. AI models must be designed with input from diverse stakeholders, including marginalized communities, urban planners, policymakers, and civil engineers, ensuring that the voices of historically underserved groups are reflected in AI-driven decisions. One method for incorporating participatory governance in AI-driven infrastructure planning is the weighted community feedback model, formulated as:

$$\text{Community\_Objective} = \sum_{i=1}^{n} w_i \cdot f(\text{Community\_Feedback}_i)$$

where w_i, weight assigned to the input of the i-th community representative, Community _ Feedback$_i$, feedback provided by the i-th community member, $f(\cdot)$, function translating community input into actionable policy objectives.

By integrating participatory governance models into AI-driven urban planning, policymakers can ensure that infrastructure investments reflect local community needs, rather than being dictated solely by algorithmic efficiency. This fosters greater transparency, accountability, and public trust in AI-driven decision-making.

The integration of fairness-constrained optimization, participatory AI governance, and stakeholder-driven decision-making forms a comprehensive sociotechnical framework for equitable AI deployment in civil engineering and urban development. This approach ensures that AI systems not only optimize technical performance but also address historical inequalities, incorporate diverse perspectives, and contribute to long-term social equity goals.

By embedding sociotechnical considerations into AI-driven infrastructure planning, urban systems can be more inclusive, just, and responsive to the needs of all communities, ensuring that the benefits of AI technologies extend beyond privileged groups and contribute to a more equitable and sustainable future.

Framework: Sociotechnical Fairness-Driven AI for Equitable Urban Development (SFAI-EUD)

Layered AI Architecture for Fair, Inclusive, and Responsive Urban Infrastructure Planning

Layer	Main functionality	AI module (s) and components	Operational scope (Input → Processing → Output)
Layer 1	Data collection and equity integration: Ensuring data reflects diverse, equitable representation in urban planning	Data quality validation and equity enhancement engine	Input → Data from diverse urban communities → Output → Equitable dataset for AI-driven decisions
Layer 2	Sociotechnical fairness-constrained optimization: Balancing technical efficiency with social equity in infrastructure planning	Fairness-constrained optimization	Input → Data with efficiency and equity goals → Output → AI model that optimizes both technical performance and fairness
Layer 3	Weighted community feedback: Integrating community input into AI-driven decisions to reflect local needs	Participatory AI	Input → Community feedback → Output → Participatory objectives integrated into AI system
Layer 4	Multi-objective optimization: Simultaneously optimizing efficiency and equity for fair resource allocation	Multi-Objective Optimization (MOO)	Input → Multiple urban objectives (efficiency, equity) → Output → Optimized resource allocation balancing both
Layer 5	Post-AI fairness audits: Conducting fairness audits to detect and address biases in AI-driven decisions	Fairness monitoring and compliance audit system	Input → AI decisions, fairness evaluation → Output → Fairness audit reports and compliance checks

Figure 5.1 highlights the key ethical challenges of integrating AI into civil engineering, urban development, and governance, focusing on three main areas: Fairness and Equity, Accountability, and Sociotechnical Context. "Fairness and Equity" addresses issues of bias in data, fair resource distribution, and prioritizing underserved communities. "Accountability" includes governance frameworks, regular fairness audits, and the use of explainable AI to ensure transparency and responsibility in AI decisions. "Sociotechnical Context" explores AI's role in social dynamics, emphasizing the importance of balancing efficiency with community needs and aligning AI practices with social justice goals.

Fairness and equity are critical ethical considerations in the deployment of AI systems in civil engineering, urban development, and governance. Addressing these challenges requires a multifaceted approach that includes mitigating bias in data and algorithms, prioritizing equity in resource distribution, and implementing robust accountability and governance frameworks. By taking these steps, AI systems can be designed and deployed in a manner that promotes fairness and equity, ensuring that all communities benefit from the transformative potential of AI technologies.

Fig. 5.1 Ethical challenges of AI in civil engineering, urban development, and governance

5.2 Transparency and Explainability

Transparency and explainability are fundamental to the ethical development and deployment of AI systems, particularly in civil engineering, urban development, and governance. These principles play a crucial role in fostering public trust, ensuring accountability, and enabling stakeholders, including policymakers, engineers, and the public to scrutinize AI-driven decisions (Sargiotis 2024a). The complexity of many AI models, particularly those based on deep neural networks, often creates an inherent opacity that makes it difficult to interpret their decision-making processes (Barredo Arrieta et al. 2020). Addressing these challenges is essential for ensuring fairness, responsible governance, and ethical AI deployment in urban systems.

5.2.1 *The Need for Transparency in AI Systems*

Transparency in AI refers to the ability to understand how an AI system functions and makes decisions. In the context of civil engineering and urban development, where AI systems are increasingly used to optimize infrastructure, resource management, and urban planning, transparency is essential for fostering public trust and ensuring that decisions are made in a just and equitable manner. Without transparency, stakeholders, including policymakers, engineers, and the public, may have difficulty understanding how AI systems arrive at their conclusions, leading to skepticism and a lack of trust in AI-driven decision-making processes (Manzoor et al. 2021).

To establish trust in AI-based urban systems, transparency must be embedded at every stage of AI model development and deployment. This requires explainability techniques that make AI decisions interpretable, allowing stakeholders to assess the fairness, accuracy, and reliability of AI-driven recommendations. Explainable AI (XAI) provides such a methodological basis by clarifying how models process data and generate predictions. For this purpose, Shapley value-based feature attribution is especially useful because it quantifies the contribution of different input variables to an AI system's decisions. The Shapley value-based transparency model is formulated as:

$$\phi_i(f) = \sum_{S \subseteq N \setminus \{i\}} \frac{|S|!(|N| - |S| - 1)!}{|N|!} [f(S \cup \{i\}) - f(S)]$$

where $\phi_i(f)$, Shapley value measuring the impact of feature i on the AI model's decision, $f(S)$, function output for a subset S of features, N, total number of features, $|S|$, number of elements in subset S.

By computing Shapley values, AI-driven urban planning models can provide clear justifications for their decisions, allowing stakeholders to verify whether AI-generated recommendations align with fair and equitable urban policies.

Beyond explainability, algorithmic transparency also requires model interpretability audits, ensuring that AI systems remain comprehensible and free from unintended biases. One method for conducting interpretability audits is the model confidence scoring approach, which evaluates how well an AI system justifies its decisions relative to its level of certainty. The confidence scoring model is formulated as:

$$C(x) = \frac{1}{1 + e^{-\sigma(f(x) - \mu)}}$$

where $C(x)$, confidence score of the AI model's decision on input x, $f(x)$, predicted output for input x, μ, threshold for model certainty, σ, scaling factor controlling sensitivity to deviations.

This model provides an explicit confidence level for each AI-driven decision, allowing urban planners and policymakers to identify unreliable or low-certainty AI outputs before acting on them.

Moreover, transparency in AI-driven urban systems must extend beyond technical mechanisms to institutional governance, ensuring that AI decision-making processes are accessible and accountable to all stakeholders. This requires the adoption of participatory AI frameworks, where communities and policymakers play an active role in reviewing AI-driven urban planning. A formal participatory AI governance model can be formulated as:

$$\text{Transparency\_Objective} = \sum_{i=1}^{n} w_i \cdot f(\text{Stakeholder\_Feedback}_i)$$

where w_i, weight assigned to the feedback of stakeholder i, Stakeholder _ Feedback$_i$, feedback provided by the i-th policymaker, engineer, or community member, $f(\cdot)$, function incorporating stakeholder input into AI transparency metrics.

This participatory AI governance model ensures that AI-driven urban systems do not operate in opaque, black-box environments, but rather in a collaborative, accountable manner where the public and policymakers can evaluate and refine AI-driven recommendations.

The combination of XAI techniques, confidence scoring, and participatory governance mechanisms forms a comprehensive AI transparency framework for urban planning and infrastructure development. This framework ensures that AI models are interpretable, verifiable, and accountable, fostering public trust while mitigating the risks of opaque, unaccountable AI-driven decision-making.

By embedding transparency at every stage of AI system design and deployment, urban AI governance can enhance civic trust, promote equitable decision-making, and ensure that AI technologies remain aligned with the broader goals of social justice and inclusive urban development.

Framework: AI Transparency and Governance for Urban Systems (AI-TGUS)

Layered AI Architecture for Transparent, Explainable, and Accountable Urban Planning and Infrastructure

Layer	Main functionality	AI module (s) and components	Operational scope (Input → Processing → Output)
Layer 1	Data collection and transparency integration: Ensuring transparency in data collection and usage practices	Data provenance and transparency management system	Input → Urban data → Output → Transparent and accountable data usage protocols
Layer 2	Shapley value-based explainability: Quantifying the contribution of each feature to AI-driven decisions	Shapley Values (XAI)	Input → AI decision outputs, feature data → Output → Transparent feature attributions for decision-making
Layer 3	Confidence scoring: Evaluating the model's certainty and justifiability of its decisions	Uncertainty quantification and confidence evaluation engine	Input → AI decision outputs → Output → Confidence score indicating model certainty
Layer 4	Participatory AI governance: Incorporating feedback from stakeholders to improve transparency and accountability	Stakeholder feedback integration platform	Input → Stakeholder feedback → Output → Actionable transparency objectives from community input

(continued)

Layer	Main functionality	AI module (s) and components	Operational scope (Input → Processing → Output)
Layer 5	Post-AI transparency audits: Conducting audits to ensure fairness, transparency, and compliance with governance standards	Governance transparency and compliance audit platform	Input → AI decisions, transparency metrics → Output → Audited reports for transparency and fairness

One of the greatest challenges to achieving transparency in AI-driven urban systems is the complexity of modern machine learning models. Deep learning algorithms, in particular, are often highly intricate, functioning as black-box systems that generate predictions without offering clear insights into their internal decision-making processes (Barredo Arrieta et al. 2020). This lack of interpretability is a significant concern, especially in high-stakes applications such as public resource allocation, infrastructure planning, and transportation management, where transparency is essential for accountability, fairness, and trust.

The opacity of deep learning models can lead to several risks, including biased decision-making, lack of public trust, and difficulties in challenging unfair AI-driven outcomes. For example, if an AI system assigns public transportation routes based on historical demand patterns but fails to account for underserved communities, the lack of interpretability could prevent stakeholders from understanding why these areas receive inadequate service. To address this issue, AI models must incorporate explainability techniques that allow engineers, policymakers, and the public to interpret and validate AI-driven decisions.

One effective approach to improving AI transparency is through Layer-Wise Relevance Propagation (LRP), a method that assigns important scores to different input features, helping to explain which factors most strongly influenced an AI model's decision. The LRP-based explainability model is formulated as:

$$R_i = \sum_j \frac{a_i w_{ij}}{\sum_k a_k w_{kj}} R_j$$

where R_i, relevance score of input feature i, a_i, activation of neuron i, w_{ij}, weight connecting neuron i to neuron j, R_j, relevance propagated from neuron j, $\sum_k a_k w_{kj}$, normalization factor ensuring the sum of relevances remains conserved.

This model provides a layer-by-layer breakdown of how an AI system arrives at its decisions, offering interpretable insights into which data features played the most significant role. By using LRP, AI-driven urban planning models can explain why certain neighborhoods receive infrastructure investments while others do not, ensuring that AI systems remain fair, accountable, and free from hidden biases.

In addition to interpretability techniques like LRP, another method for improving transparency is decision rule extraction, which converts complex AI models into simpler rule-based systems that mimic AI decision-making using human-understandable logic. A widely used technique for rule extraction is decision tree approximation, formulated as:

$$\text{Decision\_Rule} = \arg \max_{y} P(y|X)$$

where Decision _ Rule, extracted rule approximating the AI system's decision, $P(y|X)$, probability of output y given input features X.

By converting deep learning predictions into decision rules, stakeholders can gain a clear, logical explanation of how an AI system distributes public resources or prioritizes urban infrastructure projects.

However, improving transparency in AI systems is not only a technical challenge but also a governance issue. To ensure AI transparency remains a fundamental principle in urban development, cities must adopt AI transparency regulations that require AI-driven decisions to be auditable, explainable, and accountable. A formal AI governance framework for transparency can be expressed as:

$$\text{Transparency\_Objective} = \sum_{i=1}^{n} w_i \cdot f(\text{Audit\_Log}_i)$$

where w_i, weight assigned to the importance of each audit record, Audit _ Log_i, record of AI-driven decisions and justifications, $f(\cdot)$, function mapping audit logs to transparency metrics.

By implementing auditable AI decision logs, urban governance frameworks can ensure that all AI-driven decisions are recorded, reviewed, and validated for fairness and accountability.

The integration of LRP-based explainability, decision rule extraction, and AI transparency audits creates a comprehensive transparency framework for AI-driven urban systems. This approach ensures that AI models used in civil engineering, resource management, and public service allocation are interpretable, accountable, and aligned with societal goals.

By harnessing transparency techniques into AI models and governance policies, cities can enhance public trust, prevent discriminatory outcomes, and ensure that AI-driven decision-making remains fair and explainable in urban development and civil engineering.

Framework: AI Transparency and Governance for Urban Systems (AI-TGUS II)

Layered AI Architecture for Transparent, Explainable, and Accountable Urban Planning and Infrastructure

Layer	Main functionality	AI module (s) and components	Operational scope (Input → Processing → Output)
Layer 1	Data collection and transparency integration: Ensuring transparency in data collection and usage practices	Data provenance and transparency management system	Input → Urban data → Output → Transparent and accountable data usage protocols

(continued)

Layer	Main functionality	AI module (s) and components	Operational scope (Input → Processing → Output)
Layer 2	Layer-wise relevance propagation: Explaining which features contribute most to AI-driven decisions	LRP (Explainable AI)	Input → AI decision outputs → Output → Relevance scores for each feature
Layer 3	Decision rule extraction: Converting AI predictions into human-understandable decision rules	Decision rule extraction (Decision Trees)	Input → AI model predictions → Output → Human-readable decision rules
Layer 4	AI transparency governance: Implementing auditable logs to ensure accountability in decision-making	Audit logging and accountability engine	Input → AI-driven decisions, audit logs → Output → Transparency metrics and audit reports
Layer 5	Post-AI transparency audits: Conducting audits to ensure fairness, transparency, and compliance with governance standards	Governance transparency and compliance audit platform	Input → AI decisions, transparency metrics → Output → Audited reports for transparency and fairness

5.2.2 *Explainability and Its Role in AI Governance*

Explainability is a fundamental aspect of AI transparency, referring to an AI system's ability to provide meaningful and interpretable justifications for its decisions. In urban governance and civil engineering, where AI is increasingly employed to manage public services, optimize traffic flow, and allocate infrastructure resources, explainability is critical for ensuring that AI-driven decisions are fair, just, and accountable (Sargiotis 2024b). Without a clear understanding of how AI systems arrive at their conclusions, stakeholders, including city planners, policymakers, and the public, may struggle to assess the validity and fairness of AI-driven recommendations, particularly when these decisions affect large segments of the population (Mökander 2023).

A core challenge of AI explainability in urban governance lies in the complexity of modern machine learning models, which often operate as black-box systems, making it difficult to trace the reasoning behind their decisions. To address this issue, AI models must incorporate explainability techniques that provide interpretable insights into their decision-making processes. One effective method for supporting explainability is Local Interpretable Model-agnostic Explanations (LIME), which approximates complex AI models with simpler, interpretable models that explain why certain predictions were made. The LIME-based explanation model is formulated as:

$$\min_{g \in G} \mathbb{E}_{x \sim D}[L(f, g, \pi_x)] + \Omega(g)$$

where f, complex AI model, g, interpretable approximation of f, G, class of interpretable models, D, dataset used for explanation, $L(f, g, \pi_x)$, loss function measuring the difference between the original model and the approximation, πxπx, proximity function assigning importance to local samples, $\Omega(g)$, regularization function ensuring interpretability.

This model allows urban planners and policymakers to understand how specific input features influence AI-driven decisions, ensuring that AI models do not inadvertently reinforce biases or make discriminatory resource allocation decisions.

Another approach to enhancing explainability is counterfactual reasoning, which provides insight into how an AI model's decision would change if certain input variables were altered. This technique is particularly useful in urban governance, where policymakers need to simulate different planning scenarios to evaluate AI-driven recommendations. The counterfactual explanation model is formulated as:

$$\arg \min_{\tilde{x}} \lambda d(x, \tilde{x}) + L(f(\tilde{x}), y^*)$$

where $\tilde{x}$, counterfactual input instance, x, original input instance, λ, regularization parameter balancing similarity and plausibility, $d(x, \tilde{x})$, distance function measuring similarity between original and counterfactual input, $L(f(\tilde{x}), y^*)$, loss function ensuring the counterfactual leads to the desired alternative outcome y^*.

By using counterfactual explanations, AI systems can demonstrate alternative urban planning outcomes based on different policy inputs, helping decision-makers to explore fairer and more equitable resource allocation strategies.

In addition to technical explainability methods, AI systems must be governed by institutional frameworks that enforce explainability as a regulatory requirement. This necessitates explainability audits, ensuring that AI models in urban governance are systematically evaluated for interpretability and fairness. The explainability audit framework can be mathematically expressed as:

$$\text{Explainability\_Score} = \sum_{i=1}^{n} w_i \cdot f(\text{Interpretability\_Metric}_i)$$

where w_i, weight assigned to each interpretability metric, Interpretability _ Metric$_i$, score measuring how explainable AI decisions are based on predefined evaluation criteria, $f(\cdot)$, function aggregating the weighted interpretability scores into an overall explainability assessment.

Through leveraging LIME-based explanations, counterfactual reasoning, and explainability audits, AI-driven urban governance systems can ensure that decision-making processes remain transparent, accountable, and just. These techniques allow policymakers, engineers, and the public to understand, challenge, and refine AI-driven recommendations, ensuring that urban planning and civil engineering projects align with societal goals and ethical principles.

A well-structured explainability framework ensures that AI is not only technically efficient but also interpretable and trustworthy, fostering public confidence and ethical governance in AI-driven urban development.

Framework: AI Explainability for Urban Systems (AI-EXUS)

Layered AI Architecture for Transparent, Explainable, and Accountable Urban Planning and Infrastructure

Layer	Main functionality	AI module (s) and components	Operational scope (Input → Processing → Output)
Layer 1	Data collection and transparency integration: Ensuring transparency in data collection and usage practices	Data provenance and transparency management system	Input → Urban data → Output → Transparent and accountable data usage protocols
Layer 2	Lime-based explanation: Approximating complex AI models with simpler, interpretable models	LIME (Explainable AI)	Input → AI decision outputs, feature data → Output → Interpretable model approximating AI decisions
Layer 3	Counterfactual reasoning: Generating alternative outcomes to explain AI decisions	Counterfactual Reasoning	Input → AI decision outputs, input features → Output → Counterfactual explanations for alternative decisions
Layer 4	Explainability audits: Implementing audits to assess and ensure the interpretability of AI-driven decisions	Interpretability evaluation and audit engine	Input → AI decisions, interpretability metrics → Output → Transparency scores and audit reports
Layer 5	Post-AI explainability and governance oversight: Evaluating AI models for fairness, transparency, and accountability	Governance transparency and compliance audit platform	Input → AI decisions, transparency and fairness checks → Output → Audited explainability reports and governance oversight

The growing complexity of machine learning models has created significant challenges in understanding how AI systems make decisions, particularly in domains such as civil engineering and urban governance, where AI-driven recommendations impact infrastructure development, resource allocation, and public policy. To address these challenges, Explainable AI (XAI) techniques have been developed to enhance transparency, interpretability, and trust in AI-driven decision-making (Barredo Arrieta et al. 2020). XAI methods provide insights into the internal workings of AI models, making it possible for engineers, policymakers, and stakeholders to evaluate and manage AI systems effectively.

Explainability is especially crucial in high-stakes applications where AI-driven decisions carry significant social and economic consequences. In civil engineering, AI is used to optimize traffic systems, predict structural failures, and allocate public resources, yet without proper explainability, these models may lead to unintended biases, unfair outcomes, or inefficient planning strategies. XAI techniques ensure that AI models do not function as black-box systems but instead provide clear, interpretable justifications for their recommendations, fostering accountability and public trust in AI-assisted urban governance.

A foundational approach in XAI is the use of Shapley value-based feature attribution, which assigns contribution scores to input variables, allowing stakeholders to understand how AI models prioritize different factors. The Shapley value-based explainability model is formulated as:

$$\phi_i(f) = \sum_{S \subseteq N\setminus\{i\}} \frac{|S|!(|N|-|S|-1)!}{|N|!}[f(S \cup \{i\}) - f(S)]$$

where $\phi_i(f)$, Shapley value quantifying the contribution of input feature i to the model's decision, $f(S)$, model output for a subset S of features, N, total set of features, $|S|$, number of elements in subset S. This model enables urban planners to assess the factors driving AI-driven decisions, ensuring that policies, such as infrastructure investments or resource allocations, are based on fair and justifiable reasoning rather than opaque model outputs.

Another widely used XAI method is saliency mapping, which highlights which parts of an input dataset most strongly influence the AI model's decision. In urban governance, saliency maps can be applied to geospatial AI models to determine which city zones receive higher AI-assigned priorities for development projects. The saliency-based explainability model is formulated as:

$$S_{i,j} = \left| \frac{\partial f(X)}{\partial X_{i,j}} \right|$$

where $S_{i,j}$, saliency score for input feature at position (i,j), $f(X)$, AI model's output given input X, $\frac{\partial f(X)}{\partial X_{i,j}}$, gradient of the model's output with respect to input feature $X_{i,j}$, measuring how sensitive the model is to change in this variable. This technique allows urban planners and civil engineers to visually interpret how AI systems prioritize different areas for infrastructure investment, public service distribution, or urban development initiatives, ensuring fair and unbiased decision-making.

To further reinforce XAI-driven accountability, governance frameworks should incorporate explainability audits, ensuring that AI systems comply with fairness and transparency regulations. A formal XAI audit model can be expressed as:

$$\text{XAI\_Compliance\_Score} = \sum_{i=1}^{n} w_i \cdot f(\text{Explainability\_Metric}_i)$$

where w_i, weight assigned to each explainability metric, Explainability _ Metric$_i$, score measuring AI model transparency based on predefined evaluation criteria, $f(\cdot)$, function aggregating the weighted explainability scores into an overall compliance assessment.

Through the integration of Shapley value-based feature attribution, saliency mapping, and explainability audits, AI-driven urban governance systems can operate with higher levels of trust, fairness, and public accountability. These techniques enhance oversight and scrutiny of AI systems, ensuring that decisions affecting public infrastructure, resource distribution, and civil engineering projects are well-justified, interpretable, and equitable.

A well-designed XAI framework ensures that AI-driven decision-making in urban governance remains transparent, auditable, and aligned with ethical principles, fostering greater trust and public confidence in AI-assisted urban planning and infrastructure development.

Framework: AI Explainability for Urban Systems (AI-EXFUS)

Layered AI Architecture for Transparent, Explainable, and Accountable Urban Planning and Infrastructure

Layer	Main functionality	AI module (s) and components	Operational scope (Input → Processing → Output)
Layer 1	Data collection and transparency integration: Ensuring transparency in data collection and usage practices	Data provenance and transparency management system	Input → Urban data → Output → Transparent and accountable data usage protocols
Layer 2	Shapley value-based explanation: Quantifying the contribution of each feature to AI-driven decisions	Shapley Values (XAI)	Input → AI decision outputs, feature data → Output → Transparent feature attributions for decision-making
Layer 3	Saliency mapping: Highlighting influential features in AI-driven decisions	Saliency Mapping (XAI)	Input → AI decision outputs, input features → Output → Visualized influence of features on decision-making
Layer 4	Counterfactual reasoning: Generating alternative outcomes to explain AI decisions	Counterfactual Reasoning (XAI)	Input → AI decision outputs, input features → Output → Counterfactual explanations for alternative decisions
Layer 5	Explainability audits: Implementing audits to assess and ensure the interpretability of AI-driven decisions	Interpretability evaluation and audit engine	Input → AI decisions, interpretability metrics → Output → Transparency scores and audit reports

Explainability in AI is not only a technical requirement but also a fundamental ethical necessity, ensuring that AI systems align with fairness, equity, and justice. In urban governance and civil engineering, where AI-driven decisions influence public resource distribution, infrastructure planning, and social services, explainability plays a crucial role in identifying and mitigating biases that may lead to discriminatory outcomes (Floridi et al. 2018). By making AI decision-making processes transparent and interpretable, explainability allows policymakers, engineers, and stakeholders to evaluate whether AI-driven recommendations are fair and equitable, particularly in cases affecting vulnerable or marginalized populations.

One of the primary risks associated with black-box AI models is that they can inadvertently reinforce historical inequalities by learning from biased data. For example, an AI system trained on historical infrastructure investments may prioritize wealthier districts for development while neglecting low-income communities. To prevent such biases, AI models must be designed with explainability constraints, ensuring that decision-making processes can be scrutinized for fairness. One approach to operationalizing this requirement is the Fairness-Constrained Explainability Model (FCEM), formulated as:

$$\min_{\mathrm{w}} L(\mathrm{w}, D) + \lambda \sum_{i=1}^{n} \text{Bias\_Penalty}(y_i, \widehat{y_i})$$

where $L(\mathrm{w}, D)$, standard loss function of the AI model, w, model parameters, λ, fairness regularization term, $\text{Bias\_Penalty}(y_i, \widehat{y_i})$, fairness constraint penalizing biased predictions. This model ensures that AI-driven urban planning decisions are not only optimized for efficiency but also audited for fairness, allowing for early detection of potentially discriminatory patterns.

In addition to fairness-constrained optimization, counterfactual explainability is another powerful tool for evaluating AI decisions in urban governance. Counterfactual explanations allow decision-makers to explore alternative AI-driven outcomes, helping to determine whether changes in input variables would have led to different results, thereby identifying potential biases. The counterfactual fairness model is formulated as:

$$\arg \min_{\tilde{x}} \lambda d(x, \tilde{x}) + L(f(\tilde{x}), y^*)$$

where $\tilde{x}$, counterfactual input, x, original input, λ, regularization parameter, $d(x, \tilde{x})$, distance function measuring similarity between original and counterfactual input, $L(f(\tilde{x}), y^*)$, loss function ensuring the counterfactual results in a fairer alternative outcome y^*. By applying counterfactual fairness analysis, policymakers can evaluate whether marginalized communities receive equitable AI-driven recommendations and adjust policies accordingly.

To reinforce explainability and ethical AI practices, urban governance frameworks must also institutionalize AI fairness audits. These audits should regularly evaluate AI models to ensure compliance with fairness and equity principles. A formal fairness audit model can be expressed as:

$$\text{Fairness\_Audit\_Score} = \sum_{i=1}^{n} w_i \cdot f(\text{Equity\_Metric}_i)$$

where w_i, weight assigned to each fairness metric, Equity _ Metric$_i$, evaluation score measuring AI fairness, f (·), function aggregating fairness metrics into an overall audit score.

The integration of fairness-constrained optimization, counterfactual fairness analysis, and institutional fairness audits forms a comprehensive ethical AI governance framework, ensuring that AI-driven urban systems operate in an accountable, transparent, and just manner. These techniques provide the necessary oversight to prevent AI from reinforcing existing inequalities, ensuring that technological advancements benefit all communities equitably.

By embedding explainability within AI-driven decision-making, urban governance can enhance public trust, prevent discriminatory AI outcomes, and ensure that AI aligns with broader ethical principles of fairness and justice.

Framework: Fairness-Constrained Urban AI Governance Framework (FCUAGF)

Layered AI Architecture for Urban Infrastructure and Governance

Layer	Main functionality	AI module(s) and components	Operational scope
Layer 1	Data collection: Gathering data from public urban infrastructure	Data quality validation and bias correction engine	Input → Raw data (e.g., sensor streams, demographic data) → Output → Preprocessed data for AI models
Layer 2	Learning: Fairness-Constrained Explainability Model (FCEM) for decision-making	Fairness-Constrained Explainability Model (FCEM)	Input → Historical urban data → Output → Fairness-optimized infrastructure decisions
Layer 3	Counterfactual explainability: Evaluate alternative urban outcomes for fairness	Counterfactual fairness model	Input → Original urban data → Output → Counterfactual predictions ensuring fairness
Layer 4	Ethical AI audits: Fairness audits of AI-driven urban systems	Fairness monitoring and compliance audit system	Input → Audit data, fairness metrics → Output → Fairness scores and audit reports
Layer 5	Governance: Post-AI fairness evaluation and feedback	Governance feedback and policy adaptation engine	Input → Feedback from AI audits → Output → Policy adjustments based on fairness findings

5.2.3 Challenges to Achieving Transparency and Explainability

While transparency and explainability are fundamental to ensuring ethical AI systems, their implementation presents significant challenges. Modern AI models, particularly those based on deep learning, operate using vast datasets and highly complex mathematical architectures, making it difficult to provide clear and intuitive explanations for their outputs (Barredo Arrieta et al. 2020). This lack of interpretability is particularly concerning in urban governance and civil engineering, where AI-driven decisions influence infrastructure planning, public services, and resource allocation.

A primary challenge to explainability is the black-box nature of deep learning models. Unlike traditional rule-based systems, deep neural networks process data through multiple hidden layers, each applying nonlinear transformations that are difficult to trace back to a human-understandable decision logic. This opacity reduces stakeholder trust and makes it challenging for policymakers and engineers to validate AI-generated recommendations.

Additionally, there is often a trade-off between model performance and interpretability. More accurate AI models, including ensemble learning approaches, deep neural networks, and transformer-based architectures, tend to be less interpretable, whereas simpler models, such as decision trees and linear regression, are easier to understand but may lack predictive power and robustness (Mökander 2023). Balancing these competing demands is one of the most pressing challenges faced by AI researchers and developers.

To mitigate this challenge, AI systems must integrate hybrid models that balance interpretability and performance. One approach is the Performance-Explainability Trade-off Model (PETM), which introduces a regularization constraint to optimize both accuracy and interpretability. The model is formulated as:

$$\min_{\mathrm{w}} \left[\alpha L_{\text{performance}}(\mathrm{w}, D) + \beta L_{\text{interpretability}}(\mathrm{w}, D)\right]$$

where $L_{\text{performance}}(\mathrm{w}, D)$, loss function optimizing AI accuracy, $L_{\text{interpretability}}(\mathrm{w}, D)$, loss function penalizing excessive complexity, α, weight assigned to performance, β, weight assigned to interpretability.

This model ensures that AI-driven urban governance systems achieve high performance without sacrificing explainability, making decisions more accessible to policymakers and stakeholders.

Another challenge to explainability is the computational cost associated with generating interpretable explanations for deep learning models. Post hoc explainability techniques, such as Local Interpretable Model-Agnostic Explanations (LIME) and Shapley values, require extensive computational resources to approximate AI model behavior. To address this, an Efficient Approximate Explainability Model (EAEM) can be employed, reducing computational complexity while maintaining interpretability. The EAEM model is formulated as:

$$\hat{f}(X) = \sum_{i=1}^{m} w_i g_i(X)$$

where $\hat{f}(X)$, approximated explainable model, w_i, weight assigned to simple model $g_i(X)$, m, number of interpretable models combined. This approach replaces fully opaque deep learning models with simplified, interpretable approximations, allowing AI systems to provide transparent decisions without significant performance losses.

Beyond computational constraints, a key governance challenge is ensuring that explainability is not treated as an afterthought but rather as a core requirement in AI development. This necessitates regulatory frameworks that enforce explainability audits at each stage of AI deployment. A formal AI Explainability Audit Model is expressed as:

$$\text{Explainability\_Compliance\_Score} = \sum_{i=1}^{n} w_i \cdot f(\text{Transparency\_Metric}_i)$$

where w_i, weight assigned to each transparency metric, Transparency _ Metric$_i$, score measuring explainability, $f(\cdot)$, function aggregating transparency evaluations.

This governance framework ensures that AI models in urban development are systematically evaluated for transparency, preventing black-box decision-making that could result in unjust or biased outcomes.

By employing performance-explainability trade-off models, efficient approximation techniques, and regulatory audits, AI-driven urban governance systems can achieve greater transparency while maintaining accuracy and efficiency. This approach ensures that AI remains both a powerful and accountable tool in infrastructure planning, public services, and urban decision-making.

Framework: Performance-Explainability Hybrid AI Governance Framework (PEHAGF)

Layered AI Architecture for Urban Infrastructure and Governance Transparency

Layer	Main functionality	AI module (s) and components	Operational scope
Layer 1	Data collection: Gather urban infrastructure and service data	Data provenance and quality management system	Input → raw urban data (e.g., sensor data, demographic information) → Output → Preprocessed data for AI models
Layer 2	Learning: Performance-Explainability Trade-off Model (PETM) for balancing accuracy and interpretability	Performance-Explainability Trade-off Model (PETM)	Input → Raw data, model weights → Output → Optimized urban decisions balancing accuracy and interpretability
Layer 3	Approximation: Efficient Approximate Explainability Model (EAEM) for reducing computational cost	Efficient Approximate Explainability Model (EAEM)	Input → Complex deep learning models → Output → Approximated interpretable models

(continued)

Layer	Main functionality	AI module (s) and components	Operational scope
Layer 4	Governance: AI explainability audit and compliance	Explainability compliance and audit engine	Input → Transparency metrics, audit data → Output → Transparency compliance score and audit reports
Layer 5	Post-AI feedback: Continuous governance and feedback for explainability and performance	Governance feedback and adaptive optimization platform	Input → Feedback data, transparency reports → Output → Refined urban governance strategies

Another critical challenge in ensuring transparent and accessible AI systems is the need for context-specific explainability, namely the ability to tailor explanations to different stakeholders according to their role in the decision-making process. In urban governance and civil engineering, AI-driven decisions affect multiple actors, from city planners and policymakers to engineers and the general public. Each group requires different levels of explanation to effectively understand, evaluate, and engage with AI-driven decision-making processes (Floridi et al. 2018).

For instance, city planners and engineers require technical and model-specific explanations to assess how an AI system optimizes traffic flow, allocates public resources, or designs infrastructure projects. These explanations often include detailed algorithmic insights, statistical justifications, and model performance metrics. Conversely, members of the public may require a high-level summary of how the AI system operates and why certain decisions were made, allowing them to assess whether the system aligns with community needs and fairness principles.

Developing multi-tiered explainability frameworks that provide appropriate levels of detail to different audiences is therefore a fundamental challenge in AI-driven urban governance. One approach is the Adaptive Explainability Framework (AEF), which dynamically adjusts the level of explanation provided according to the needs of each stakeholder group. This framework is mathematically formulated as:

$$E_i^* = \arg \underset{E_i}{min} \left[\alpha L_{\text{complexity}}(E_i) + \beta L_{\text{informativeness}}(E_i)\right]$$

where E_i^*, optimal explanation for stakeholder i, E_i, candidate explanations, $L_{\text{complexity}}(E_i)$, complexity penalty ensuring that non-technical audiences receive simplified explanations, $L_{\text{informativeness}}(E_i)$, penalty for overly vague explanations ensuring that technical stakeholders receive sufficient detail, α, β, weight parameters balancing complexity and informativeness. This model ensures that technical audiences receive high-detail justifications while non-technical audiences receive simplified yet meaningful summaries of AI decisions.

Beyond adaptive explainability, AI systems must also support multimodal explanations, delivering information through text, visualizations, and interactive dashboards to improve accessibility. One technique for achieving this is explainability through visualization, which helps non-expert stakeholders interpret AI-generated decisions in an intuitive manner. The Visualization-Based Explainability Model (VBEM) is formulated as:

$$V_i^* = \arg\max_{V_i} [U(V_i, S_i)]$$

where V_i^*, optimal visualization method for stakeholder i, V_i, candidate visual representations (e.g., heatmaps, bar charts, geospatial maps), $U(V_i, S_i)$, utility function measuring interpretability of visualization V_i for stakeholder S_i. This framework ensures that policy reports, public consultations, and urban planning documents can incorporate AI-driven visual explanations, making AI outputs more accessible to diverse audiences.

To further institutionalize context-specific explainability, governance frameworks must include Explainability Access Controls (EACs), which automatically tailor AI-generated reports based on the stakeholder's role. The EAC model is formulated as:

$$A_i^* = \arg\max_{A_i} [P(A_i|R_i)]$$

where A_i^*, optimal explanation access level for stakeholder i, A_i, available explanation levels, R_i, role of stakeholder i, $P(A_i| R_i)$, probability distribution ensuring that explanations are appropriate for the stakeholder's role and expertise.

By taking advantage of adaptive explainability, visualization-based explanations, and role-specific access controls, AI-driven urban systems can ensure greater transparency, accountability, and stakeholder engagement. These techniques enhance public trust, enable informed policymaking, and foster responsible AI governance, ensuring that AI systems are understood, trusted, and effectively integrated into urban decision-making.

Framework: Context-Specific AI Explainability Governance Framework (CSAEGF)

Layered AI Architecture for Adaptive, Visualization, and Role-Specific Explainability

Layer	Main functionality	AI module (s) and components	Operational scope
Layer 1	Data collection: Gather diverse data from urban infrastructure	Data provenance and quality management system	Input → Raw urban data (e.g., sensor data, public reports) → Output → Preprocessed data for AI models
Layer 2	Learning: Adaptive Explainability Framework (AEF) for role-based explanations	Adaptive Explainability Framework (AEF)	Input → Raw data, stakeholder roles → Output → Tailored explanations balancing complexity and informativeness

(continued)

Layer	Main functionality	AI module (s) and components	Operational scope
Layer 3	Visualization: Visualization-Based Explainability Model (VBEM) for intuitive understanding	Visualization-Based Explainability Model (VBEM)	Input → AI-generated decision data → Output → Visual representations such as heatmaps, bar charts, geospatial maps
Layer 4	Access Control: Explainability Access Controls (EAC) for tailored report distribution	Explainability Access Controls (EAC)	Input → Stakeholder roles, available explanations → Output → Role-specific explanation access levels
Layer 5	Governance: Post-AI transparency audits and feedback for continuous improvement	Transparency audit and governance feedback platform	Input → Feedback data, transparency evaluations → Output → Refined governance strategies ensuring continuous explainability

Figure 5.2 presents the essential components of transparency and explainability in AI, organized into three main areas: the Need for Transparency, the Role of Explainability, and Challenges to Transparency and Explainability. "Need for Transparency" emphasizes understanding AI decisions, especially in high-stakes areas like infrastructure and resource allocation, to prevent discriminatory outcomes. "Role of Explainability" focuses on Explainable AI (XAI) techniques, enhancing accountability, and aligning AI with ethical principles to address biases and promote

Fig. 5.2 Transparency and explainability in AI

fairness. "Challenges to Transparency and Explainability" highlights obstacles like the complexity of deep learning models, the trade-off between accuracy and interpretability, and the need for context-specific explanations tailored to different stakeholders, such as planners and the public.

Transparency and explainability are essential for promoting the ethical use of AI in civil engineering, urban development, and governance. These principles are critical for building trust, ensuring accountability, and enabling stakeholders to understand and scrutinize AI-driven decisions. While achieving transparency and explainability presents significant challenges, particularly in the context of complex machine learning models, the development of explainable AI techniques offers promising solutions for enhancing the interpretability and accountability of AI systems. By prioritizing transparency and explainability, AI developers and policymakers can ensure that AI systems are aligned with ethical principles and contribute to the fair and equitable development of urban environments.

5.3 Accountability and Responsibility

The integration of AI systems into civil engineering, urban development, and governance brings forward the critical need for accountability and responsibility. These principles are essential for ensuring that AI-driven decisions are not only technically sound but also ethically grounded and aligned with societal values. Without clear mechanisms for accountability, the deployment of AI systems risks exacerbating inequalities, perpetuating biases, and causing harm without appropriate redress (Sargiotis 2024d).

5.3.1 Challenges in Defining Accountability

In the context of AI systems used in civil engineering and urban development, accountability refers to the mechanisms through which individuals or institutions are held responsible for decisions made by AI systems, especially when those decisions have significant impacts on public infrastructure, safety, and resource allocation (Floridi et al. 2018). One of the most prominent challenges in AI governance is the lack of direct human intervention in the decision-making process, as autonomous AI models can make decisions without direct human approval. This characteristic becomes particularly problematic when such decisions influence high-stakes areas such as urban planning and public services, where the effects of AI systems are far-reaching and cannot easily be traced back to a single source or actor (Manzoor et al. 2021).

This problem is further complicated by the distributed nature of responsibility within AI-driven urban systems. These systems are typically integrated with data from various sources, including sensor networks, transportation grids, and energy systems, making it challenging to define clearly who is accountable when AI systems fail to deliver fair or just outcomes. When AI models are developed and deployed by multiple actors, including government agencies, private contractors, and technology vendors, responsibility becomes diffused. This diffusion often creates ambiguity in identifying the party responsible for unintended consequences resulting from AI-driven decisions.

To resolve these challenges, we propose a framework for AI accountability that ensures AI-driven decisions are traceable, auditable, and attributable to the responsible stakeholders. A well-defined approach is necessary to distribute accountability across the various stakeholders based on their role in AI development, deployment, and oversight. The AI Decision Attribution Model (AIDAM) introduces a weighted responsibility matrix, which assigns a proportional accountability score to each stakeholder based on their influence on the AI system's decisions.

This model allows for the quantification of accountability by measuring the extent of responsibility of each stakeholder involved in the decision-making process. The equation is formulated as:

$$R_i = \sum_{j=1}^{n} w_j \cdot C_{ij}$$

where R_i, accountability score, w_j, weight, C_{ij}, responsibility attribution matrix. This function ensures the precise attribution of accountability to each stakeholder, allowing for the fair distribution of responsibility based on their influence on AI decision-making.

As we move toward a more comprehensive framework, another significant challenge arises due to the absence of direct human oversight in autonomous systems. Traditional decision-making processes often involve explicit human accountability, but in autonomous AI models, responsibility becomes obscured, as the system operates independently. Therefore, to preserve accountability, it is critical to establish a Human-in-the-Loop (HITL) framework that enables human oversight at key decision-making junctures. The HITL model integrates human judgment with AI decision-making, ensuring that human decision-makers can intervene when necessary, particularly in high-stakes scenarios.

$$D_t^* = \lambda D_t^{\text{AI}} + (1 - \lambda) D_t^{\text{Human}}$$

where D_t^*, final AI-driven decision at time t, D_t^{AI}, AI-generated decision, D_t^{Human}, human override decision, λ, control parameter balancing AI autonomy and human intervention. This formulation ensures that human oversight remains an integral component of AI governance, preventing fully autonomous AI systems from making high-impact urban planning or civil engineering decisions without human validation.

Another challenge in defining AI accountability arises in the legal domain, where traditional liability frameworks do not fully accommodate the autonomous decision-making of AI systems. When AI-driven decisions lead to harmful outcomes, such as infrastructure failures or biased resource distribution, determining legal liability becomes increasingly complex. In such cases, it is necessary to develop a model that accounts for the legal and regulatory aspects of accountability.

The AI Liability Model (ALM) introduces a framework to categorize liability into three components: technical, operational, and regulatory. This model allows for the dynamic adjustment of liability weights based on the specific application of the AI system and the severity of the consequences. The equation is:

$$L = \alpha L_{\text{technical}} + \beta L_{\text{operational}} + \gamma L_{\text{regulatory}}$$

where L, overall liability, $L_{\text{technical}}$, technical responsibility, $L_{\text{operational}}$, operational responsibility, $L_{\text{regulatory}}$, regulatory responsibility, α, β, γ, adjustable parameters. This function assigns liability for AI-driven decisions, distributing responsibility among the developers, deployers, and regulators based on their role in the AI decision-making process.

Next, a transparency metric is calculated using Shapley values, a method derived from cooperative game theory that assigns each input feature a quantitative value based on its contribution to the final decision. This allows for an interpretable breakdown of AI decisions, ensuring clarity and public accountability. The Shapley value-based transparency model is formulated as follows:

$$\phi_i(f) = \sum_{S \subseteq N \setminus \{i\}} \frac{|S|!(|N| - |S| - 1)!}{|N|!} [f(S \cup \{i\}) - f(S)]$$

where $\phi_i(f)$, Shapley value for feature i, $f(S)$, function output for subset S, N, total set of features. This function measures the impact of each feature on the AI system's decision, providing an interpretable explanation for stakeholders. This function computes the contribution of each feature to the AI model's decision, offering insights into the reasoning behind the model's predictions and making the decision-making process transparent. This method is vital for both public trust and for internal audits, ensuring that AI systems are both interpretable and accountable. By understanding the contribution of each feature, stakeholders are assured that AI-driven decisions are based on fair, transparent, and traceable methodologies.

Following the need for transparency, the regulatory compliance aspect ensures that AI decisions adhere to both legal and ethical standards. In urban environments, where AI decisions can have significant societal impacts, ensuring compliance with local, national, and international laws is essential. AI-driven systems must not only comply with established regulations but also be able to demonstrate that their actions align with ethical guidelines. This process is supported by Compliance Auditing Models (CAM), which provide a comprehensive mechanism for auditing AI decisions to ensure that all decisions are in compliance with legal frameworks.

To establish this compliance, the Compliance Auditing Model (CAM) evaluates the extent to which AI decisions meet regulatory standards by producing a compliance score. This score is a weighted summation of various regulatory metrics, with each metric reflecting the degree to which a particular decision adheres to legal or ethical standards. The compliance score formula is expressed as:

$$\text{Compliance Score} = \sum_{i=1}^{n} w_i \cdot f(\text{Regulatory Metric}_i)$$

where w_i, weight for each regulatory metric, Regulatory Metric$_i$, evaluation score for each metric.

This model ensures that AI-driven urban systems not only meet performance goals but also comply with necessary legal and ethical requirements. Compliance score allows for systematic assessment of AI systems, providing a tangible measure of their compliance with regulatory frameworks. By incorporating compliance auditing into the decision-making process, we ensure that AI systems remain responsible and within the boundaries of established laws, preventing any legal or ethical violations in their operation.

Thus, the combination of Post-AI reporting & transparency and regulatory compliance frameworks ensures that AI systems in urban environments are transparent, auditable, and compliant with the law. These frameworks guarantee that AI decisions are not only efficient and effective but also ethical, fair, and justifiable in the eyes of the public and regulatory bodies, fostering a robust governance structure for AI-driven urban systems.

Framework: AI Accountability and Governance Framework (AIAGF)

Layered AI Architecture for Accountability, Responsibility, and Transparency in Urban Decision-Making

Layer	Main functionality	AI module(s) and components	Operational scope
Layer 1	Data collection	Data provenance and quality management system	Input → Raw sensor data (e.g., traffic, energy consumption, demographic data) → Output → Preprocessed input ready for AI models
Layer 2	Stakeholder influence and responsibility	Stakeholder responsibility mapping engine	Input → Stakeholder data, influence metrics → Output → Responsibility scores for each stakeholder
Layer 3	AI decision attribution	AIDAM	Input → Responsibility scores, decision logs → Output → Accountability scores based on influence over decisions
Layer 4	Human-in-the-Loop integration	HITL	Input → AI-generated decisions, human feedback → Output → Final decision adjusted by human feedback
Layer 5	Dynamic liability adjustment	AI Liability Model (ALM)	Input → Decision outcomes, failure logs → Output → Liability assignments for each stakeholder based on role
Layer 6		Shapley values	

(continued)

Layer	Main functionality	AI module(s) and components	Operational scope
	Post-AI reporting and transparency		Input → AI decision logs, accountability scores → Output → Public-facing transparency reports on decision-making
Layer 7	Regulatory compliance	Compliance Auditing Models (CAM)	Input → Regulatory guidelines, decision logs → Output → Compliance scores and corrective actions

One of the most complex issues in AI governance is determining who should be held accountable for decisions made by AI systems. In traditional decision-making processes, responsibility is clearly assigned to individuals or institutions that make the final decision. However, in AI-driven urban governance and civil engineering, responsibility becomes more diffuse, as multiple actors contribute to the design, deployment, and oversight of AI systems. This diffusion of responsibility complicates efforts to ensure that AI-driven decisions remain ethical and aligned with public interest (Mökander 2023).

In AI-driven urban systems, there are several possible points of accountability. One perspective is that responsibility should lie with the engineers and data scientists who design and develop the AI models. These individuals define the model's structure, select the training data, and establish optimization objectives, all of which influence the model's decision-making behavior. If an AI system produces biased or unjust outcomes, should the responsibility be attributed to the technical choices made by its developers?

Alternatively, one could argue that accountability should rest with the organizations that deploy AI systems in urban infrastructure projects, public services, and transportation networks. These organizations, whether public institutions or private companies, decide how AI models are integrated into decision-making processes. They determine whether AI-generated recommendations will be followed automatically or whether human oversight will be involved. If AI-driven policies lead to social inequalities or resource misallocation, should the deploying organizations bear the responsibility for those outcomes?

A third perspective suggests that accountability should be placed on policymakers and regulatory bodies that oversee the use of AI in urban governance. Governments and regulatory institutions establish the legal frameworks that define acceptable AI applications, ethical guidelines, and compliance standards. If AI-driven urban planning results in unfair distribution of resources or infrastructure bias, should policymakers be held accountable for failing to impose adequate safeguards?

To translate these accountability concerns into decision-level governance, we propose a Fairness-Constrained Reinforcement Learning (FCRL) model. While accountability requires institutional attribution of responsibility across developers, deploying organizations, and regulatory bodies, it also requires that the AI decision process itself be constrained against unfair or socially harmful outcomes. The FCRL model addresses this second requirement by integrating fairness directly into the policy optimization process, ensuring that AI decisions, especially those affecting urban planning, resource allocation, and infrastructure development, are equitable, reviewable, and ethically aligned.

The primary objective of the FCRL model is to minimize a cost function $C(\pi)$, subject to fairness constraints represented by $F(\pi)$, ensuring that AI decisions are both efficient and equitable. This problem is formulated as:

$$\text{minimize} J(\pi) = \mathbb{E}[C(\pi)] + \lambda \mathbb{E}[F(\pi)]$$

where π, policy learned by the reinforcement learning agent dictating the decision-making process, $C(\pi)$, cost function representing the operational efficiency of the AI system, $F(\pi)$, fairness constraint ensuring that decisions are equitably distributed across different demographic groups, λ, hyperparameter balancing the trade-off between optimizing costs and ensuring fairness.

Building upon the foundational integration of the accountability & traceability module and Fairness-Constrained Reinforcement Learning (FCRL) Module, we now extend the AI framework to address other critical aspects of governance, responsibility, and decision-making transparency in AI-driven urban systems. These modules, designed to ensure not only technical effectiveness but also fairness and accountability, must be complemented by further enhancements that encapsulate regulatory oversight, human involvement, and dynamic adaptability in an ever-changing environment.

The following AI modules contribute to a holistic, ethically grounded framework in urban governance:

$$R_i = \sum_{j=1}^{n} w_j \cdot C_{ij}$$

where R_i represents the responsibility assigned to entity i, w_j is the weight representing the level of control over decision component j, and C_{ij} is a matrix defining the contributions of each entity to the AI system's decisions.

This formulation ensures that responsibility is distributed based on the degree of influence each actor has on the AI decision-making process, allowing for a more precise allocation of accountability.

Another key challenge is the issue of legal liability in AI-driven urban systems. Traditional legal frameworks were designed for human-led decision-making and do not always account for autonomous decision-making processes. If an AI system deployed in urban planning or civil engineering causes harm such as an infrastructure failure due to flawed predictive modeling, determining legal liability becomes complex. Should liability fall on the AI's creators, the agency that implemented the model, or the policymakers who allowed its deployment?

To navigate this challenge, a structured liability model must be developed to distinguish between technical, operational, and regulatory accountability. This model can be expressed mathematically as:

$$\text{Compliance}(t) = \mathbb{E}[\text{Violation}(t)] - \alpha \cdot \mathbb{E}[\text{Compliance Indicator}(t)]$$

where Compliance(t), measure of the system's compliance at time t, Violation(t), violation of ethical or legal standards at time t, α, a regularization factor governing the trade-off between compliance and violation, Compliance Indicator(t), a function representing adherence to legal frameworks and policies at time t. This function monitors the AI's adherence to the set regulatory standards, ensuring that the AI's decisions are legally acceptable and ethically sound. By introducing the compliance indicator and violation terms, this module ensures that AI-driven actions in urban governance remain within the bounds of predefined frameworks, ensuring accountability to society.

While AI can significantly enhance decision-making processes, human judgment remains crucial in complex, ethically sensitive domains such as urban governance. The Human-in-the-Loop (HITL) Module allows human oversight to influence AI decisions, ensuring that AI-generated recommendations are validated and adjusted when necessary, by human experts. The HITL model can be expressed as follows:

$$H(s_t, a_t) = \beta \cdot R(s_t, a_t) + (1 - \beta) \cdot \text{Human Input}(t)$$

where $H(s_t, a_t)$, final decision at time t for state-action pair (s_t, a_t), $R(s_t, a_t)$, reward function guiding the AI's decision-making process, β, weight assigned to the AI's recommendation, Human Input(t), input provided by human oversight at time t. This function models the collaborative process where the AI's decision is adjusted by human feedback. The parameter β controls how much weight is given to the AI-generated decision relative to human input, allowing for adaptive decision-making in dynamic urban governance scenarios.

Transparency and interpretability are core components of responsible AI deployment, particularly in urban governance, where decisions impact public welfare. The Explainable AI (XAI) Module ensures that AI decisions are not only transparent but also understandable by non-experts, including policymakers, engineers, and the general public. The XAI framework can be mathematically expressed as:

$$\text{Explained Output} = f(\pi) + \gamma \cdot \text{Interpretability}(f(\pi))$$

where Explained Output, final human-readable explanation of the AI's decision, $f(\pi)$, the output generated by the AI system based on its policy π, γ, weight assigned to the interpretability of the decision, Interpretability($f(\pi)$), function representing how well the AI's decision can be explained to human stakeholders. This function ensures that the AI system provides clear, interpretable justifications for its decisions, increasing trust and enabling better engagement from stakeholders in urban governance. The parameter γ regulates the trade-off between making decisions that are both optimized and explainable.

Urban systems are dynamic, requiring AI models to adapt in real time to changing conditions. The dynamic adaptation and learning module ensures that the AI continuously learns from new data and feedback, improving over time while maintaining its fairness and compliance characteristics. The dynamic learning process can be modeled by:

$$\Delta\theta(t) = \eta \cdot \nabla_{\theta}\mathcal{L}(\theta) + \lambda \cdot \mathbb{E}[\text{Fairness Penalty}(t)]$$

where $\Delta\theta(t)$, change in model parameters θ at time t, η, learning rate for the model updates, $\nabla_{\theta}\mathcal{L}(\theta)$, gradient of the loss function $\mathcal{L}(\theta)$ with respect to the model parameters, λ, regularization factor, Fairness Penalty(t), fairness penalty ensuring that new learning does not violate previously established fairness constraints. This update rule ensures that the AI system learns continuously from new data while incorporating fairness constraints. By updating model parameters over time, this module allows the AI to remain adaptive to evolving urban conditions without compromising fairness or compliance.

The proposed AI framework addresses the complex issue of accountability in urban governance by ensuring that all AI-driven decisions are traceable, fair, and aligned with both legal and societal expectations. The Accountability & Traceability Module and FCRL serve as the foundation of this framework, with the additional modules ensuring compliance, transparency, adaptability, and human involvement. This multifaceted approach ensures that AI in urban systems operates not only with technical efficacy but also with ethical responsibility, paving the way for more just, equitable, and accountable AI deployments in future cities.

To mathematically formalize this multidimensional attribution of responsibility, we define the overall liability L as the weighted sum of its technical, operational, and regulatory components, each reflecting the distinct role of stakeholders in AI governance:

$$L = \alpha L_{\text{technical}} + \beta L_{\text{operational}} + \gamma L_{\text{regulatory}}$$

where L represents overall liability, $L_{\text{technical}}$ accounts for the responsibility of AI developers, $L_{\text{operational}}$ covers the responsibility of deploying organizations, and $L_{\text{regulatory}}$ refers to the accountability of policymakers. The parameters α, β, γ adjust the relative weight of each responsibility component depending on the AI system's specific application.

By instituting a rigorously structured approach to responsibility attribution and liability, AI-driven urban systems can operate with heightened ethical accountability, strengthen legal clarity, and reinforce public trust. Such a framework ensures that all stakeholders—engineers, deploying organizations, and policymakers—clearly understand and fulfill their respective roles in safeguarding fairness and transparency within AI-assisted decision-making processes. Establishing a well-defined accountability architecture is imperative to pre-empt ethical blind spots in AI governance, guaranteeing that these systems remain instruments of public good while steadfastly upholding fairness, transparency, and responsible innovation in civil engineering and urban development.

Framework: Fairness-Constrained Urban AI Governance (FCUAG II)

Layered AI Architecture for Decentralized Urban Intelligence

Layer	Main functionality	AI module(s) and components	Operational scope (Input → Processing → Output)
Layer 1	Identify socio-political constraints and inequality mapping	Socio-political data analysis and fairness mapping engine	Input: Historical data, socio-political context → Processing: Analyze biases, inequalities → Output: Ethical constraints, fairness measures
Layer 2	Define AI decision-making process	Fairness-Constrained Reinforcement Learning (FCRL)	Input: Training data, reward functions, fairness constraints → Processing: Optimize cost and fairness in decision-making → Output: Policy that minimizes cost while ensuring fairness
Layer 3	Assign responsibility based on contributions	Responsibility assignment model	Input: Entity contributions, decision matrix → Processing: Weight responsibility by contributions → Output: Responsibility matrix for accountability assignment
Layer 4	Monitor regulatory compliance	Regulatory compliance monitoring model	Input: Legal frameworks, ethical standards → Processing: Track adherence to laws and regulations → Output: Compliance reports, violation indicators
Layer 5	Incorporate human feedback into decisions	Human-in-the-Loop (HITL) model	Input: AI recommendations → Processing: Human adjustment and validation → Output: Final decision incorporating human oversight
Layer 6	Enhance transparency of AI decisions	Explainable AI (XAI)	Input: AI output → Processing: Interpretability models and transparency functions → Output: Human-readable explanation of AI decisions
Layer 7	Adapt to new data and feedback	Dynamic adaptation and learning model	Input: New data, feedback → Processing: Update AI parameters to adapt to changes → Output: Updated, optimized policy maintaining fairness constraints
Layer 8	Monitoring, evaluation, and regulatory auditing	Continuous monitoring and regulatory audit platform	Input: Real-world outcomes, system performance → Processing: Continuous monitoring, performance audits → Output: Evaluated outcomes, system improvements

5.3.2 Governance Frameworks for Accountability

Ensuring accountability in AI-driven decision-making requires structured governance frameworks that integrate ethical principles, regulatory oversight, and technical mechanisms for monitoring AI behavior. Without clearly defined governance structures, AI systems risk operating in opaque and weakly regulated environments, making it difficult to trace responsibility when decisions lead to unintended or harmful outcomes. In civil engineering and urban development, where AI systems influence infrastructure planning, resource distribution, and public services, responsibility for AI-driven decisions must be clearly assigned, auditable, and enforceable (Barredo Arrieta et al. 2020).

A key component of AI governance is responsibility assignment, which establishes who is accountable for the decisions AI systems make. This requires clear roles for AI developers, deploying organizations, and regulatory authorities. One approach to formalizing responsibility within AI governance is a hierarchical accountability structure, through which decisions remain traceable and oversight mechanisms are embedded across levels of responsibility. A mathematical representation of an AI governance hierarchy can be formulated as:

$$G = \sum_{i=1}^{n} A_i \cdot R_i$$

where G represents the overall governance accountability score, A_i represents the assigned accountability weight for stakeholder i, and R_i represents the responsibility level of that stakeholder in AI deployment. This model ensures that accountability is distributed in proportion to the level of influence each stakeholder has over the AI system, making it easier to identify who should be held responsible in cases of bias, errors, or failures.

Another essential aspect of AI governance is AI auditing, which involves systematically evaluating AI models to ensure they operate fairly, transparently, and in compliance with ethical and regulatory standards. Auditing mechanisms help detect biases in training data, algorithmic decision-making, and model outputs, allowing stakeholders to take corrective action before AI-driven policies negatively impact communities. The AI Auditing Compliance Model (AACM) is formulated as:

$$C = \sum_{j=1}^{m} w_j \cdot M_j$$

where C represents the overall AI compliance score, w_j is the weight assigned to compliance metric j, and M_j represents the evaluation metric for bias detection, transparency assessment, or fairness audit. This model allows regulatory bodies to quantify AI system compliance, ensuring that urban governance applications of AI adhere to ethical and legal standards.

Beyond auditing, explainable AI (XAI) techniques play a crucial role in strengthening accountability by making AI systems more interpretable. When AI models operate in high-stakes urban environments, where decisions affect transportation, energy, and public services, it is essential that engineers, policymakers, and the public understand how AI-driven recommendations are made. One method to enhance explainability is the Shapley-based explainability framework, which ensures that each input factor's contribution to the AI model's output is clearly quantified. This method is formulated as:

$$\phi_i = \sum_{S \subseteq N \setminus \{i\}} \frac{|S|!(|N| - |S| - 1)!}{|N|!} [f(S \cup \{i\}) - f(S)]$$

where ϕ_i represents the importance score of input factor i, $f(S)$ is the function output for subset S of features, and N is the total set of features.

By incorporating explainability techniques such as Shapley values, model interpretability audits, and visualization-based transparency tools, AI governance frameworks can ensure that AI-driven decisions are not only accurate but also comprehensible to stakeholders.

The integration of responsibility assignment structures, auditing mechanisms, and explainability frameworks forms a comprehensive AI governance model that enhances accountability, trust, and ethical compliance in urban decision-making. These mechanisms ensure that AI serves as a transparent and equitable tool for urban planning and civil engineering, rather than operating as an opaque system with unclear responsibility.

By embedding governance principles into AI system design and deployment, policymakers and engineers can ensure that AI-driven urban solutions align with societal values, regulatory standards, and ethical imperatives, fostering trustworthy and responsible AI development in the built environment.

Framework: AI Governance and Accountability Framework (AIGAF)

Layered AI Architecture for Responsibility, Auditing, and Explainability in AI-Driven Urban Systems

Layer	Main functionality	AI module (s) and components	Operational scope
Layer 1	Data collection: Gathering data for AI-driven urban decision-making	Data provenance and quality management system	Input → Raw data (urban systems, sensor networks) → Output → Preprocessed data for AI systems
Layer 2	Responsibility assignment: Hierarchical accountability structure for AI governance	AI governance hierarchy model	Input → AI decision data, stakeholder roles → Output → Overall governance accountability score

(continued)

Layer	Main functionality	AI module (s) and components	Operational scope
Layer 3	AI auditing: Systematic evaluation of AI compliance with ethical standards	AI Auditing Compliance Model (AACM)	Input → AI system metrics, compliance data → Output → AI compliance score
Layer 4	Explainability: Shapley-based framework for AI model interpretability	Shapley-based explainability framework	Input → AI model data, feature set → Output → Importance scores of input features
Layer 5	Operational governance: Embedding governance principles into AI deployment	Governance integration and policy enforcement engine	Input → Governance frameworks, policy data → Output → Transparent, accountable AI-driven urban solutions

Governance frameworks for AI-driven urban systems must include mechanisms for redress, allowing individuals or communities negatively impacted by AI-driven decisions to challenge those decisions and seek remedies. As AI systems increasingly influence resource allocation, infrastructure development, and public safety, the absence of clear redress mechanisms poses a serious ethical and governance risk (Floridi et al. 2018). Without formal processes for reviewing AI decisions and providing corrective actions, AI systems may perpetuate harm without accountability, disproportionately affecting marginalized populations.

A robust redress framework must allow for effective review, intervention, and correction of AI-driven decisions. This requires three key components: the right of affected individuals or communities to challenge AI-generated outcomes, an independent review procedure capable of assessing the fairness and ethical validity of contested decisions, and a corrective mechanism through which harmful or biased outcomes can be revised. A structured AI redress mechanism therefore ensures that affected stakeholders have access to appeals, review, and remediation processes. One approach is to implement a review function, where contested AI decisions are evaluated according to fairness and ethical considerations. This can be formulated as:

$$R_t = \alpha H_t + \beta A_t + \gamma C_t$$

where R_t represents the final redress outcome at time t, H_t is the human-reviewed decision, A_t is the AI system's original decision, and C_t is a corrective adjustment factor. The parameters α, β, γ ensure that human intervention and corrective processes appropriately balance automated decision-making and ethical considerations. This approach ensures that AI-generated decisions are not absolute but remain subject to human oversight and legal evaluation.

To prevent unjust AI-driven decisions from requiring frequent legal intervention, governance frameworks must include proactive bias correction mechanisms. Bias audits ensure that AI systems do not systematically disadvantage certain communities and allow for real-time adjustments to prevent discriminatory outcomes before they escalate. A bias-aware AI model correction function can be represented as:

$$\min_{w} L(\mathrm{w}, D) + \delta \sum_{i=1}^{n} \mathrm{Bias\_Adjustment}(y_i, \widehat{y_i})$$

where $L(\mathrm{w}, D)$ is the AI system's loss function, ww represents model parameters, and $\mathrm{Bias\_Adjustment}(y_i, \widehat{y_i})$ quantifies the necessary correction to align AI outputs with fairness principles. The term δ serves as a regulatory constraint, ensuring that fair adjustments are integrated into AI training and decision-making processes.

For redress mechanisms to be effective, AI governance frameworks must include institutional bodies responsible for enforcing accountability. Independent AI ethics boards, ombudsman offices, and regulatory commissions should be established to review grievances and impose corrective measures when AI systems fail to operate fairly. The formal governance accountability function can be structured as:

$$O = \sum_{j=1}^{m} w_j \cdot P_j$$

where O represents the overall oversight score, w_j is the weight assigned to each governance policy, and P_j represents policy enforcement metrics ensuring AI compliance. This model ensures that AI governance is not merely advisory but has enforceable mechanisms to ensure fairness and accountability.

The integration of structured appeals processes, algorithmic fairness audits, and independent oversight bodies creates a comprehensive redress framework that ensures AI-driven urban systems remain accountable, fair, and responsive to community needs. By embedding these governance structures into AI policy, urban decision-making can align with ethical standards, legal compliance, and public trust, ensuring that AI serves as a tool for equitable urban development rather than an unchecked decision-making entity.

Framework: AI Redress and Accountability Framework (AIRAF)

Layered AI Architecture for Redress Mechanisms, Bias Correction, and Governance in AI-Driven Urban Systems

Layer	Main functionality	AI module (s) and components	Operational scope
Layer 1	Data collection: Gathering data for contested AI-driven decisions	Data provenance and quality management system	Input → Raw data (urban decisions, impacted community feedback) → Output → Preprocessed data for review functions

(continued)

Layer	Main functionality	AI module (s) and components	Operational scope
Layer 2	Redress mechanism: Structured review and correction of AI decisions	Redress review function	Input → AI decision data, human review data → Output → Final redress outcome (R_t) with corrective adjustments
Layer 3	Bias correction: Proactive bias correction to ensure fairness in AI-driven decisions	Bias-aware correction function	Input → AI model data, fairness constraints → Output → Bias-adjusted predictions
Layer 4	Governance accountability: Institutional enforcement of accountability in AI systems	Governance accountability function	Input → Governance policies, AI compliance data → Output → Overall oversight score (O) ensuring fairness
Layer 5	Institutional oversight: Independent bodies ensuring fairness and transparency in AI decision-making	Independent oversight and compliance enforcement platform	Input → Grievances, oversight reports → Output → Enforcement of corrective actions and accountability

5.3.3 *The Role of Auditing in AI Accountability*

Auditing is a fundamental aspect of AI accountability, providing a systematic approach to evaluating AI-driven decision-making processes. Regular audits help identify biases, errors, and unintended consequences, allowing for corrective measures to be implemented before AI-driven policies negatively impact communities (Mökander 2023). In civil engineering and urban development, where AI systems are deployed for infrastructure design, resource management, and public service optimization, auditing ensures that AI functions as intended, adheres to ethical principles, and remains aligned with regulatory standards.

Auditing AI systems is particularly crucial in high-stakes applications, where flawed AI-driven decisions can compromise safety, reinforce inequalities, or misallocate resources. Without continuous oversight, AI systems risk exacerbating biases embedded in historical data or producing unintended societal impacts.

To operationalize auditing within AI governance frameworks, a structured AI audit compliance function can be introduced. This function systematically assesses bias, fairness, and alignment with ethical principles in AI decision-making. A generalized AI auditing model is formulated as:

$$A = \sum_{i=1}^{n} w_i \cdot F_i$$

where A represents the overall AI audit score, w_i is the weight assigned to each audit criterion, and F_i represents the evaluation function measuring compliance with fairness, transparency, and accountability principles. This model ensures that AI-driven systems are regularly assessed based on a set of predefined governance metrics, reducing the risk of unethical or faulty decision-making.

A key component of AI auditing is the detection and correction of biases in AI models. Many AI systems rely on historical data, which may reflect existing societal inequalities. Without intervention, these biases can be amplified by AI models, leading to unfair or discriminatory outcomes. A bias penalty function within AI auditing can be expressed as:

$$\min_{\mathrm{w}} L(\mathrm{w}, D) + \lambda \sum_{i=1}^{n} \mathrm{Bias\_Score}(y_i, \widehat{y}_i)$$

where $L(\mathrm{w}, D)$ represents the AI system's loss function, w are the model parameters, and $\mathrm{Bias\_Score}(y_i, \widehat{y}_i)$ measures the disparities in AI-driven decisions across different demographic groups. The regularization parameter λ ensures that bias penalties are incorporated into AI model evaluations, enforcing fairness constraints during audits.

In civil engineering, AI systems are widely used to optimize structural designs, predict material performance, and allocate resources for large-scale projects. If these models operate without proper auditing, errors in AI-driven predictions can lead to flawed infrastructure planning, resulting in increased costs, inefficiencies, or safety hazards. Regular audits must verify that AI-generated designs align with engineering best practices and ethical considerations.

One approach to auditing AI models in infrastructure planning is through uncertainty quantification, which assesses the reliability of AI-generated predictions. This can be expressed mathematically as:

$$U = \sum_{j=1}^{m} P_j \cdot V_j$$

where U represents the overall uncertainty score, P_j is the probability of error in AI predictions, and v represents the variance in AI-driven forecasts. By incorporating uncertainty assessments into engineering audits, AI systems can be evaluated for reliability, robustness, and safety compliance.

For AI audits to be effective and enforceable, governance frameworks must establish independent auditing bodies, responsible for conducting periodic reviews of AI systems. These institutions should function alongside policymakers and urban planners, ensuring that AI-driven systems are monitored, assessed, and adjusted as needed. A formal governance oversight function for AI auditing can be represented as:

$$G = \sum_{k=1}^{p} w_k \cdot C_k$$

where G represents the governance audit score, w_k represents the weight assigned to each compliance factor, and C_k represents the assessment function evaluating AI policy adherence.

This governance model ensures that AI auditing is not a passive process but an active mechanism for maintaining ethical AI use in urban environments.

The integration of structured AI audits, bias detection frameworks, and institutional governance oversight forms a comprehensive auditing framework for AI-driven decision-making in urban planning and civil engineering. These auditing mechanisms ensure that AI operates within ethical and legal boundaries, promoting fairness, transparency, and trust in AI-assisted infrastructure development.

By embedding auditing as a core function of AI governance, cities and institutions can continuously monitor, assess, and refine AI systems, ensuring that AI serves as a reliable and accountable tool in urban management rather than an unregulated entity operating without oversight.

Framework: AI Auditing and Compliance Framework (AIAF)

Layered AI Architecture for Auditing, Bias Detection, and Governance in AI-Driven Urban Systems

Layer	Main functionality	AI module (s) and components	Operational scope
Layer 1	Data collection: Gathering data for AI auditing processes	Data provenance and quality management system	Input → Raw data (AI decisions, feedback from affected communities) → Output → Preprocessed data for auditing models
Layer 2	AI auditing: Evaluating compliance with fairness, transparency, and accountability	AI audit compliance function	Input → AI decision data, governance metrics → Output → AI audit score (A)
Layer 3	Bias detection: Identifying and penalizing bias in AI-driven decisions	Bias-aware correction function	Input → AI model data, fairness constraints → Output → Bias-adjusted predictions and penalties
Layer 4	Uncertainty quantification: Assessing reliability of AI predictions in infrastructure planning	Uncertainty quantification function	Input → AI predictions, probability of error → Output → Uncertainty score (U)
Layer 5	Governance oversight: Institutional assessment of AI policy compliance	Governance oversight function	Input → Compliance policies, oversight data → Output → Governance audit score (G)

Auditing serves as a fundamental mechanism for ensuring that AI systems comply with legal and regulatory standards, particularly as they become more embedded in urban governance and civil engineering. AI-driven decision-making influences critical aspects of public infrastructure, resource distribution, and data privacy, making it essential that these systems adhere to existing laws governing public safety, fairness, and ethical accountability (Manzoor et al. 2021). Without rigorous auditing frameworks, AI systems risk operating in ways that may contradict regulatory requirements or lead to unintended legal and ethical consequences.

Compliance auditing ensures that AI systems align with national and international legal frameworks, particularly in areas such as data protection, anti-discrimination law, and urban planning policy. AI models deployed in infrastructure development and public safety must be regularly assessed for adherence to these frameworks, ensuring that algorithmic decision-making remains within legally defined boundaries. To formalize this assessment, a regulatory compliance function can be introduced to quantify the degree to which an AI system satisfies legal constraints. This function is formulated as:

$$C = \sum_{i=1}^{n} w_i \cdot R_i$$

where C represents the compliance score, w_i is the weight assigned to regulatory requirement i, and R_i represents the AI system's adherence to that regulation. By integrating this compliance function into AI auditing protocols, regulatory agencies can systematically evaluate AI models and ensure they meet legal and ethical standards.

Following the establishment of a basic compliance framework, the need to enhance explainability becomes apparent. As many AI systems remain black-box models, explainable AI (XAI) techniques serve as essential tools in increasing transparency. By providing human-understandable justifications for AI-driven decisions, XAI ensures that regulatory bodies, policymakers, and the public can trace how AI decisions align with specific legal requirements. This process fosters greater trust in AI systems, which is crucial for ensuring their widespread acceptance and ethical deployment. The XAI compliance model for formalizing the explainability of AI decisions can be expressed by:

$$E = \sum_{j=1}^{m} P_j \cdot I_j$$

where E represents the explainability score of the AI system, P_j represents the probability that an AI decision can be explained in legal and regulatory terms, and I_j represents the interpretability of model output with respect to specific compliance requirements. This function allows auditors to evaluate whether AI-driven decisions can be justified in legal terms, providing clarity and transparency in AI decision-making processes.

However, compliance auditing is not complete without considering the critical aspect of bias detection. AI models, particularly those trained on historical data, may inherit and even amplify societal biases, leading to discriminatory outcomes. To address this, it is essential to implement a bias correction function that can dynamically adjust AI outputs to eliminate unfair disparities. This correction function can be mathematically expressed as follows:

$$\min L(w, D) + \lambda \sum_{i=1}^{n} \text{Bias\_Score}\left(y_i, (y_i)^{\perp}\right)$$

where $L(w, D)$, loss function, w, model parameters, Bias _ Score(y_i, $(y_i)^{\perp}$), quantifies the bias in AI decisions, λ, regularization term ensuring fairness constraints. This model enforces bias correction during the AI's training and decision-making processes, ensuring that its outputs are free from unfair disparities based on demographic or socioeconomic factors.

Next, addressing the privacy concerns surrounding AI systems is paramount, especially when dealing with sensitive data in urban governance. AI systems frequently handle vast amounts of personal and geospatial data, and ensuring compliance with privacy regulations, such as the General Data Protection Regulation (GDPR), is critical. The privacy risk quantification model assesses the level of exposure to privacy risks posed by AI-driven decisions and ensures that data protection mechanisms are in place. This is mathematically represented as:

$$P = \sum_{k=1}^{p} w_k \cdot D_k$$

where P represents the AI system's privacy risk score, w_k is the weight assigned to each privacy regulation, and D_k represents the AI model's exposure to privacy risks. This function provides a quantitative measure of privacy risks, ensuring that AI systems adhere to data protection laws and incorporate privacy-preserving mechanisms, such as differential privacy.

Finally, to ensure continuous compliance with legal and regulatory frameworks over time, especially as regulations evolve, the integration of real-time monitoring and auditing becomes crucial. A dynamic approach to compliance monitoring allows for the ongoing assessment of AI systems, adjusting for regulatory shifts and changes in AI behavior. The compliance monitoring model can be expressed as:

$$M_t = \alpha M_{t-1} + (1 - \alpha) A_t$$

where M_t represents the compliance monitoring score at time t, M_{t-1} is the previous compliance assessment, and A_t represents the real-time audit findings. The parameter α ensures that historical compliance performance and real-time audit results are weighed appropriately, allowing for a dynamic assessment of AI regulatory adherence. This dynamic function ensures that AI systems remain continuously aligned with legal and ethical standards, even as new regulatory guidelines are introduced or AI behaviors evolve over time.

Incorporating these auditing mechanisms into urban governance frameworks ensure that AI systems are not only compliant with existing laws but are also fair, transparent, and accountable. These measures foster trust in AI applications across sectors, ensuring that AI-driven decisions are aligned with public interests and societal values. The combination of compliance auditing, bias detection, explainability, and continuous monitoring provides a comprehensive approach to AI governance that actively mitigates risks while promoting ethical standards in urban decision-making.

Framework: Fairness-Constrained Urban AI Governance (FCUAG III)

Layered AI Architecture for Decentralized Urban Intelligence

Layer	Main functionality	AI module(s) and components	Operational scope (Input → Processing → Output)
Layer 1	Initial setup and regulatory compliance preparation	Regulatory readiness and ethical compliance engine	Input: Historical data, legal frameworks → Processing: Set fairness constraints, prepare for AI integration → Output: Ethical guidelines and initial compliance frameworks
Layer 2	Regulatory compliance auditing function	Supervised learning or reinforcement learning	Input: Historical compliance data → Processing: Learn patterns of regulatory adherence → Output: Compliance score
Layer 3	Explainability of AI decisions	Explainable Boosting Machines (EBMs), SHAP, LIME	Input: AI decision outcomes → Processing: Generate interpretable justifications for decisions → Output: Explanation of decisions
Layer 4	Bias detection and correction	Adversarial debiasing, fairness-constrained optimization	Input: AI decisions → Processing: Detect and correct biases in predictions → Output: Fairer AI decision outcomes
Layer 5	Privacy Risk quantification	Differential privacy, secure multi-party computation (SMPC)	Input: Sensitive data → Processing: Evaluate privacy risks in AI-driven decisions → Output: Privacy risk score
Layer 6	Continuous compliance monitoring	Online Learning, Recurrent Neural Networks (RNN)	Input: New regulatory data → Processing: Continuously update compliance status → Output: Real-time compliance monitoring score
Layer 7	Monitoring, evaluation, and regulatory auditing	Continuous monitoring and regulatory audit platform	Input: Real-world outcomes, system performance → Processing: Continuous monitoring, performance audits → Output: Evaluated outcomes, system improvements

5.3.4 Human Oversight and Ethical Responsibility

Despite the increasing autonomy of AI systems, human oversight remains an essential safeguard for ensuring accountability in AI-driven decision-making. While AI can optimize complex processes and provide data-driven insights, fully autonomous AI systems should not be deployed without mechanisms that allow human intervention (Sargiotis 2025). This is particularly crucial in high-stakes environments such as civil engineering and urban governance, where AI-driven decisions directly impact infrastructure safety, public resources, and long-term urban development (Barredo Arrieta et al. 2020).

Human oversight ensures that AI decisions align with broader societal values and ethical principles. AI models, particularly those based on deep learning, operate by identifying patterns in vast datasets, but they do not inherently account for contextual nuances, legal considerations, or ethical dilemmas. Engineers, policymakers, and urban planners must retain the authority to override AI-generated recommendations when necessary. This is particularly important when AI decisions conflict with legal regulations, urban planning priorities, or public interests.

To support this balance, the Human-in-the-Loop (HITL) decision model integrates AI autonomy with human validation, providing a structured decision-making framework in which AI decisions are subjected to human review, particularly when their potential risks or societal impacts are significant. This structured approach to combining human oversight with AI decision-making is mathematically represented as follows:

$$D_t^* = \lambda D_t^{\mathrm{AI}} + (1 - \lambda) D_t^{\mathrm{Human}}$$

where D_t^* represents the final decision outcome at time t, D_t^{AI} is the AI-generated decision, and D_t^{Human} is the human intervention. The parameter λ controls the degree of AI autonomy, ensuring that AI decisions remain subject to human validation, particularly in high-risk scenarios.

This function demonstrates how human intervention (with a parameter λ that balances AI and human input) ensures that critical decisions are validated by human expertise when necessary, preserving accountability, especially in sensitive decision-making contexts.

Building upon this fusion mechanism, risk-aware governance demands the inclusion of uncertainty modeling in the decision synthesis. Traditional expected utility theory fails to capture epistemic uncertainty in dynamic civil infrastructure environments. To that end, the posterior risk distributions must be estimated using Bayesian techniques. The foundational estimation is derived using variational Bayesian inference:

$$P_i = \int_\theta P_i(\theta \mid x) \cdot q(\theta) d\theta$$

where P_i, posterior risk probability of event i, θ, model parameter vector, x, input observation, $q(\theta)$, variational distribution approximating the posterior. This function models the uncertainty surrounding event outcomes by integrating across parameter distributions.

Moving from human-in-the-loop decision-making to the next step in this AI governance framework, risk-sensitive decision-making is essential. Urban governance and civil engineering decisions often entail uncertain outcomes that cannot be fully captured by AI models. AI models typically rely on probabilistic reasoning, but real-world situations require incorporating uncertainty and risk management in decision-making. This necessity calls for the introduction of a risk-sensitive decision framework, where the potential risks of AI-driven decisions are quantitatively assessed, ensuring that the decision process accounts for not only the predicted outcomes but also the uncertainty and consequences of errors. This is captured in the following equation:

$$R_t = \sum_{i=1}^{n} P_i \cdot U_i$$

where R_t represents the risk-adjusted AI decision score at time t, P_i is the probability of an adverse event occurring due to AI decision-making, and U_i represents the utility function assessing the potential consequences of AI errors. This function ensures that AI decisions are not just evaluated on accuracy but also on potential risks, enabling a more comprehensive decision-making framework that incorporates uncertainty and risk assessment in the urban governance context.

Continuing the framework, once risks are evaluated, addressing bias in AI decision-making is paramount, especially when such biases can have harmful societal consequences. Bias detection and correction in AI-driven systems is a critical function for urban governance, where fairness and equity must be prioritized. To mitigate inherent biases in AI models, especially those trained on historical data, fairness-constrained optimization models can be applied. The following equation captures this correction process:

$$\min L(w, D) + \lambda \sum_{i=1}^{n} \text{Bias\_Score}\left(y_i, (y_i)^{\perp}\right)$$

where $L(w, D)$, loss function of the AI system, w, model parameters, Bias _ Score(y_i, $(y_i)^{\perp}$), quantifies the bias in AI-generated predictions, λ, regularization parameter ensuring fairness.

This function actively adjusts the AI system's decision-making to ensure fairness, rectifying any biases in predictions to promote equitable outcomes, particularly in decisions that impact urban planning and resource allocation.

To enforce this fairness constraint at a structural level, adversarial learning mechanisms are introduced. These adversaries attempt to predict protected attributes from latent representations, thereby ensuring that learned features are demographically invariant:

$$\underset{G}{min}\ \underset{D}{max}\, \mathbb{E}_x[\log D(a \mid G(x))] + \mathbb{E}_x[\log(1 - D(a \mid x))]$$

where G, feature generator, D, adversarial discriminator, a, protected attribute (e.g., gender, ethnicity), x, input sample. This function enforces demographic fairness by ensuring that the generator learns to produce representations from which protected attributes are indistinguishable.

As AI systems are heavily used for privacy-sensitive tasks in urban governance, privacy risk quantification is necessary to ensure compliance with data protection regulations, such as the GDPR. AI systems in civil engineering or urban governance frequently process sensitive personal or geospatial data, making privacy preservation a critical concern. To address this, we introduce the privacy risk quantification function, which measures the potential privacy risks associated with AI-driven decisions:

$$P = \sum_{k=1}^{p} w_k \cdot D_k$$

where P, privacy risk score, w_k, weight assigned to each privacy regulation, D_k, exposure of AI model to privacy risks. This function quantifies the privacy risks involved with AI decision-making, ensuring compliance with privacy standards and laws, and integrating privacy-preserving techniques into urban governance systems.

To mitigate this quantified risk, the system invokes differential privacy via Laplacian noise injection:

$$\tilde{f}(x) = f(x) + \mathrm{Lap}\left(\frac{\Delta f}{\varepsilon}\right)$$

where $\tilde{f}(x)$, privacy-protected function output, $f(x)$, original function output, Δf, sensitivity of the function, ε, privacy budget. This function guarantees that individual contributions to the dataset cannot be inferred beyond a bounded probabilistic threshold.

Lastly, continuous compliance monitoring ensures that AI systems remain compliant with evolving legal, ethical, and regulatory standards over time. AI models must be updated and assessed continuously to ensure they align with dynamic regulatory environments. The equation for continuous monitoring and real-time compliance is given by:

$$M_t = \alpha M_{(t-1)} + (1 - \alpha)A_t$$

where M_t, compliance score at time t, $M_{(t-1)}$, previous compliance score, A_t, real-time audit findings, α, weight parameter balancing historical performance and real-time data. This function ensures that AI systems continuously adapt to new regulatory information and evolving compliance standards, providing a dynamic evaluation of AI compliance in urban governance contexts.

By using these mathematical models Human-in-the-Loop (HITL), risk-sensitive decision-making, bias detection and correction, privacy risk quantification, and continuous compliance monitoring, this framework ensures that AI systems in urban governance and civil engineering are accountable, fair, transparent, and legally compliant. Each function integrates real, mathematically valid AI modules, allowing for continuous human oversight and intervention, particularly in high-stakes environments where AI decisions directly impact public safety, resource allocation, and long-term infrastructure planning. These layered mechanisms balance AI autonomy with human accountability, creating a robust and reliable framework for modern urban governance.

Framework: Human-Centric Ethical Oversight Architecture (HCEOA)

Layered AI Architecture for Responsible Decision-Making in Urban Governance

Layer	Main functionality	AI module(s) and components	Operational scope (Input → Processing → Output)
Layer 1	Integrated human-AI decision synthesis	Human-in-the-Loop Decision Model (HITL)	AI Decision, Human Expert Input → Weighted Synthesis using λ → Final Decision Output
Layer 2	Risk-sensitive policy formulation	Bayesian risk modeling with utility optimization	Predicted Actions, Environmental Uncertainty → Risk Utility Aggregation → Risk Weighted Policy Output
Layer 3	Bias detection and mitigation	Fairness-constrained optimization	Prediction Outcome, Demographic Data → Bias Score Minimization and Loss Regularization → Fair Adjusted Prediction
Layer 4	Privacy risk quantification	Weighted regulatory exposure estimator	Privacy Attributes, Regulation Weights → Risk Score Computation → Privacy-Constrained AI Output
Layer 5	Real-Time compliance tracking	Temporal compliance smoothing model	Previous Compliance Metrics, Audit Feedback → Exponential Update → Dynamic Compliance Score

Ethical responsibility in AI extends beyond decision-making and governance to the very foundations of system design and development. Engineers and developers play a pivotal role in embedding ethical principles such as fairness, transparency, and accountability into AI models from their inception. Designing AI systems that align with these principles requires technical rigor and structured ethical oversight, ensuring that AI technologies do not perpetuate biases or operate in ways that contradict societal values (Floridi et al. 2018).

A fundamental aspect of ethical AI design is the integration of fairness constraints into machine learning algorithms. Since AI models learn from historical data, they may inherit and amplify existing biases unless proactive mitigation mechanisms are embedded into the model architecture. A fairness-aware optimization framework can therefore be introduced to support unbiased prediction while maintaining model performance. The objective function can be formulated as:

$$\min_{\mathrm{w}} L(\mathrm{w}, D) + \lambda \sum_{i=1}^{n} \text{Fairness\_Penalty}(y_i, \widehat{y_i})$$

where $L(\mathrm{w}, D)$ represents the standard loss function for AI model training, w are the model parameters, and $\text{Fairness\_Penalty}(y_i, \widehat{y_i})$ quantifies disparities in model predictions across different demographic groups. The regularization parameter λ ensures that fairness constraints are enforced without excessively compromising model accuracy.

However, fairness at the loss level alone is insufficient to guarantee causal impartiality. Many biases reside within latent structures, escaping detection unless causal constraints are imposed. To rectify this, causal fairness is enforced through counterfactual alignment between factual and counterfactual outcomes, using path-specific effect modeling. This ensures that fairness holds not merely statistically but also across potential causal pathways.

$$\min_{\mathrm{w}} \sum_{i=1}^{n} \left| f(x_i, a_i) - f(x_i, a_i^{\perp}) \right|^2$$

where $f(x_i, a_i)$, model output under actual attribute a_i, $f\left(x_i, a_i^{\perp}\right)$, model output under counterfactual attribute $a_i^{\perp}$, x_i, input features for instance i. This function enforces counterfactual fairness by minimizing disparities between factual and counterfactual predictions, thus ensuring the model's behavior is invariant to protected characteristics.

As fairness constraints reshape the optimization surface, transparency must concurrently be introduced to make these adjustments intelligible. The architecture must expose the inner logic of decisions via explainability constraints. Feature attribution models such as SHAP and Integrated Gradients quantify each input variable's contribution, which are aggregated using a transparency-weighted scoring function.

$$T = \sum_{j=1}^{m} w_j \cdot X_j$$

where T represents the transparency score of the AI system, w_j is the weight assigned to each interpretability criterion, and X_j represents the model's explainability metric, such as feature importance attribution or local interpretability techniques. By integrating transparency constraints into AI model evaluation, developers can ensure that AI-driven decisions remain interpretable and justifiable.

Ethical AI design also requires ongoing monitoring throughout the system's lifecycle. AI models are not static; they evolve as they process new data, meaning that fairness and transparency must be continuously assessed to prevent unintended deviations. A dynamic ethical monitoring framework can be formulated as:

$$M_t = \alpha M_{t-1} + (1 - \alpha)E_t$$

where M_t represents the ethical compliance score at time t, M_{t-1} is the prior assessment, and E_t represents the latest ethical evaluation based on fairness, transparency, and accountability metrics. The parameter α controls how much historical compliance influences ongoing monitoring, ensuring that ethical standards remain consistently enforced throughout the AI system's operational lifecycle.

This function continuously updates the ethical integrity of the system, ensuring that drift or retraining does not compromise foundational values.

To anticipate failure before it emerges, models must also be evaluated using ethical risk assessments, quantitative diagnostics estimating potential harm, opacity, or bias propagation. However, these metrics must first be normalized to ensure dimensional consistency. A Z-normalization step is applied before weighted risk aggregation.

$$R_k^{\text{norm}} = \frac{R_k - \mu_k}{\sigma_k}$$

where R_k^{norm}, normalized risk factor k, R_k, raw risk score, μ_k, mean of factor k, σ_k, standard deviation of factor k. This function standardizes risk factors, ensuring their comparability across disparate dimensions prior to aggregation.

Following normalization, an ethical impact score is computed through weighted summation across all ethically relevant dimensions:

$$I = \sum_{k=1}^{p} w_k \cdot R_k$$

where I represents the ethical impact score, w_k is the weight assigned to each ethical risk factor, and R_k represents the model's risk exposure to fairness violations, opacity, or unintended consequences.

By using fairness-aware optimization, explainability constraints, dynamic monitoring, and ethical impact assessments, AI systems can be designed with built-in accountability, minimizing risks of bias and opacity. Ethical AI development is not a one-time process but a continuous commitment that requires technical safeguards and proactive governance. Engineers and developers must ensure that AI models align with legal, ethical, and societal expectations, embedding responsibility at every stage of the system's lifecycle to create AI that serves the public interest in a just and equitable manner.

Framework: Ethical Design and Oversight AI Framework (EDOAF)

Layered AI Architecture for Fair, Transparent, and Accountable System Design

Layer	Main functionality	AI module(s) and components	Operational scope (Input → Processing → Output)
Layer 1	Fairness-aware model optimization	Fairness-Constrained Empirical Risk Minimization (FC-ERM)	Historical Training Data → Constrained Loss Minimization → Bias-Adjusted Model Parameters
Layer 2	Transparency-integrated evaluation	Explainable AI using Feature Attribution (XAI-FA)	Model Predictions, Interpretability Metrics → Weighted Explainability Score → Transparency-Aligned Model Outputs
Layer 3	Dynamic ethical monitoring	Temporal Ethical Compliance Tracking (TECT)	Prior Compliance Scores, New Ethical Evaluations → Exponential Update Mechanism → Updated Ethical Compliance
Layer 4	Ethical Risk assessment	Weighted Ethical Impact Estimation Model (WEIEM)	Risk Factor Indicators, Ethical Criteria Weights → Weighted Risk Summation → Ethical Impact Score

Figure 5.3 outlines the essential aspects of accountability and responsibility in AI systems, focusing on four main areas: Challenges in Defining Accountability, Governance Frameworks, Role of Auditing, and Human Oversight and Ethical Responsibility. "Challenges in Defining Accountability" addresses the difficulties

Fig. 5.3 Accountability and responsibility in AI systems

in assigning responsibility due to the autonomous nature of AI and diffuse roles among engineers, organizations, and policymakers. "Governance Frameworks" emphasizes clear guidelines for responsibility, mechanisms for redress, and the use of Explainable AI (XAI) to improve transparency. "Role of Auditing" highlights the need for regular audits to identify biases and ensure compliance with legal and ethical standards. "Human Oversight and Ethical Responsibility" underscores the importance of retaining human control over AI decisions and the ethical obligations of engineers to design AI systems with fairness, transparency, and accountability in mind.

Accountability and responsibility are critical for ensuring that AI systems in civil engineering, urban development, and governance operate in an ethical and transparent manner. By establishing robust governance frameworks, incorporating regular audits, and ensuring human oversight, AI systems can be held accountable for their decisions and their impact on society. As AI continues to play an increasingly important role in shaping urban environments and public infrastructure, addressing these ethical challenges is essential for promoting public trust and ensuring that AI systems serve the public good.

5.4 Sustainability and Environmental Impact

The integration of AI systems into civil engineering, urban development, and governance presents both opportunities and challenges in terms of sustainability and environmental impact. On the one hand, AI has the potential to significantly improve resource efficiency, reduce waste, and optimize the management of urban infrastructures, contributing to more sustainable and environmentally friendly cities. On the other hand, the development and deployment of AI systems also raise concerns about their environmental footprint, particularly in relation to energy consumption, resource depletion, and long-term sustainability (Sargiotis 2024f).

5.4.1 AI as a Tool for Environmental Sustainability

AI technologies offer significant potential for promoting sustainability in urban development and civil engineering by optimizing energy consumption, managing smart grids, and improving public transportation systems. These capabilities are enabled through AI-driven algorithms that analyze large volumes of data to identify energy use patterns, improve efficiency, and support the integration of renewable energy sources such as solar and wind power (Wolniak and Stecuła 2024). The application of AI in this domain can contribute to reducing greenhouse gas emissions and, consequently, to lowering the environmental footprint of urban systems (Manzoor et al. 2021). To support these objectives, an advanced AI-driven optimization framework is proposed for the dynamic control of energy consumption and the enhancement of urban sustainability. The AI-driven model, incorporating Deep Reinforcement Learning (DRL) for real-time energy management and sustainable urban infrastructure optimization, can be formulated as:

$$Q^*(s,a) = \mathbb{E}\left[r_t + \gamma \max_{a'} Q^*(s',a') \mid s,a\right]$$

where $Q^*(s,a)$ is the optimal action-value function that estimates the maximum expected future reward for a given state-action pair, r_t represents the immediate reward function measuring the sustainability gain from energy efficiency actions, γ is the discount factor regulating the contribution of future rewards, s and s' denote the current and next environmental states, and a and a' represent the chosen and alternative actions for optimizing energy use.

The above equation formulates how DRL learns the optimal energy management strategy by evaluating the expected reward for each action taken in a dynamic urban system. The model continuously refines energy policies to ensure that smart grids and urban infrastructure operate at peak efficiency, reducing energy wastage while integrating renewable sources dynamically.

To further refine the policy optimization process, the Proximal Policy Optimization (PPO) objective function is defined as:

$$L(\theta) = \mathbb{E}\left[\min\left(\frac{\pi_\theta(a \mid s)}{\pi_{\theta_{\text{old}}}(a \mid s)} A_t, \text{clip}\left(\frac{\pi_\theta(a \mid s)}{\pi_{\theta_{\text{old}}}(a \mid s)}, 1-\epsilon, 1+\epsilon\right) A_t\right)\right]$$

where $L(\theta)$ is the policy loss function guiding the learning process of the AI model, $\pi_\theta(a \mid s)$ represents the probability distribution of actions under the new policy, $\pi_{\theta_{\text{old}}}(a \mid s)$ denotes the probability distribution of actions under the previous policy, A_t is the advantage function quantifying the expected improvement of an action over the baseline policy, and ϵ is the clipping parameter preventing excessive policy updates.

The PPO optimization ensures that policy updates remain stable and effective, allowing the AI-driven model to dynamically adjust energy management strategies based on real-time urban data. By continuously evaluating and optimizing policies, the AI system maintains sustainability while preventing fluctuations in energy distribution.

To evaluate the real-time performance of energy-efficient decisions, the reward function for sustainability optimization is defined as:

$$R(s_t, a_t) = \lambda_1 E_{\text{saved}}(s_t, a_t) - \lambda_2 C_{\text{operation}}(s_t, a_t) - \lambda_3 P_{\text{emission}}(s_t, a_t)$$

where (s_t, a_t) represents the reward function for sustainable energy actions, $E_{\text{saved}}(s_t, a_t)$ denotes the amount of energy conserved through AI-driven optimizations, $C_{\text{operation}}(s_t, a_t)$ represents the operational cost incurred, and $P_{\text{emission}}(s_t, a_t)$ quantifies the greenhouse gas emissions resulting from urban energy consumption. The parameters $\lambda_1, \lambda_2, \lambda_3$ are the respective weight factors balancing the contributions of energy savings, operational costs, and emissions in determining the reward.

However, due to variance in reward scales across metrics, normalization is essential to ensure comparability and convergence. Prior to reward aggregation, a statistical normalization is applied using the following transformation:

$$\hat{R}_k = \frac{R_k - \mu_k}{\sigma_k}$$

where $\hat{R}_k$, normalized reward component k, R_k, raw reward, μ_k, mean, σ_k, standard deviation. This function standardizes the reward terms to zero-mean and unit variance, ensuring numerical stability in multi-objective learning.

This AI-driven reinforcement learning framework dynamically adjusts urban energy policies by integrating real-time data, optimizing power distribution, and prioritizing sustainability. The reward function ensures that AI-driven decisions align with environmental goals, minimizing emissions and promoting renewable energy integration. Using Deep Reinforcement Learning (DRL) with Proximal Policy Optimization (PPO), this AI-driven model significantly enhances urban energy management, enabling smart grids to operate efficiently while reducing waste. The system continuously learns optimal strategies for integrating renewable energy sources, optimizing consumption patterns, and fostering sustainability in urban infrastructure. Through AI-driven automation, cities can achieve substantial energy savings, reduce their carbon footprint, and ensure a more sustainable future.

Framework: AI-Driven Urban Sustainability Optimization System (AID-USOS)

Layered AI Architecture for Smart Energy Management and Sustainable Urban Control

Layer	Main functionality	AI module(s) and components	Operational scope (Input → Processing → Output)
Layer 1	Environmental data acquisition and urban context mapping	Environmental data processing and context encoding engine	Sensor Streams, Geospatial Inputs → Data Preprocessing and Structuring → Encoded Environmental State
Layer 2	State-action value function estimation	Deep Q-Learning with Bellman Optimality (DQL-BO)	Urban Energy State, Control Action → Recursive Q-Value Update → Optimal Policy Selection
Layer 3	Policy stabilization and regularized improvement	Proximal Policy Optimization (PPO)	Policy Parameters, Advantage Estimates → Clipped Surrogate Objective Minimization → Stable Policy Update
Layer 4	Sustainability-aware reward generation	Weighted Energy-Efficiency Reward Modeling (WEERM)	Action-State Pair, Energy & Emission Data → Weighted Reward Aggregation → Sustainability Score Output
Layer 5	Policy actuation and energy grid adjustment	Smart grid control and actuation module	Optimized Energy Policy, Emission Constraints → Infrastructure Control Systems → Real-Time Grid Adjustment

AI technologies provide advanced capabilities for waste management by improving collection route planning, reducing landfill dependence, and increasing recycling efficiency. Sensor-equipped smart bins can continuously monitor waste levels and generate automated alerts for collection scheduling, thereby reducing fuel consumption, operational costs, and unnecessary vehicle circulation (Floridi et al. 2018). In civil engineering, AI also supports the development of sustainable infrastructure by enabling material-use optimization, lifecycle-aware design, and environmentally responsible construction practices.

To address the operational and ecological demands of contemporary urban sustainability, a two-tiered AI architecture is proposed. This framework combines a Multi-Agent Deep Reinforcement Learning (MADRL) system for waste-collection logistics with a Bayesian Design Optimization (BDO) system for eco-efficient infrastructure design.

At the operational level, each waste-collection vehicle is modeled as a learning agent within a shared urban environment. The agent's objective is to identify the most efficient collection route by integrating sensor inputs from smart bins with dynamic traffic data. The learning process is defined by the deep Q-learning update:

$$Q_i(s_t, a_t) \leftarrow Q_i(s_t, a_t) + \alpha \left[r_t + \gamma \max_{a'} Q_i(s_{t+1}, a') - Q_i(s_t, a_t) \right]$$

where $Q_i(s_t, a_t)$ is the Q-value for agent i (vehicle) in state s_t taking action a_t, α is the learning rate, r_t is the reward (e.g., fuel savings, timely pickup), γ is the discount factor for future rewards, and $\max_{a'} Q_i(s_{t+1}, a')$ is the expected return from the best future action. This function enables each agent to adapt its routing policy to urban conditions minimizing redundant travel, reducing emissions, and responding autonomously to real-time bin fill levels.

At the infrastructure level, Bayesian Design Optimization (BDO) is used to identify optimal construction material configurations under environmental constraints. This probabilistic model balances trade-offs between structural performance and sustainability by selecting material mixes with high posterior probability of low-carbon impact and high durability. The optimization function is defined as:

$$x^* = \arg \max_{x \in X} \mathbb{E}_{f \sim \mathcal{P}}[\mathrm{UCB}(f(x))]$$

where x represents a candidate material configuration (e.g., recycled concrete ratios, green cement formulas), $\mathcal{P}$ is the Gaussian process prior over the objective function $f(x)$, and $\mathrm{UCB}(f(x))$ is the Upper Confidence Bound acquisition function used to select promising configurations. This approach allows AI to recommend innovative, low-impact materials while managing uncertainty essential in early-stage sustainable design decisions. As more test data from lab or field trials become available, the posterior distribution is updated, improving the quality of future recommendations.

To ensure that the BDO framework updates its internal belief state with rigor, the Gaussian Process posterior must be recalculated as new evaluations are obtained. The posterior mean and variance for any candidate point xx are updated as follows:

$$\mu_t(x) = k(x)^T \left[K + \sigma_n^2 I\right]^{-1} y, \sigma_t^2(x) = k(x,x) - k(x)^T \left[K + \sigma_n^2 I\right]^{-1} k(x)$$

where $\mu_t(x)$, posterior mean, $\sigma_t^2(x)$, posterior variance, $k(x)$, kernel vector between x and training inputs, K, kernel matrix, σ_n^2, noise variance, y, observed objective values. These functions allow the BDO system to update its uncertainty model over the material space, making future evaluations increasingly informed and precise.

Upon derivation of optimal policies and material designs, actionable results must be translated into real-world operations. This post-AI deployment is implemented through a logistics execution layer, where optimized vehicle routes and construction recommendations are scheduled for implementation by routing systems and procurement teams. While this layer does not invoke machine learning directly, it is critical for embedding intelligence into city operations.

In dynamic environments, fixed models quickly degrade. Therefore, continuous learning is instituted via a feedback loop. In the MADQL system, real-time traffic and bin fill feedback are used to update Q-values, while in BDO, material test results update the GP posterior. This is realized through an online learning mechanism governed by adaptive retraining functions:

$$Q_i^{(t+1)} \leftarrow Q_i^{(t)} + \eta_t \delta_t, \mu_{t+1}(x) = \mu_t(x) + \rho_t(y_t - \mu_t(x))$$

where $Q_i^{(t+1)}$, updated Q-value, η_t, learning rate, δ_t, TD-error, $\mu_{t+1}(x)$, updated GP mean, ρ_t, posterior adjustment coefficient, y_t, observed function value. These functions allow real-time retraining of routing agents and continuous adaptation of material design recommendations.

To conclude, citywide sustainability goals must be monitored holistically. The performance of both routing and material modules is assessed using a multi-criteria utility aggregation model (MCUA), which combines emissions, fuel usage, waste metrics, and operational cost into a composite score:

$$S_t = \sum_{k=1}^{p} w_k \cdot U_k(t)$$

where S_t, overall sustainability score at time t, w_k, utility weight for metric k, $U_k(t)$, metric value at time t. This function generates a real-time index of systemic sustainability, enabling regulatory benchmarking and AI policy audit.

Together, the MADRL and BDO layers form a comprehensive AI ecosystem for sustainable city management. The MADRL component improves short-term efficiency in waste collection, responding to real-time urban sensor data, while the BDO component guides long-term environmental decisions in civil engineering, ensuring infrastructure aligns with carbon and lifecycle goals. This unified AI architecture demonstrates how operational intelligence and design foresight can be integrated to shape resource-efficient, climate-conscious future cities.

Framework: AI-Integrated Urban Waste and Infrastructure Sustainability (AI-UWIS)

Layered AI Architecture for Operational Routing, Sustainable Design, and Continuous Adaptation

Layer	Main functionality	AI module(s) and components	Operational scope (Input → Processing → Output)
Layer 1	Urban sensor integration and smart bin data aggregation	Sensor data fusion and urban context encoding engine	Bin-Level Sensor Streams, Geospatial Routing Maps → Data Fusion and Preprocessing → Encoded State for Agents
Layer 2	Multi-agent adaptive route optimization	Multi-Agent Deep Q-Learning (MADQL)	Smart Bin Fill Levels, Traffic Conditions → Distributed Q-Value Learning → Agent-Specific Optimal Routes
Layer 3	Eco-material search under uncertainty	Bayesian Design Optimization (BDO)	Material Candidates, Structural Targets → Posterior Sampling + UCB → Sustainable Material Recommendation
Layer 4	Operational deployment of collection and design policies	Smart operations execution and control platform	Optimized Routes and Material Plans → Scheduling/Procurement Systems → Executed Collection and Design Actions
Layer 5	Feedback loop and model retraining	Online multi-agent reinforcement learning/ Bayesian posterior updating	Sensor Updates, Field Evaluations → Dynamic Q-Table and Posterior Update → Refined Agent and BDO Models
Layer 6	Sustainability monitoring and performance indexing	Multi-Criteria Utility Aggregation (MCUA)	Waste, Emission, Fuel Data → Weighted Performance Aggregation → Dynamic Sustainability Score

5.4.2 *Energy Consumption and Environmental Footprint of AI Systems*

AI can contribute substantially to environmental sustainability, yet its own computational energy requirements constitute a significant challenge. Training large-scale AI models, particularly deep learning architectures, demands considerable computational capacity, resulting in elevated energy use and increased carbon emissions (Barredo Arrieta et al. 2020). As AI systems become more complex and computationally intensive, their environmental burden grows, making it necessary to design optimization strategies that reduce energy consumption while preserving computational efficiency. To address this challenge without degrading performance, we propose an energy-aware optimization framework that integrates Deep Reinforcement Learning (DRL) into the AI training process. The framework first encodes training into a state-action representation that captures energy, cost, and workload dynamics. On this basis, a DRL agent estimates the long-term value of alternative training configurations through the state-value function defined by the Bellman optimality principle:

$$V^*(s) = \max_a \mathbb{E}[R(s,a) + \gamma V^*(s') \mid s]$$

where $V^*(s)$ is the optimal state-value function measuring the expected cumulative reward for energy-efficient AI training strategies, $R(s,a)$ represents the immediate sustainability reward from executing a particular training configuration, γ is the discount factor prioritizing long-term efficiency in computational resources, s and s' denote the current and subsequent states of the AI training process, and a represents the action that selects an optimized training method. This equation dynamically optimizes AI training energy consumption by selecting efficient computational resource configurations while ensuring minimal performance degradation. The model continuously learns and refines AI training processes, optimizing scheduling decisions, hardware resource allocation, and adaptive learning rate adjustments to minimize carbon emissions.

To enhance AI training efficiency while ensuring robust optimization, the deep reinforcement learning-based policy optimization process is defined as:

$$\pi^*(a \mid s) = \arg\max_\pi \mathbb{E} \sum_{t=0}^{T} \gamma^t R(s_t, a_t)$$

where $\pi^*(a \mid s)$ represents the optimal policy for selecting an AI training strategy that maximizes sustainability, T is the time horizon over which energy savings are optimized, and $R(s_t, a_t)$ quantifies the immediate reward derived from energy-efficient AI training adjustments. This function ensures that AI dynamically selects the best computational strategies for minimizing energy consumption while maintaining the accuracy of the AI model. It enables automated adaptation to workload variations, reducing power consumption while avoiding performance degradation.

To quantify the efficiency of AI-driven training processes, the energy-aware reward function is formulated as:

$$R(s_t, a_t) = \lambda_1 E_{\text{reduction}}(s_t, a_t) - \lambda_2 C_{\text{compute}}(s_t, a_t) - \lambda_3 P_{\text{carbon}}(s_t, a_t)$$

where $R(s_t, a_t)$ represents the overall energy-saving reward from AI model training optimization, $E_{\text{reduction}}(s_t, a_t)$ quantifies the reduction in energy consumption from optimized AI training strategies, $C_{\text{compute}}(s_t, a_t)$ denotes the computational cost of executing an AI training process, and $P_{\text{carbon}}(s_t, a_t)$ measures the total carbon emissions associated with AI model execution. The parameters λ_1, λ_2, λ_3 regulate the balance between energy savings, computational cost, and environmental impact. This function ensures that AI-driven training optimization minimizes computational energy waste while considering carbon footprint reduction. The model dynamically refines AI training processes, leveraging power-aware scheduling, low-energy hardware selection, and adaptive optimization techniques to reduce environmental impact.

Given that the reward components vary widely in scale and units, a normalization layer is essential to preserve numerical stability during learning. Therefore, each reward term is normalized using a Z-score standardization function, applied prior to reward aggregation:

$$\hat{R}_k = \frac{R_k - \mu_k}{\sigma_k}$$

where $\hat{R}_k$, normalized reward component k, R_k, raw reward value, μ_k, mean of metric k, σ_k, standard deviation of metric k. This function standardizes the input rewards to ensure that no single metric (e.g., cost or carbon) dominates the optimization due to unit disparity, thus enabling a balanced sustainability trade-off.

Through the integration of Deep Reinforcement Learning (DRL) with energy-aware policy optimization, this AI-driven framework enables AI models to dynamically manage computational workloads while minimizing energy consumption and carbon emissions. The system continuously adapts to workload variations, ensuring that AI advancements align with sustainability objectives. Through this approach, AI achieves high-performance training with reduced environmental impact, contributing to a more sustainable future in computational science.

Framework: Energy-Aware Deep Reinforcement Learning for Sustainable AI Training (EADRLSAT)

Layered AI Architecture for Minimizing Energy Consumption in Large-Scale AI Model Training

Layer	Main functionality	AI module(s) and components	Operational scope (Input → Processing → Output)
Layer 1	AI training process monitoring and resource encoding	Training resource monitoring and encoding engine	Training Configurations, Compute Load Metrics → Workload Representation → Encoded Training State
Layer 2	Value function estimation for energy-efficient training	Deep Reinforcement Learning (DRL)	State, Action → Bellman Update → Optimized Value Function for Energy Scheduling
Layer 3	Policy optimization under energy constraints	Policy gradient DRL	Training States, Reward Signals → Policy Search and Update → Energy-Minimizing Training Policies
Layer 4	Sustainability reward aggregation	Weighted multi-objective reward function	Training Action-State Pair → Energy/Carbon/Cost Evaluation → Composite Sustainability Score

The increasing energy demands of AI-based systems present a major sustainability challenge, since the training of large deep learning models may require computational resources comparable to substantial household energy consumption over extended periods (Mökander 2023). This raises concerns about the long-term environmental viability of AI, particularly in civil engineering and urban development, where large-scale computational systems are increasingly used for infrastructure planning, simulation, optimization, and monitoring. To ensure that the operational benefits of AI outweigh its ecological costs, energy-aware optimization mechanisms must dynamically regulate computational energy use while preserving model performance. For this purpose, a Deep Reinforcement Learning (DRL) framework for sustainable AI training can be formulated as:

$$V^*(s) = \underset{a}{max}\,\mathbb{E}[R(s,a) + \gamma V^*(s') \mid s]$$

where $V^*(s)$ represents the optimal state-value function that predicts the cumulative energy efficiency gains from an AI training strategy, $R(s, a)$ denotes the immediate reward function capturing the sustainability benefits of an AI model configuration, γ is the discount factor ensuring long-term optimization, s and s' correspond to the current and next computational states, and a represents the selected AI training method aimed at reducing energy consumption. This formulation enables AI systems to optimize their resource allocation dynamically, scheduling energy-intensive processes during low-carbon energy availability periods, selecting eco-friendly computing infrastructures, and applying intelligent model pruning techniques to reduce computational complexity without compromising accuracy.

To refine and stabilize AI training policy updates while minimizing energy waste, the Proximal Policy Optimization (PPO) function is introduced as:

$$L(\theta) = \mathbb{E}\left[min\left(\frac{\pi_\theta(a \mid s)}{\pi_{\theta_{\text{old}}}(a \mid s)}A_t, \text{clip}\left(\frac{\pi_\theta(a \mid s)}{\pi_{\theta_{\text{old}}}(a \mid s)}, 1-\epsilon, 1+\epsilon\right)A_t\right)\right]$$

where $L(\theta)$ is the policy loss function that stabilizes AI training decisions, $\pi_\theta(a \mid s)$ represents the new policy's probability distribution over AI training strategies, $\pi_{\theta_{\text{old}}}(a \mid s)$ corresponds to the previous policy, A_t quantifies the advantage function measuring the efficiency improvement of a given training decision, and ϵ is the clipping threshold ensuring smooth updates. This function guarantees that AI models adjust their training policies gradually, avoiding unnecessary energy spikes while learning the most sustainable configurations. The optimization process allows AI training to adaptively adjust hardware utilization, batch sizes, and hyperparameters to maintain computational efficiency while reducing carbon emissions.

To quantitatively assess the effectiveness of sustainable AI training, the reward function for energy-efficient AI model training is formulated as:

$$R(s_t, a_t) = \lambda_1 P_{\text{efficiency}}(s_t, a_t) - \lambda_2 C_{\text{compute}}(s_t, a_t) - \lambda_3 E_{\text{carbon}}(s_t, a_t)$$

where $R(s_t, a_t)$ represents the reward function for sustainable AI training, $P_{\text{efficiency}}(s_t, a_t)$ measures improvements in computational efficiency resulting from optimized AI training strategies, $C_{\text{compute}}(s_t, a_t)$ quantifies the energy cost of executing a given AI training task, and $E_{\text{carbon}}(s_t, a_t)$ represents the carbon emissions associated with the selected computational configuration. The parameters λ_1, λ_2, λ_3 control the balance between energy efficiency, operational cost, and environmental impact. This reinforcement learning-based optimization strategy ensures that AI-driven models autonomously learn to minimize their carbon footprint while maintaining computational robustness. By dynamically allocating computational resources based on energy availability and environmental impact, AI systems can significantly reduce their contribution to climate change while continuing to power innovative applications in civil engineering and urban development.

By unifying Deep Reinforcement Learning (DRL) with Proximal Policy Optimization (PPO), this AI-driven framework achieves real-time adaptive energy management for AI training. The system continuously refines AI training strategies, leveraging energy-efficient computing, model compression, and scheduling optimization to minimize environmental costs. This approach ensures that AI's rapid expansion aligns with sustainability goals, preventing its ecological footprint from undermining its long-term societal benefits.

Framework: Energy-Aware Deep Reinforcement Learning for Sustainable AI Training (EADRLSAT2)

Layered AI Architecture for Minimizing Energy Consumption in Large-Scale AI Model Training

Layer	AI module	Mathematical model/framework	Functional flow
Layer 1	Monitoring and pre-AI encoding	Training resource monitoring and encoding engine	System Metrics, Model Metadata → Computational State Encoding → Initial State for DRL Training
Layer 2	Value function estimation for sustainable AI training	Deep Reinforcement Learning (DRL)	Training State, Model Configuration → Bellman Update → Optimized Energy-Aware Value Function
Layer 3	Stable policy refinement with energy awareness	Proximal Policy Optimization (PPO)	Policy Probabilities, Efficiency Advantages → Clipped Policy Update → Smooth Adaptive Training Policy
Layer 4	Sustainability-aware reward construction	Multi-factor reward modeling	Efficiency, Emission, Compute Data → Weighted Trade-Off Evaluation → Composite Reward Signal
Layer 5	Dynamic energy-conscious scheduling	Energy-aware orchestration and scheduling engine	Reward Signals, State Encodings → Adaptive Resource Allocation → Sustainable Training Strategy

5.4.3 Balancing AI Innovation with Sustainability Goals

Sustainability in AI-enabled urban development and civil engineering requires a balanced approach that maximizes the environmental benefits of intelligent systems while reducing their computational energy demands. AI technologies can improve resource utilization, reduce waste, and support the integration of renewable energy sources into urban infrastructures. However, if left unmanaged, the energy-intensive character of advanced AI models may partially offset these benefits, making it necessary to develop optimization frameworks that maintain performance while improving energy efficiency (Manzoor et al. 2021). To address this balance, an advanced Deep Reinforcement Learning (DRL) model with dynamic energy management is introduced to align computational efficiency with urban sustainability objectives. The model can be formulated as:

$$Q^*(s,a) = \mathbb{E}\left[r_t + \gamma \underset{a'}{max}\, Q^*(s',a') \mid s,a\right]$$

where $Q^*(s,a)$ represents the optimal action-value function predicting the long-term sustainability gain from AI-driven urban planning strategies, r_t quantifies the immediate environmental and energy efficiency rewards from implementing AI-based optimizations, γ is the discount factor prioritizing long-term benefits, s and s' denote the current and next urban system states, and a and a' represent different AI-driven decision-making strategies. This model ensures that AI-driven urban solutions dynamically adjust energy usage, optimize infrastructure efficiency, and integrate renewable energy sources while mitigating excessive energy consumption. By continuously evaluating the impact of AI-driven actions, the system refines urban planning strategies to achieve maximum environmental benefits.

To stabilize policy updates and ensure optimal decision-making, the Proximal Policy Optimization (PPO) function is defined as:

$$L(\theta) = \mathbb{E}\left[min\left(\pi_\theta\left(a_{\theta_{\text{old}}}\left(\frac{a_\theta(a \mid s)}{\pi_{\theta_{\text{old}}}(a \mid s), 1-\epsilon, 1+\epsilon}\right)A_t\right)\right)\right]$$

where $L(\theta)$ is the policy loss function ensuring stable and gradual learning updates, $\pi_\theta(a| s)$ represents the probability distribution of AI-driven sustainability strategies under the new policy, $\pi_{\theta_{\text{old}}}(a \mid s)$ corresponds to the probability under the previous policy, A_t denotes the advantage function measuring the improvement of a selected sustainability strategy, and ϵ is the clipping parameter that prevents instability in policy updates.

By leveraging Proximal Policy Optimization (PPO), AI-driven urban sustainability strategies are continuously optimized, ensuring that energy efficiency and resource conservation measures align with broader sustainability goals. This process enables AI models to dynamically adapt to shifting urban conditions, improving decision-making over time.

To quantitatively assess the effectiveness of sustainable AI-driven urban planning, the sustainability reward function is formulated as:

$$R(s_t, a_t) = \lambda_1 S_{\text{efficiency}}(s_t, a_t) - \lambda_2 E_{\text{AI}}(s_t, a_t) - \lambda_3 C_{\text{operation}}(s_t, a_t)$$

where $R(s_t, a_t)$ represents the sustainability reward for AI-driven urban optimization, $S_{\text{efficiency}}(s_t, a_t)$ measures the improvement in resource efficiency achieved through AI-driven solutions, $E_{\text{AI}}(s_t, a_t)$ quantifies the energy consumption of AI systems used in urban development, and $C_{\text{operation}}(s_t, a_t)$ represents the overall operational cost incurred. The parameters λ_1, λ_2, λ_3 control the balance between urban efficiency, AI energy consumption, and economic sustainability. This function ensures that AI-driven urban strategies prioritize environmental sustainability by optimizing resource use and energy efficiency while preventing excessive computational energy consumption.

Building upon the structured reward landscape established in the preceding level, the system now requires a policy that can act optimally in real-time environments by dynamically responding to fluctuations in urban states and energy profiles. This transition from reward structuring to actionable strategy deployment is operationalized through a policy derivation mechanism rooted in the optimal Q-function previously defined via Bellman's formulation. The optimal policy function selects, for each state, the action that maximizes long-term sustainability-adjusted reward:

$$\pi^*(s) = \arg \underset{a}{max}\, Q^*(s, a)$$

where $\pi^*(s)$ is the optimal policy mapping from state to action, s denotes the current urban state, a represents possible actions within the sustainability domain, $Q^*(s, a)$ is the optimal action-value function capturing expected cumulative reward. This function executes real-time strategy deployment by selecting, for each urban condition, the AI-driven decision that maximizes long-term sustainability performance while accounting for prior reward formulations.

Via the combination of Deep Reinforcement Learning (DRL) with Proximal Policy Optimization (PPO), this AI-driven framework achieves a sustainable equilibrium between AI-enabled urban advancements and energy-efficient computing. The system dynamically regulates AI energy usage, optimizes urban resource management, and ensures that AI-driven solutions contribute to long-term environmental sustainability rather than exacerbating energy consumption challenges. Through this intelligent approach, AI enhances urban development while aligning with global sustainability goals.

Framework: Sustainable Deep Reinforcement Urban Intelligence Framework (SDRUIF)

Layered AI Architecture for Sustainable and Energy-Efficient Urban Development

Layer	Main functionality	AI module(s) and components	Operational scope (Input → Processing → Output)
Layer 1	Urban sustainability state encoding	Urban data encoding and sustainability mapping engine	Urban metrics, environmental indicators → Structured urban state representation → DRL-ready state vector
Layer 2	Policy optimization via action-value estimation	Deep Reinforcement Learning (DRL)	Urban state, AI-driven action → Bellman update → Optimal sustainable planning decision
Layer 3	Stability-constrained policy refinement	Proximal Policy Optimization (PPO)	Previous and current policy π, advantage estimates → Clipped surrogate loss → Stable sustainability policy update
Layer 4	Energy-constrained reward structuring	Energy-Aware Reward Modeling Engine	AI energy usage, efficiency gain, operational cost → Weighted sustainability reward → Sustainability-adjusted feedback
Layer 5	Policy deployment and adaptive scheduling	Deep Reinforcement Learning (DRL)	Sustainability-optimized strategy, dynamic environment data → Adaptive control scheduling → Real-time AI-based policy execution
Layer 6	Post-deployment monitoring and strategy adjustment	Continuous Monitoring and Adaptive Feedback System	Policy outcomes, environmental response → Performance evaluation → Updated metrics for retraining feedback

To address the sustainability challenges created by the computational demands of AI, researchers are developing energy-efficient models capable of maintaining strong performance while reducing power consumption. These advances are particularly important in energy-intensive domains such as civil engineering and urban governance, where intelligent systems increasingly support infrastructure optimization and resource management. The integration of AI workloads with renewable energy sources also offers a viable pathway for lowering the carbon footprint of computation, ensuring that training and inference processes are powered, where possible, by sustainable energy rather than fossil fuels (Floridi et al. 2018). To support these objectives, an energy-aware Deep Reinforcement Learning (DRL) framework is introduced to improve computational efficiency while coordinating the use of renewable energy resources. The model can be formulated as:

$$V^*(s) = \max_a \mathbb{E}[R(s,a) + \gamma V^*(s') \mid s]$$

where $V^*(s)$ represents the optimal state-value function predicting the long-term sustainability gains from selecting an energy-efficient AI algorithm, $R(s,a)$ denotes the immediate reward derived from computational energy savings and renewable energy integration, γ is the discount factor prioritizing long-term efficiency improvements, s and v denote the current and next system states, and a represents the choice of AI optimization strategies that balance performance and energy consumption. This equation allows AI models to autonomously select energy-efficient configurations, such as adaptive model pruning, quantization techniques, and lightweight neural architectures, ensuring reduced power consumption without compromising accuracy. It also facilitates dynamic adjustments that prioritize computation during peak renewable energy availability, further minimizing AI's carbon footprint.

To refine energy-efficient AI decision-making, the Proximal Policy Optimization (PPO) function is defined as:

$$L(\theta) = \mathbb{E}\left[min\left(\frac{\pi_\theta(a \mid s)}{\pi_{\theta_{\text{old}}}(a \mid s)}A_t, \text{clip}\left(\frac{\pi_\theta(a \mid s)}{\pi_{\theta_{\text{old}}}(a \mid s)}, 1 - \epsilon, 1 + \epsilon\right)A_t\right)\right]$$

where $L(\theta)$ is the policy loss function ensuring stable updates to AI model configurations, $\pi_\theta(a \mid s)$ represents the probability distribution of AI optimization strategies under the new policy, $\pi_{\theta_{\text{old}}}(a \mid s)$ corresponds to the probability distribution under the previous policy, A_t is the advantage function measuring the energy efficiency improvement of a selected AI model, and ϵ is the clipping threshold preventing excessive adjustments. This optimization process enables AI-driven models to dynamically refine their computational processes, reducing unnecessary energy consumption while continuously improving efficiency. It ensures that AI systems transition toward more sustainable computing strategies, promoting adaptive and resource-conscious AI development.

To quantitatively assess AI sustainability improvements, the reward function for energy-aware AI optimization is formulated as:

$$R(s_t, a_t) = \lambda_1 P_{\text{optimization}}(s_t, a_t) - \lambda_2 C_{\text{compute}}(s_t, a_t) - \lambda_3 E_{\text{renewable}}(s_t, a_t)$$

where $R(s_t, a_t)$ represents the sustainability reward for AI energy efficiency, $P_{\text{optimization}}(s_t, a_t)$ quantifies the performance-to-energy ratio improvements achieved through AI model compression and adaptive computing, $C_{\text{compute}}(s_t, a_t)$ denotes the energy cost of executing a selected AI training process, and $E_{\text{renewable}}(s_t, a_t)$ represents the proportion of renewable energy used in AI computation. The parameters λ_1, λ_2, λ_3 regulate the trade-offs between performance optimization, energy consumption, and renewable energy integration. This function ensures that AI-driven systems continuously refine their energy efficiency, applying intelligent strategies such as computational offloading to low-power processors, leveraging edge AI for decentralized processing, and dynamically adjusting model complexity to suit available renewable energy resources.

Following the synthesis of energy-aware rewards established in the previous level, the model must now translate these quantified sustainability trade-offs into actionable decision policies. The transition from reward structuring to real-time adaptive control necessitates a mechanism capable of selecting, among all viable

AI configurations, the one that maximizes expected long-term energy efficiency. This is achieved through the derivation of the optimal policy from the previously defined action-value function, aligning energy-aware reinforcement strategies with performance-aware deployment:

$$\pi^*(s) = \arg \underset{a}{max}\, Q^*(s, a)$$

where $\pi^*(s)$ is the optimal policy mapping from state to action, s denotes the current system state capturing energy and operational variables, a represents the set of available AI strategies, $Q^*(s, a)$ is the optimal action-value function encoding expected cumulative sustainability-adjusted reward. This function enables the autonomous selection of AI configurations that maximize long-term computational sustainability, dynamically balancing performance optimization with energy and environmental constraints.

Using Deep Reinforcement Learning (DRL) with Proximal Policy Optimization (PPO), this AI-driven framework enables AI models to operate with significantly reduced energy consumption while leveraging renewable energy for computational sustainability. The system dynamically refines AI decision-making to balance computational efficiency, model performance, and energy resource availability. This approach ensures that AI-powered urban governance and civil engineering applications align with sustainability principles, preventing excessive energy waste while supporting long-term environmental objectives.

Framework: Energy-Aware Reinforcement Optimization Framework (EAROF)

Layered AI Architecture for Renewable-Efficient AI Computing

Layer	Main functionality	AI module(s) and components	Operational scope (Input → Processing → Output)
Layer 1	Energy-aware system state representation	Energy state encoding and resource mapping engine	Hardware load, energy consumption rate, renewable availability → Structured environment vector → DRL input state
Layer 2	Energy-efficient value function optimization	Deep Reinforcement Learning (DRL)	System state, AI configuration strategy → Bellman update → Long-term efficiency-oriented value function
Layer 3	Stability-constrained policy refinement	Proximal Policy Optimization (PPO)	Current and past policy, energy-aware advantage → Clipped policy loss → Stable AI energy optimization policy
Layer 4	Energy-constrained sustainability reward structuring	Sustainability-aware reward modeling engine	Performance-energy ratios, compute cost, renewable usage → Composite sustainability reward → Reward feedback for optimization

(continued)

Layer	Main functionality	AI module(s) and components	Operational scope (Input → Processing → Output)
Layer 5	Adaptive AI policy deployment	Deep Reinforcement Learning (DRL)	Optimized policy, real-time energy dynamics → Adaptive computation control → Energy-aligned AI system execution
Layer 6	Post-deployment monitoring and adaptation	Continuous monitoring and adaptive optimization system	System performance logs, energy metrics → Evaluation and adaptation trigger → Feedback for retraining and policy update

Figure 5.4 outlines the sustainability and environmental impacts of AI, structured into three main areas: "AI as a Tool for Environmental Sustainability," "Energy Use and Environmental Footprint," and "Balancing AI with Sustainability Goals." "AI as a Tool for Environmental Sustainability" highlights the potential for AI to optimize resource use, improve waste management, and promote sustainable infrastructure. "Energy Use and Environmental Footprint" addresses the high energy demands of AI systems and the need to manage their carbon emissions. "Balancing AI with Sustainability Goals" emphasizes developing energy-efficient AI models and integrating renewable energy sources, such as solar and wind, to support sustainability in AI applications.

Fig. 5.4 Sustainability and environmental impact of AI

The deployment of AI systems in civil engineering, urban development, and governance has the potential to significantly enhance environmental sustainability by optimizing resource use, reducing waste, and promoting the integration of renewable energy. However, the energy consumption and environmental impact of AI systems themselves must be carefully managed to ensure that the benefits of AI-driven innovations are not overshadowed by their contribution to carbon emissions and environmental degradation. Achieving sustainability in AI systems requires ongoing efforts to develop energy-efficient algorithms, integrate renewable energy sources, and adopt a balanced approach to AI deployment that prioritizes long-term environmental sustainability.

5.5 Privacy and Data Governance

As artificial intelligence (AI) becomes increasingly integrated into civil engineering, urban development, and governance, issues surrounding privacy and data governance are becoming more urgent. AI systems rely on vast amounts of data, often including personal, sensitive, and sometimes confidential information. These systems are used for decision-making in public infrastructure, urban planning, and service delivery, necessitating robust frameworks to protect individual privacy and ensure ethical data use. Effective data governance is crucial to managing the collection, storage, and use of data in AI applications, balancing innovation with privacy protection and compliance with legal standards (Sargiotis 2024a, d).

5.5.1 Privacy Challenges in AI Systems

AI plays a pivotal role in smart cities and civil engineering by leveraging large volumes of data from diverse sources, including sensors, surveillance systems, and personal devices (Barredo Arrieta et al. 2020; Sargiotis 2024a, g). These datasets enable intelligent systems to improve urban infrastructure, enhance traffic management, and strengthen public service delivery. However, the extensive collection of such information raises critical privacy concerns, particularly when sensitive data related to individual movements, behaviors, and service interactions is involved. Preserving privacy while maintaining operational efficiency therefore requires a robust computational framework that integrates Federated Learning (FL) with Differential Privacy (DP). This combined approach supports decentralized model training while limiting the exposure of sensitive personal data in smart-city environments. The privacy-preserving model for urban data processing can be formulated as:

$$\theta^t = \sum_{i=1}^{N} w_i \theta_i^t + \mathcal{N}(0, \sigma^2)$$

where θ^t represents the globally aggregated model parameters at time t, w_i is the weight assigned to each local model θ_i^t based on its contribution to the overall training, $\mathcal{N}$ denotes the number of local devices or urban data collection points participating in the federated learning process, and $\mathcal{N}(0, \sigma^2)$ represents the Gaussian noise introduced to ensure differential privacy.

This formulation enables privacy-preserving AI training by ensuring that sensitive urban data remains decentralized, processed locally on individual devices or edge nodes rather than being transmitted to centralized cloud servers. By applying differential privacy, the model guarantees that even if an adversary accesses AI training data, individual-level details remain obfuscated, protecting citizens from potential surveillance risks.

To further refine privacy-preserving optimization, the Differential Privacy (DP) mechanism for limiting data exposure is defined as:

$$P(M(D) \in S) \leq e^{\epsilon} P(M(D') \in S) + \delta$$

where $P(M(D) \in S)$ is the probability that a randomized algorithm M produces an output S given dataset D, $P(M(D')$ is the probability of obtaining the same output from an adjacent dataset D' differing by a single individual's record, ϵ represents the privacy budget controlling the trade-off between privacy and utility, and δ is a relaxation parameter allowing for minimal privacy leakage. This mechanism ensures that individual contributions to AI models remain indistinguishable, even if an adversary attempts to infer information from urban datasets. By regulating the privacy budget $\epsilon\epsilon$, city planners can maintain an optimal balance between data utility and privacy protection, ensuring ethical AI deployment in civil engineering applications.

To evaluate the effectiveness of privacy-preserving AI systems in urban environments, the privacy-aware reward function is defined as:

$$R(s_t, a_t) = \lambda_1 U_{\text{optimization}}(s_t, a_t) - \lambda_2 L_{\text{privacy}}(s_t, a_t) - \lambda_3 C_{\text{compliance}}(s_t, a_t)$$

where $R(s_t, a_t)$ represents the reward function for privacy-aware AI optimization in smart cities, $U_{\text{optimization}}(s_t, a_t)$ quantifies the urban service improvements derived from AI-driven data analytics, $L_{\text{privacy}}(s_t, a_t)$ measures the privacy risks associated with a particular AI deployment, and $C_{\text{compliance}}(s_t, a_t)$ represents the regulatory compliance cost of ensuring legal adherence to data protection frameworks such as GDPR or CCPA. The parameters λ_1, λ_2, λ_3 regulate the trade-offs between AI optimization, privacy preservation, and legal constraints. This formulation enables AI-driven smart-city applications to dynamically adjust privacy parameters based on real-time urban data demands, ensuring compliance with privacy laws while maintaining the effectiveness of AI-driven optimizations.

Following the formulation of the privacy-constrained reward function, which elegantly balances utility maximization, privacy preservation, and regulatory compliance, the system proceeds to translate these multidimensional trade-offs into an operational policy that governs AI-driven decisions across distributed smart-city environments. Within the federated learning paradigm, this policy derivation operates atop the decentralized training architecture previously established. By optimizing over the learned action-value function, the model ensures that the selected AI strategies align with the highest expected reward under strict privacy guarantees:

$$\pi^*(s) = \arg\max_{a} Q^*(s, a)$$

where $\pi^*(s)$ is the optimal decision policy derived under federated conditions, s denotes the privacy-structured system state at deployment, a represents available AI-driven urban decision actions, $Q^*(s, a)$ is the privacy-informed action-value function representing cumulative reward expectations under the federated privacy constraints. This function ensures that urban AI services autonomously adopt strategies that maximize operational utility while conforming to established privacy and compliance thresholds, thereby executing trustworthy and adaptive decisions in real-time smart-city applications.

By employing Federated Learning (FL) with Differential Privacy (DP), this AI-driven framework ensures that smart-city applications and civil engineering projects leverage data-driven insights without compromising individual privacy. The system dynamically regulates data sharing mechanisms, applying secure aggregation techniques to prevent unauthorized access while optimizing AI-driven urban management. Through this approach, AI enhances smart-city services while maintaining public trust and regulatory compliance, ensuring that privacy remains a central pillar of AI-enabled urban governance.

Framework: Federated Privacy-Aware Smart Urban Framework (FPSUF)

Layered AI Architecture for Privacy-Preserving Smart-City Optimization

Layer	Main functionality	AI module(s) and components	Operational scope (Input → Processing → Output)
Layer 1	Urban data decentralization and local processing	Edge data processing and decentralized encoding engine	Sensor inputs, surveillance feeds, device-level logs → Localized feature extraction → Edge-processed training input
Layer 2	Federated aggregation with privacy noise	Federated Learning (FL) with Differential Privacy (DP)	Local model updates, secure aggregation → Weighted global model with noise → Privacy-preserving model parameters
Layer 3	Formal privacy budget mechanism enforcement	Differential Privacy (DP)	Urban datasets, randomized mechanisms → Privacy constraints enforcement → Provable privacy guarantees

(continued)

Layer	Main functionality	AI module(s) and components	Operational scope (Input → Processing → Output)
Layer 4	Privacy-constrained reward formulation	Privacy-aware reward modeling engine	Urban service utility, privacy loss, compliance costs → Composite reward → Feedback to learning process
Layer 5	Policy deployment and privacy-aware adaptation	Federated Learning (FL)	Privacy-tuned model, real-time urban data → Adapted AI-driven decisions → Smart-city service execution
Layer 6	Compliance monitoring and data trust assurance	Compliance monitoring and trust audit system	Deployment reports, access patterns → Risk assessment and update triggers → Accountability and transparency records

The reliance of AI systems on cloud-based platforms and extensive data networks introduces vulnerabilities to cyberattacks, exposing sensitive personal information to unauthorized access (Floridi et al. 2018). Data breaches can lead to identity theft, financial loss, and reputational damage, making robust security mechanisms necessary for protecting user privacy. As AI systems process large volumes of sensitive information, data integrity and confidentiality become critical conditions for ethical and secure deployment. To address these risks, an advanced framework integrating Blockchain Technology with Federated Learning (FL) and Homomorphic Encryption (HE) is introduced to strengthen data security and improve resilience against cyber threats.

The secure data processing model based on blockchain, federated learning, and homomorphic encryption can be formulated as:

$$\mathcal{H}(D_i) = \sum_{i=1}^{N} \mathrm{Enc}\left(\theta_i^t\right) + \mathcal{B}(h_i, t)$$

where $\mathcal{H}(D_i)$ represents the encrypted and securely stored data for user i, $\mathrm{Enc}\left(\theta_i^t\right)$ denotes the homomorphic encryption of the local AI model parameters θ_i^t, N is the number of federated learning participants contributing to the AI training, and $\mathcal{B}(h_i, t)$ represents the blockchain-secured hash function applied to the data transaction history at time t. This function enables secure parameter exchange in hostile network environments, ensuring that model updates are cryptographically shielded throughout the training cycle.

Proceeding from encrypted local computation, the system invokes Federated Learning (FL) to aggregate secure model updates across the distributed network. The privacy-preserving aggregation mechanism accommodates the encrypted contributions of each node, while maintaining differential privacy through Gaussian noise injection. This is formally expressed through the federated aggregation function:

$$\theta^t = \sum_{i=1}^{N} w_i \theta_i^t + \mathcal{N}(0, \sigma^2)$$

where θ^t is the globally aggregated model at round t, w_i is the weight of client i's contribution, θ_i^t is the local encrypted model, $\mathcal{N}(0, \sigma^2)$ is Gaussian noise for differential privacy. This function ensures that global learning proceeds without exposing individual records or violating privacy guarantees.

To guarantee transaction integrity, the framework integrates blockchain technology for cryptographic anchoring of training events. Each contribution is hashed and stored on a distributed ledger, offering tamper-proof traceability. The blockchain hash formulation ensures that the data origin and authenticity are verifiable:

$$B(h_i, t) = \text{Hash}\left(D_i^t \parallel T_i^t\right)$$

where $B(h_i, t)$ is the blockchain hash for participant i at time t, D_i^t is the participant's encrypted data, T_i^t is the associated timestamp, $\parallel$ denotes concatenation. This function immutably records data transactions, thereby enforcing ledger-based integrity in federated AI operations.

To prevent manipulation and ensure consensus validation across network nodes, the architecture adopts a blockchain consensus mechanism, wherein authorized nodes cryptographically vote to validate each AI transaction block. The formal consensus validation function is expressed as:

$$P(T_t) = \arg \max_{h_i} \sum_{i=1}^{M} \sigma(h_i, t)$$

where $P(T_t)$ represents the validated AI data transaction at time t, M is the total number of blockchain nodes participating in verification, and $\sigma(h_i, t)$ is the digital signature function ensuring cryptographic validation of each AI data interaction. This mechanism ensures that all AI transactions are authenticated using blockchain consensus, mitigating risks of data breaches by preventing unauthorized alterations. The model verifies AI operations through decentralized validation, making it highly resilient to cyber threats targeting centralized cloud platforms.

To quantify the effectiveness of AI-driven cybersecurity measures, the data security reward function is formulated as:

$$R(s_t, a_t) = \lambda_1 I_{\text{integrity}}(s_t, a_t) - \lambda_2 C_{\text{encryption}}(s_t, a_t) - \lambda_3 R_{\text{breach}}(s_t, a_t)$$

where $R(s_t, a_t)$ represents the security reinforcement reward for AI-driven data protection, $I_{\text{integrity}}(s_t, a_t)$ quantifies improvements in data security achieved through AI-driven cryptographic techniques, $C_{\text{encryption}}(s_t, a_t)$ represents the computational overhead required for secure AI data encryption, and $R_{\text{breach}}(s_t, a_t)$ measures the potential risk of AI system vulnerabilities being exploited. The parameters λ_1, λ_2, λ_3

regulate the trade-offs between security enhancement, computational efficiency, and privacy risk mitigation.

Lastly, in accordance with this reward signal, the model deploys an optimized policy that dynamically balances security-performance trade-offs. Derived from the Q-function encoded with encrypted data flows, the adaptive policy execution function is expressed as:

$$\pi^*(s) = \arg \underset{a}{max}\, Q^*(s, a)$$

where $\pi^*(s)$ is the optimal federated policy under secure computation, s is the encrypted system state, a is the possible AI strategy, $Q^*(s, a)$ is the expected secure reward function. This function enables the final decision mechanism to select the most secure, privacy-preserving AI action, effectively closing the reinforcement loop in federated, blockchain-integrated, homomorphically encrypted AI systems.

This AI-driven reinforcement learning framework dynamically strengthens AI data security, integrating blockchain authentication, homomorphic encryption, and federated learning techniques to prevent unauthorized access while maintaining AI system performance.

By incorporating Blockchain Technology with Federated Learning (FL) and Homomorphic Encryption (HE), this AI-driven model ensures that sensitive data remains protected throughout its lifecycle. AI computations can be performed on encrypted data, preventing unauthorized access while maintaining the integrity of urban and civil engineering applications. The decentralized nature of blockchain further reinforces data protection, eliminating single points of failure and making AI-driven systems resilient to cyberattacks. Through this intelligent security framework, AI continues to enhance urban operations while ensuring the highest level of privacy and protection.

Framework: Secure Federated Blockchain AI Architecture (SFB-AIA)

Layered AI Architecture for Cybersecure and Privacy-Preserving AI Systems

Layer	Main functionality	AI module(s) and components	Operational scope (Input → Processing → Output)
Layer 1	Encrypted model parameter encoding	Homomorphic Encryption (HE)	Local AI parameters, privacy keys → Secure encryption → Encoded encrypted model update
Layer 2	Decentralized secure model aggregation	Federated Learning (FL)	Encrypted model updates, participant weights → Federated aggregation → Secure global AI model
Layer 3	Blockchain-authenticated data hashing	Blockchain Technology	Model transactions, encrypted hashes → Immutable ledger verification → Secured audit trail
Layer 4	Distributed transaction consensus validation	Blockchain Consensus Mechanism	Node hashes, digital signatures → Consensus voting process → Validated AI transaction block

(continued)

Layer	Main functionality	AI module(s) and components	Operational scope (Input → Processing → Output)
Layer 5	Security-constrained reward structuring	Cybersecurity-Aware Reward Modeling Engine	Integrity indicators, encryption costs, breach probabilities → Composite cybersecurity reward → Feedback to reinforcement loop
Layer 6	Adaptive policy deployment for threat resilience	Federated Learning (FL)	Reinforced secure model, threat analytics → Updated privacy-preserving actions → Cyberattack-resilient AI execution

The large-scale deployment of AI in urban development involves extensive data collection, often without explicit consent from individuals. Citizens frequently remain unaware of how their data is gathered, processed, and used, raising ethical concerns regarding transparency and control over personal information (Manzoor et al. 2021). To address these issues, AI-enabled urban systems must integrate transparent data governance frameworks and consent-driven mechanisms that empower individuals to control their data. For this purpose, a framework combining Federated Learning (FL), Explainable AI (XAI), and smart contract-based consent systems is introduced to strengthen transparency and support ethical AI deployment.

$$\theta^t = \sum_{i=1}^{N} w_i \theta_i^t + \mathcal{SC}(c_i, t)$$

where θ^t represents the aggregated model parameters at time t, w_i denotes the weight assigned to each local model θ_i^t in the federated learning framework, N is the number of participating decentralized nodes in AI-driven urban data processing, and $\mathcal{SC}(c_i, t)$ represents the smart contract ensuring user-consent validation for data utilization at time t. This formulation ensures that AI models operate with decentralized data processing while implementing transparent consent mechanisms via blockchain-based smart contracts. These contracts verify individual user agreements before any data is processed, enabling an ethical framework where citizens actively control their information.

To enhance explainability in AI driven urban systems and ensure transparency, the Explainable AI (XAI) decision function is defined as:

$$X(s, a) = \arg \max_a \mathbb{E}[R(s, a) \mid \text{interpretability constraints}]$$

where $X(s, a)$ represents the optimized AI decision-making process subject to interpretability constraints, ensuring that urban AI actions remain explainable, $R(s, a)$ denotes the expected reward for a transparent AI-driven urban action, and the interpretability constraints ensure that AI decisions are comprehensible to human users. This function guarantees that AI models operating in urban environments remain transparent, allowing citizens to understand how their data influences urban planning, traffic optimization, and infrastructure management.

To assess the effectiveness of AI-driven consent mechanisms and ethical governance, the transparency-aware reward function is defined as:

$$R(s_t, a_t) = \lambda_1 T_{\text{governance}}(s_t, a_t) - \lambda_2 C_{\text{processing}}(s_t, a_t) - \lambda_3 L_{\text{consent}}(s_t, a_t)$$

where $R(s, a)$ represents the reward function for ethical AI data governance, $T_{\text{governance}}(s_t, a_t)$ quantifies the transparency level achieved in AI-driven urban data management, $C_{\text{processing}}(s_t, a_t)$ measures the computational cost incurred for executing consent verification and privacy-preserving AI processes, and $L_{\text{consent}}(s_t, a_t)$ represents the risk of processing data without explicit user approval. The parameters λ_1, λ_2, λ_3 regulate the balance between AI transparency, efficiency, and privacy compliance. This formulation ensures that AI-driven urban solutions continuously adapt to enforce transparency, actively verifying consent-based interactions through decentralized governance mechanisms. The model dynamically refines AI operations to comply with ethical AI principles while ensuring that data-driven urban optimizations remain effective.

Following the formulation of the transparency-constrained reward function, the model transitions toward operationalizing ethical AI behavior through policy execution. At this stage, the decision-making mechanism must align with both federated learning outcomes and interpretability guarantees, ensuring that each action deployed in the urban environment maximizes expected utility under ethical constraints. The optimal policy is thus derived from the previously optimized action-value function, incorporating governance-informed feedback into the federated decision architecture:

$$\pi^*(s) = \arg \underset{a}{max}\, Q^*(s, a)$$

where $\pi^*(s)$ is the optimal policy selecting the best AI-driven urban action, s denotes the current state shaped by consent and transparency dynamics, a represents the available explainable AI-driven decisions, $Q^*(s, a)$ is the action-value function capturing the expected ethical and operational reward. This function ensures that each AI deployment instance adheres to ethical data governance principles while delivering effective, privacy-compliant services within federated urban environments.

Through the integration of Federated Learning (FL) with Explainable AI (XAI) and smart contract-based consent systems, this AI-driven framework ensures that urban AI applications operate with full transparency and accountability. Citizens retain control over their data through automated blockchain-based consent validation, while AI-driven decision-making remains interpretable and ethically sound. This approach guarantees that AI enhances urban development while maintaining strict compliance with privacy and ethical standards, fostering public trust in AI-powered governance.

Framework: Transparent Federated Urban Ethics Framework (TFUEF)

Layered AI Architecture for Ethical Data Governance and Explainable Smart Consent Systems

Layer	Main functionality	AI module(s) and components	Operational scope (Input → Processing → Output)
Layer 1	Consent-aware data initialization	Smart contract mechanism	Citizen identifiers, access requests → Consent validation → Permissioned data access
Layer 2	Federated learning with consent enforcement	Federated Learning (FL) with Smart Contracts	Local model updates, user-consent policies → Aggregated model parameters with compliance trace → Secure federated output
Layer 3	Explainable urban decision computation	Explainable AI (XAI)	Urban AI predictions, interpretability thresholds → Constrained optimization → Transparent decision outcome
Layer 4	Ethical reward structuring for governance	Ethics-aware reward modeling engine	Transparency score, processing load, consent violations → Governance-aware reward function → Optimization feedback signal
Layer 5	Policy execution with explainability guarantees	Federated Learning (FL)	Explainable policy, validated consent structure → Privacy-compliant urban execution → Transparent service optimization
Layer 6	Governance feedback and audit assurance	Governance audit and trust assurance system	Urban audit logs, citizen feedback → Governance metric evaluation → Ethical model refinement and trust reporting

5.5.2 Data Governance Frameworks

Data governance establishes the foundational policies and processes that regulate data collection, storage, management, and use, ensuring that AI systems operate ethically, securely, and in compliance with legal frameworks (Mökander 2023; Sargiotis 2024c, g). A robust data governance structure is essential for balancing the operational value of AI-enabled innovation with the need to protect individual privacy and maintain regulatory compliance. In addition, AI models depend on high-quality data for accurate decision-making. Poor-quality data, including incomplete, outdated, or biased datasets, can lead to erroneous predictions, inequitable outcomes, and systemic discrimination (Floridi et al. 2018). To address these challenges, a computational governance framework incorporating Graph Neural Networks (GNNs), Federated Learning (FL), and Fairness Constraints (FC) is introduced to strengthen data integrity, fairness, and compliance.

The data governance model based on graph neural networks, federated learning, and fairness constraints can be formulated as:

$$\theta^t = \sum_{i=1}^{N} w_i \theta_i^t + \mathcal{F}(b_i, t)$$

where θ^t represents the aggregated AI model parameters at time t, w_i is the weight assigned to each local model θ_i^t in the federated learning process, N denotes the number of decentralized nodes contributing to AI training, and $\mathcal{F}(b_i, t)$ represents the fairness constraint ensuring that biases in the dataset are dynamically adjusted at time t. This formulation enables AI-driven data governance by ensuring that decentralized models operate with privacy-preserving learning while dynamically adjusting fairness constraints to mitigate biases. By leveraging federated learning, AI systems can maintain compliance with legal standards without exposing sensitive data, while fairness constraints proactively correct imbalances in training data to prevent discrimination.

To further enhance data integrity and transparency, the Graph Neural Network (GNN)-based data validation function is defined as:

$$G(s, a) = \sum_{j=1}^{M} \sigma\left(W_j h_j + b_j\right)$$

where $G(s, a)$ represents the AI-driven function validating data integrity and relational consistency in structured datasets, M is the number of graph nodes representing data relationships, W_j is the weight matrix that adjusts node connections based on data quality, h_j represents the node feature vector containing data attributes, b_j is the bias term regulating adjustments in data classification, and σ $(\cdot)$ is the activation function ensuring non-linearity in AI-driven pattern recognition. This function allows AI systems to validate data quality dynamically, ensuring that only high-integrity datasets contribute to AI model training. The GNN-based validation framework identifies missing, outdated, or inconsistent records while preserving fairness in AI-driven decision-making.

To assess the effectiveness of AI-driven data governance, the fairness-aware reward function is defined as:

$$R(s_t, a_t) = \lambda_1 I_{\text{data}}(s_t, a_t) - \lambda_2 B_{\text{bias}}(s_t, a_t) - \lambda_3 C_{\text{compliance}}(s_t, a_t)$$

where $R(s_t, a_t)$ represents the reward function optimizing AI-driven data governance, $I_{\text{data}}(s_t, a_t)$ quantifies improvements in data integrity and representativeness, $B_{\text{bias}}(s_t, a_t)$ measures the bias detected in datasets and AI models, and $C_{\text{compliance}}(s_t, a_t)$ accounts for regulatory compliance costs associated with ensuring fairness and privacy standards. The parameters λ_1, λ_2, λ_3 regulate the trade-offs between data integrity, bias mitigation, and legal adherence.

Following the bias-aware reward formulation, which quantifies integrity, fairness deviation, and compliance cost, the system must now transition into an actionable decision mechanism that operationalizes these ethical objectives within the federated architecture. This is achieved through the optimization of a policy function derived from the federated Q-value estimator, which is continuously updated by the fairness-constrained reward signal. The ethical policy, therefore, encapsulates the system's ability to prioritize high-integrity, bias-mitigated, and regulation-compliant strategies:

$$\pi^*(s) = \arg \max_a Q^*(s, a)$$

where $\pi^*(s)$ is the optimal policy function over the federated system state s, a denotes the space of admissible decision actions available to the AI agent, $Q^*(s, a)$ is the optimal action-value function trained under bias-aware, decentralized learning dynamics. This function determines the most ethically aligned AI response by selecting the action that maximizes long-term fairness-governed utility within the federated learning infrastructure.

This AI-driven framework dynamically refines data governance policies, proactively detecting and mitigating biases, maintaining data quality, and ensuring compliance with privacy regulations. The model enables AI systems to make fair, accurate, and legally compliant decisions, safeguarding against discriminatory biases while preserving trust in AI-driven urban and civil engineering applications.

By unifying Graph Neural Networks (GNNs) with Federated Learning (FL) and Fairness Constraints (FC), this AI-driven model ensures robust data governance, enabling ethical and secure AI deployment. The system continuously optimizes data integrity, leveraging decentralized processing to protect privacy while enforcing fairness constraints to mitigate biases. This approach guarantees that AI-powered decision-making remains transparent, reliable, and aligned with ethical and legal principles, fostering trust in AI-driven innovations.

Framework: Federated Graph-Fair AI Data Governance Framework (FGF-AIDGF)

Layered AI Architecture for Bias-Constrained, Federated, and Graph-Validated Data Governance

Layer	Main functionality	AI module(s) and components	Operational scope (Input → Processing → Output)
Layer 1	Federated learning with fairness constraint injection	Federated Learning (FL) with Fairness Constraints (FC)	Local models, weighted aggregation, bias correction → Federated output model with fairness trace → Bias-mitigated model parameters
Layer 2	Graph-based data integrity verification	Graph neural networks (GNNs)	Structured datasets, relational mappings → Node-level validation and consistency checks → Integrity-assured data graph
Layer 3	Bias-aware reward structuring	Fairness-aware reward modeling engine	Data quality score, bias detection, compliance overhead → Fairness-aware reward function → Ethical model optimization signal

(continued)

Layer	Main functionality	AI module(s) and components	Operational scope (Input → Processing → Output)
Layer 4	Policy deployment under governance constraints	Federated Learning (FL)	Validated model, fairness reward gradient → Decision policy update → Bias-regulated federated execution
Layer 5	Audit-driven governance monitoring	Governance monitoring and compliance audit system	Audit logs, fairness KPIs, regulatory violations → Governance metric computation → Dynamic compliance enforcement and refinement

Data governance frameworks must incorporate mechanisms for anonymization and pseudonymization to protect individuals' privacy while enabling AI-assisted decision-making. By removing or masking personally identifiable information (PII), AI systems can leverage large datasets without compromising personal security (Manzoor et al. 2021; Sargiotis 2024c, h). However, even anonymized data can sometimes be re-identified, particularly when cross-referenced with other datasets, necessitating the implementation of advanced privacy-preserving techniques. To address these challenges, a secure data processing framework incorporating Differential Privacy (DP), Homomorphic Encryption (HE), and Secure Multi-Party Computation (SMPC) is introduced to protect data throughout AI processing.

The privacy-preserving anonymization model based on differential privacy, homomorphic encryption, and secure multi-party computation can be formulated as:

$$M(D) \approx M(D') \pm \mathcal{N}(0, \sigma^2)$$

where $M(D)$ represents the AI function processing dataset D, $M(D')$ is the function applied to a similar dataset D' that differs by a single individual's record, and $\mathcal{N}(0, \sigma^2)$ represents the noise added to prevent exact re-identification, ensuring that AI outputs remain indistinguishable between datasets with and without a specific individual's information. This formulation ensures that AI-driven models operate with differential privacy, making it computationally infeasible to re-identify individuals from anonymized data. By strategically injecting noise into AI-generated insights, differential privacy guarantees that AI models maintain utility without exposing sensitive personal attributes.

To further enhance security, the Homomorphic Encryption (HE) function for AI-driven encrypted data processing is defined as:

$$\mathcal{E}(x) + \mathcal{E}(y) = \mathcal{E}(x + y)$$

where $\mathcal{E}(x)$ and $\mathcal{E}(y)$ are the encrypted representations of two data values, and the equation ensures that encrypted computations can be performed without decrypting the underlying data. This function allows AI models to process data while keeping it encrypted, ensuring that even if an adversary gains access to stored AI inputs, the information remains indecipherable.

Next, the framework must facilitate collaborative computations across multiple distributed entities without exposing any party's raw data. This is operationalized through Secure Multi-Party Computation (SMPC), which allows multiple data holders to jointly evaluate a function over their combined inputs while preserving complete data confidentiality. The encrypted computations established in the previous layer now serve as cryptographic inputs to the secure collaborative function, ensuring zero-knowledge interoperability:

$$f(x_1, x_2, \ldots, x_n) = \bigoplus_{i=1}^{n} f_i(x_i)$$

where $f(x_1, x_2, \ldots, x_n)$ is the global function securely evaluated over all distributed inputs, x_i denotes the private data of party i, $f_i(x_i)$ is the local partial computation, $\bigoplus$ signifies the secure aggregation operator under SMPC protocols. This function enables encrypted decentralized data collaboration, allowing AI systems to derive global insights from distributed data without revealing any individual contributor's information, thereby reinforcing both privacy preservation and data utility.

To assess the effectiveness of AI-driven anonymization measures, the privacy-aware reward function is defined as:

$$R(s_t, a_t) = \lambda_1 A_{\text{privacy}}(s_t, a_t) - \lambda_2 R_{\text{re-identification}}(s_t, a_t) - \lambda_3 C_{\text{security}}(s_t, a_t)$$

where $R(s_t, a_t)$ represents the reward function optimizing AI-driven anonymization, $A_{\text{privacy}}(s_t, a_t)$ quantifies the level of privacy preservation achieved through anonymization techniques, $R_{\text{re - identification}}(s_t, a_t)$ measures the risk of individuals being re-identified in AI-driven datasets, and $C_{\text{security}}(s_t, a_t)$ accounts for the computational cost of implementing secure AI-driven anonymization protocols. The parameters λ_1, λ_2, λ_3 regulate the balance between privacy, re-identification risks, and security overhead. This AI-driven model dynamically refines anonymization policies, ensuring that AI-driven urban systems process data ethically, securely, and with full privacy compliance. By continuously adjusting security parameters, the system prevents re-identification risks while maintaining AI's analytical power for smart-city applications and civil engineering optimizations.

By harnessing Differential Privacy (DP) with Homomorphic Encryption (HE) and Secure Multi-Party Computation (SMPC), this AI-driven framework ensures that anonymized data remains untraceable while enabling AI-driven urban governance and decision-making. AI computations can be performed on encrypted data, preventing unauthorized access while ensuring compliance with global privacy regulations. This approach guarantees that AI enhances smart-city operations and civil engineering while prioritizing privacy, security, and data integrity.

Framework: Encrypted Differential Privacy AI Architecture (EDP-AIA)

Layered AI Architecture for Privacy-Preserving Data Anonymization and Secure Computation

Layer	Main functionality	AI module (s) and components	Operational scope (Input → Processing → Output)
Layer 1	Differential privacy injection for dataset anonymization	Differential Privacy (DP)	Structured datasets, personal attributes → Noise-injected output → Anonymized AI-ready data
Layer 2	Encrypted computation on sensitive data	Homomorphic Encryption (HE)	Encrypted data inputs, encrypted transformations → Secure computation → Privacy-preserving model processing
Layer 3	Collaborative privacy-preserving computation	Secure Multi-Party Computation (SMPC)	Distributed data holders, cryptographic protocols → Joint computation without data sharing → Encrypted aggregated results
Layer 4	Privacy-aware reward structuring	Privacy-Aware Reward Modeling Engine	Anonymization quality, re-identification risk, encryption cost → Privacy-driven reward → Policy feedback for optimization
Layer 5	Anonymization policy refinement and deployment	Differential Privacy (DP)	Privacy reward signal, encrypted utility metrics → Policy update and application → Secure anonymization deployment

5.5.3 *Legal and Regulatory Compliance*

Data governance frameworks for AI systems must not only address ethical considerations but also ensure compliance with legal and regulatory standards. Laws such as the General Data Protection Regulation (GDPR) in Europe impose strict requirements on the collection, processing, and storage of personal data, mandating transparency, user consent, and security measures to protect individual rights (Mökander 2023; Sargiotis 2024h, i). AI systems in civil engineering and urban development must therefore be designed to manage data lawfully, protecting personal information while maintaining operational efficiency. To support this requirement, a compliance framework incorporating Federated Learning (FL), Blockchain for Data Auditing (BDA), and Policy-Aware Reinforcement Learning (PARL) is introduced to enforce regulatory alignment within AI-enabled urban governance.

The regulatory compliance model based on federated learning, blockchain for data auditing, and policy-aware reinforcement learning can be formulated as:

$$\theta^t = \sum_{i=1}^{N} w_i \theta_i^t + \mathcal{B}(r_i, t)$$

where θ^t represents the aggregated AI model parameters at time t, w_i is the weight assigned to each local model θ_i^t in the federated learning process, N denotes the number of decentralized data processing nodes, and $\mathcal{B}(r_i, t)$ represents the blockchain-secured regulatory compliance record for AI model updates, ensuring that all AI-driven operations remain auditable and traceable at time t. This formulation enables AI-driven systems to process data in compliance with GDPR and other legal standards without centralizing sensitive information. The blockchain-based data auditing mechanism ensures that all data transactions are logged immutably, allowing authorities and regulatory bodies to verify AI model compliance in real time.

To enhance regulatory policy enforcement, the Policy-Aware Reinforcement Learning (PARL) compliance function is defined as:

$$\pi^*(a \mid s) = \arg \underset{\pi}{max} \mathbb{E} \sum_{t=0}^{T} \gamma^t R_{\text{policy}}(s_t, a_t)$$

where $\pi^*(a \mid s)$ represents the AI-driven policy that selects the optimal data governance action ensuring regulatory compliance, $R_{\text{policy}}(s_t, a_t)$ is the compliance reward function that quantifies adherence to legal standards, T is the regulatory assessment horizon, and γ^t prioritizes long-term compliance stability. This function ensures that AI-driven urban governance models dynamically adjust their data processing methods to align with evolving legal frameworks, automatically modifying operations to prevent violations of data protection laws such as GDPR

To quantitatively assess AI-driven regulatory adherence, the compliance-aware reward function is formulated as:

$$R(s_t, a_t) = \lambda_1 C_{\text{compliance}}(s_t, a_t) - \lambda_2 P_{\text{penalty}}(s_t, a_t) - \lambda_3 O_{\text{overhead}}(s_t, a_t)$$

where $R(s_t, a_t)$ represents the compliance reward function balancing legal adherence and AI operational efficiency, $C_{\text{compliance}}(s_t, a_t)$ quantifies the degree of alignment with GDPR and other regulatory frameworks, $P_{\text{penalty}}(s_t, a_t)$ represents the financial and legal penalties associated with non-compliance, and $O_{\text{overhead}}(s_t, a_t)$ measures the computational cost of integrating AI-driven compliance verification mechanisms. The parameters $\lambda_1, \lambda_2, \lambda_3$ regulate the trade-offs between compliance enforcement, legal risk mitigation, and computational efficiency.

Building upon the compliance-aware reward function that effectively integrates legal adherence, penalty risks, and overhead costs, we now move toward executing the optimal decision-making policy based on the federated and compliance-aware framework. To operationalize the dynamic policy adjustments derived from the compliance feedback, Federated Learning (FL) is used in combination with

Policy-Aware Reinforcement Learning (PARL), which ensures that the policy is not only legally compliant but also optimally tuned to maintain efficiency. The optimal policy function is determined by maximizing the action-value function $Q^*(s, a)$, which incorporates the ethical considerations embedded in the compliance reward:

$$\pi^*(s) = \arg \underset{a}{max} \, Q^*(s, a)$$

where $\pi^*(s)$ is the optimal policy function that selects the best action in the state s, s represents the current system state that is shaped by regulatory compliance considerations, a denotes the action taken by the AI agent, and $Q^*(s, a)$ is the optimal action-value function that encapsulates the expected cumulative reward, considering both operational and compliance goals. This function ensures that the AI-driven urban governance systems choose actions that are in full alignment with legal standards, ensuring compliance while preserving efficiency and maximizing the overall effectiveness of the system.

This AI-driven framework ensures that urban AI systems continuously refine compliance mechanisms, dynamically adjusting AI model policies to avoid legal infractions while maintaining efficient data governance. By proactively integrating compliance-aware AI policies, smart-city and civil engineering applications can maintain full transparency while operating within global data protection laws.

Using Federated Learning (FL) with Blockchain for Data Auditing (BDA) and Policy-Aware Reinforcement Learning (PARL), this AI-driven model ensures that data governance frameworks remain legally compliant, transparent, and resilient. AI-driven compliance mechanisms dynamically enforce GDPR and similar legal frameworks, while blockchain auditing provides an immutable record for regulatory verification. This approach guarantees that AI-powered decision-making remains lawful, privacy-preserving, and ethically sound, ensuring that civil engineering and urban AI innovations align with legal and regulatory standards while maintaining operational efficiency.

Framework: Federated Compliance-Driven AI Governance Model (FCDAI-GM)

Layered AI Architecture for Legal and Regulatory Compliance in Urban Governance

Layer	Main functionality	AI module(s) and components	Operational scope (Input → Processing → Output)
Layer 1	Federated learning with blockchain compliance injection	Federated Learning (FL) with Blockchain for Data Auditing (BDA)	Local model updates, regulatory compliance records → Blockchain-secured federated aggregation → Auditable and traceable AI model parameters
Layer 2	Regulatory policy enforcement via reinforcement learning	Policy-Aware Reinforcement Learning (PARL)	Local models, compliance reward signals → Federated policy update with compliance trace → AI-driven regulatory policy adjustments

(continued)

Layer	Main functionality	AI module(s) and components	Operational scope (Input → Processing → Output)
Layer 3	Compliance-aware reward structuring	Compliance-Aware Reward Modeling Engine	Compliance metrics, penalty assessments, computational costs → Fair compliance reward function → Feedback signal for optimization
Layer 4	Dynamic policy deployment for legal conformance	Federated Learning (FL) with Policy-Aware Reinforcement Learning (PARL)	Ethically-tuned policies, feedback from legal frameworks → Adapted execution policies → Privacy-preserving decision-making and operational efficiency

One of the foundational principles of the General Data Protection Regulation (GDPR) is data minimization, which mandates that AI systems collect and process only the minimum amount of data required for a specific purpose. This principle is particularly important for AI models that depend on large datasets for training and decision-making. Implementing data minimization practices not only reduces privacy risks but also enhances AI efficiency by eliminating redundant or unnecessary data processing. To address these challenges, a data minimization framework incorporating Federated Learning (FL), Sparse Data Encoding (SDE), and Reinforcement Learning for Data Efficiency (RLDE) is introduced to reduce unnecessary data exposure while preserving AI performance.

The data minimization model based on federated learning, sparse data encoding, and reinforcement learning for data efficiency can be formulated as:

$$\theta^t = \sum_{i=1}^{N} w_i S(\theta_i^t)$$

where θ^t represents the aggregated AI model parameters at time t, w_i denotes the weight assigned to each local model θ_i^t in the federated learning framework, N is the number of distributed AI nodes participating in privacy-preserving model training, and $S(\theta_i^t)$ represents the sparse encoding function that ensures only the essential data features are retained while minimizing redundant or unnecessary data points. This formulation allows AI models to train on decentralized, privacy-preserving datasets while automatically filtering out irrelevant or excessive data. Sparse data encoding techniques ensure that only the most critical features contribute to AI decision-making, aligning with GDPR's data minimization requirements.

To further enhance AI-driven data efficiency, the Reinforcement Learning for Data Efficiency (RLDE) function is defined as:

$$\pi^*(a \mid s) = \arg \underset{\pi}{max} \mathbb{E} \sum_{t=0}^{T} \gamma^t R_{\text{efficiency}}(s_t, a_t)$$

where $\pi^*(a \mid s)$ represents the AI-driven policy selecting the optimal data minimization strategy, $R_{\text{efficiency}}(s_t, a_t)$ is the reinforcement learning reward function quantifying improvements in AI efficiency due to reduced data requirements, T is the optimization time horizon, and γ^t prioritizes long-term efficiency improvements. This function enables AI-driven data governance systems to continuously adapt data minimization strategies in real time, ensuring that AI operates with minimal but highly relevant information.

To quantitatively assess the effectiveness of AI-driven data minimization strategies, the compliance-aware reward function is formulated as:

$$R(s_t, a_t) = \lambda_1 M_{\text{GDPR}}(s_t, a_t) - \lambda_2 O_{\text{reduction}}(s_t, a_t) - \lambda_3 P_{\text{performance}}(s_t, a_t)$$

where $R(s_t, a_t)$ represents the AI-driven reward function optimizing GDPR-compliant data minimization, $M_{\text{GDPR}}(s_t, a_t)$ measures compliance with data minimization principles, $O_{\text{reduction}}(s_t, a_t)$ quantifies the amount of redundant or unnecessary data eliminated from AI processing, and $P_{\text{performance}}(s_t, a_t)$ represents the trade-off between data reduction and AI decision-making accuracy. The parameters $\lambda_1, \lambda_2, \lambda_3$ regulate the balance between compliance enforcement, data efficiency, and AI performance optimization.

Having structured the compliance-aware reward function, which balances GDPR adherence, data reduction, and performance optimization, the system now proceeds to policy deployment. This deployment must be guided by the reinforcement learning framework to dynamically adapt the AI's decision-making behavior, ensuring compliance with legal and regulatory standards while minimizing data usage. As the reward structure influences the decision-making process, the AI now optimizes its actions based on the expected value of future rewards, leading to more efficient and compliant data usage. The policy function is defined as follows:

$$\pi^*(s) = \arg \underset{a}{max}\, Q^*(s, a)$$

where $\pi^*(s)$ is the optimal policy function that dictates the best action to take in a given state s, a represents the set of available actions, $Q^*(s, a)$ is the optimal action-value function that predicts the expected cumulative reward for taking action aa in state ss, based on previous feedback and rewards. This function ensures that the AI selects actions that maximize long-term regulatory compliance and data efficiency, while simultaneously adhering to privacy-preserving practices.

This AI-driven framework ensures that AI models operate with the minimal data required for effective decision-making, eliminating unnecessary data exposure risks while maintaining model efficiency. By dynamically adjusting AI-driven data filtering and selection mechanisms, the system ensures that AI processes remain legally compliant and privacy-conscious without compromising accuracy.

By unifying Federated Learning (FL) with Sparse Data Encoding (SDE) and Reinforcement Learning for Data Efficiency (RLDE), this AI-driven model ensures that AI-powered urban governance and civil engineering applications comply with

GDPR's data minimization requirements. The system continuously refines data selection mechanisms, leveraging reinforcement learning to optimize AI efficiency while maintaining full regulatory compliance. This approach guarantees that AI-driven decision-making remains privacy-preserving, lawful, and operationally effective, ensuring that urban development and civil engineering projects align with global data protection standards.

Framework: Federated Learning for Data Minimization (FLDM)

Layered AI Architecture for GDPR-Compliant Data Minimization

Layer	Main functionality	AI module(s) and components	Operational scope (Input → Processing → Output)
Layer 1	Federated learning with sparse data encoding	Federated Learning (FL) with Sparse Data Encoding (SDE)	Local model updates, sparse feature encoding → Federated aggregation with minimal data features → Privacy-preserving model parameters with GDPR compliance
Layer 2	Reinforcement learning for data efficiency	Reinforcement Learning for Data Efficiency (RLDE)	AI-driven data processing, compliance reward signal → Federated policy optimization with data minimization goal → Efficient decision-making with minimal data
Layer 3	Compliance-aware reward structuring	Compliance-aware reward modeling engine	GDPR compliance metrics, data reduction measures, performance trade-offs → Compliance-driven reward function → Optimization feedback for data governance
Layer 4	Dynamic data minimization policy deployment	Federated Learning (FL) with Reinforcement Learning (RLDE)	Optimized policies, dynamic data selection → Privacy-preserving AI execution → Compliant and efficient decision-making in real time

Moreover, data governance frameworks should include mechanisms for individuals to access, correct, and delete their personal data. These rights are enshrined in many data privacy laws and are essential for empowering individuals to control how their data is used. AI systems must be designed with these rights in mind, ensuring that individuals can easily exercise their rights and that data governance practices are transparent and accountable.

Privacy and data governance are critical considerations in the deployment of AI systems in civil engineering, urban development, and governance. Effective data governance frameworks must balance the need for data-driven innovation with the protection of individual privacy, ensuring that data is collected, stored, and used ethically and in compliance with legal standards. By implementing robust data governance practices, including data anonymization, consent mechanisms, and compliance with privacy laws, AI systems can be deployed in a manner that respects individual privacy and promotes public trust.

Figure 5.5 highlights the key components of privacy and data governance in AI, organized into three main areas: Privacy Challenges, Data Governance Frameworks,

Fig. 5.5 Privacy and data governance in AI

and Legal Compliance. "Privacy Challenges" focuses on issues like large-scale data collection without consent and the security risks of hacking and breaches. "Data Governance Frameworks" emphasizes maintaining data quality, ensuring fairness, and protecting personal information through anonymization. "Legal Compliance" covers adherence to privacy laws such as GDPR and the upholding of individual rights, including access, correction, and deletion of personal data. Together, these elements form a structured approach to ethical data use in AI.

The ethical challenges associated with AI in civil engineering, urban development, and governance are multifaceted, requiring careful consideration of fairness, transparency, accountability, sustainability, and privacy. Addressing these challenges is critical for ensuring that AI-driven innovations benefit society while minimizing risks. Policymakers, engineers, and technologists must work together to develop ethical frameworks and governance structures that guide the responsible use of AI in these sectors.

Limitations and Governance Implications

The frameworks presented in this chapter extend ethical AI governance into the material domain of civil engineering, urban development, and infrastructure systems, where algorithmic decisions may influence investment priorities, maintenance schedules, environmental burdens, compliance obligations, and the distribution of long-term urban risk. Fairness-Enhanced AI for Urban Governance (FEA-UG), Fairness-Enhanced AI for Equitable Urban Resource Distribution (FEA-ERD), AI-Driven Governance for Urban Systems (AI-GUS), Sociotechnical Fairness-Driven AI for Equitable Urban Development (SFAI-EUD), AI Transparency and

Governance for Urban Systems (AI-TGUS and AI-TGUS II), AI Accountability and Governance Framework (AIAGF), AI Redress and Accountability Framework (AIRAF), AI Auditing and Compliance Framework (AIAF), Human-Centric Ethical Oversight Architecture (HCEOA), AI-Driven Urban Sustainability Optimization System (AID-USOS), Federated Privacy-Aware Smart Urban Framework (FPSUF), Federated Compliance-Driven AI Governance Model (FCDAI-GM), and Federated Learning for Data Minimization (FLDM) collectively address sociotechnical fairness, responsibility attribution, compliance auditing, human-centric oversight, sustainable optimization, and privacy-aware governance. Their relevance is substantial, yet their limits must remain explicit. Infrastructure decisions are not merely computational outputs; they are public commitments with spatial, financial, environmental, and intergenerational consequences. An optimization model may identify technically efficient infrastructure priorities, a compliance module may improve audit readiness, and a federated architecture may reduce privacy exposure, but none of these mechanisms can replace professional engineering judgment, statutory responsibility, public procurement scrutiny, environmental assessment, or democratic debate over urban priorities. The governance implication is that AI in civil engineering should operate as a decision-support layer within accountable institutional practice, not as an autonomous authority over the built environment. Its outputs must remain reviewable by engineers, public authorities, affected residents, infrastructure owners, and regulatory bodies, particularly where automated recommendations risk deepening spatial inequality, obscuring liability, externalizing sustainability burdens, or weakening the civic legitimacy of infrastructure governance.

5.6 Conclusion

This chapter has demonstrated that Artificial Intelligence (AI) in civil engineering, urban development, and governance must not be viewed as a neutral technological tool, but as a profound sociotechnical construct whose ethical and institutional design determines whether cities become more equitable, transparent, and sustainable or regress into algorithmically amplified inequity. Through a systematic exploration of fairness, transparency, explainability, accountability, and sociotechnical inclusivity, the chapter established that the ethical evolution of urban AI is neither optional nor external to its technical optimization; it is the very condition of its legitimacy and public acceptance.

The analytical core of the chapter revealed that fairness and equity are not philosophical addenda but formalized constraints within the mathematical architecture of AI systems. By operationalizing fairness through Reweighed Learning, Adversarial Debiasing, and Fairness-Constrained Reinforcement Learning, the chapter converted abstract ethical imperatives into auditable computational mechanisms. These formulations quantified distributive justice as a measurable optimization variable, transforming equity from a post hoc evaluative concern into an endogenous design parameter. The introduction of multi-objective optimization further established that efficiency and equity can coexist in a Pareto-balanced equilibrium, allowing urban AI systems to sustain both operational effectiveness and moral integrity.

From a governance standpoint, the chapter showed that accountability must extend beyond compliance rhetoric to institutionalized oversight mechanisms. Frameworks such as AI-Driven Governance for Urban Systems (AI-GUS) and AI Accountability and Governance Framework (AIAGF) illustrated that responsible AI requires traceable decision attribution, human-in-the-loop oversight, and dynamically adaptive regulatory compliance. These architectures transform responsibility from a vague ethical aspiration into a quantifiable governance function, capable of assigning liability, auditing performance, and ensuring the lawful traceability of every AI-mediated decision.

Equally, transparency and explainability were reconceptualized as epistemological pillars of democratic AI governance. The incorporation of Explainable Artificial Intelligence (XAI) techniques: Shapley Values, Local Interpretable Model-Agnostic Explanations (LIME), Layer-Wise Relevance Propagation (LRP), and Counterfactual Reasoning ensured that AI decisions are intelligible not only to experts but to the public that lives under their consequences. By embedding interpretability audits and confidence scoring mechanisms within AI pipelines, the chapter made clear that transparency is both a technical achievement and a civic necessity: it renders decision-making verifiable, contestable, and therefore just.

The sociotechnical analysis advanced in this chapter elevated ethics from compliance to co-design. By aligning algorithmic efficiency with participatory governance, Sociotechnical Fairness-Driven AI for Equitable Urban Development (SFAI-EUD) demonstrated how stakeholder engagement, weighted community feedback, and participatory AI governance transform ethical oversight into a distributed collective function. This ensures that communities, especially those historically marginalized, retain agency in shaping the digital architectures that define their urban futures.

The chapter's series of layered frameworks Fairness-Enhanced AI for Urban Governance (FEA-UG), Fairness-Enhanced AI for Equitable Urban Resource Distribution (FEA-ERD), AI Transparency and Governance for Urban Systems (AI-TGUS), AI Explainability for Urban Systems (AI-EXFUS), and others constitute a comprehensive meta-architecture for ethical urban AI. Each framework is modular, interoperable, and Software-as-a-Service (SaaS) deployable, bridging theoretical ethics with scalable industrial practice. Collectively, they form a unified methodological canon for developing AI systems that are interpretable, auditable, and aligned with human values.

From an epistemological standpoint, the chapter affirmed that ethical AI in civil and urban systems is not a static end-state but an evolving equilibrium—an iterative reconciliation between computational optimization and normative governance. The continuous audit loops, fairness constraints, and adaptive learning modules described herein establish a self-corrective ethical intelligence, capable of recalibrating itself as societal expectations, data environments, and legal frameworks evolve.

In synthesis, what this chapter ultimately demonstrates is that the true measure of AI's advancement in civil engineering and urban governance lies not in its computational prowess, but in its moral architecture. An AI system that is fair, transparent, explainable, and accountable is not merely technologically superior; it is civically legitimate. The chapter thus serves as both a technical manual and an ethical charter for cities transitioning into algorithmic governance: an academic blueprint for transforming AI from a tool of prediction into an instrument of justice, sustainability, and democratic renewal.

References

Barredo Arrieta, A., Díaz-Rodríguez, N., Del Ser, J., Bennetot, A., Tabik, S., Barbado, A., Garcia, S., Gil-Lopez, S., Molina, D., & Herrera, F. (2020). Explainable artificial intelligence (XAI): Concepts, taxonomies, opportunities, and challenges toward responsible AI. *Information Fusion*, 58, 82–115. https://doi.org/10.1016/j.inffus.2019.12.012

Floridi, L., Cowls, J., Beltrametti, M., Chatila, R., Dignum, V., Luetge, C., Madelin, R., Pagallo, U., Rossi, F., Schafer, B., Valcke, P., Vayena, E. (2018). AI4People—An Ethical Framework for a Good AI Society: Opportunities, Risks, Principles, and Recommendations. *Minds and Machines*, 28, 689–707. https://doi.org/10.1007/s11023-018-9482-5

Manzoor, B., Othman, I., Durdyev, S., Ismail, S., Wahab, M. H. (2021). Influence of artificial intelligence in civil engineering toward sustainable development: A systematic literature review. *Applied System Innovation*, 4, 52. https://doi.org/10.3390/asi4030052

Mökander, J. Auditing of AI: Legal, Ethical and Technical Approaches. *DISO* 2, 49 (2023). https://doi.org/10.1007/s44206-023-00074-y

Sargiotis, D. (2024a). Ethical AI in Information Technology: Navigating Bias, Privacy, Transparency, and Accountability. *Advances in Machine Learning & Artificial Intelligence*, 5(3), 01–14. https://doi.org/10.33140/AMLAI.05.03.03

Sargiotis, D. (2024b). Key Principles of Data Governance: Building a Strong Foundation. In: Data Governance. Springer, Cham. https://doi.org/10.1007/978-3-031-67268-2_3

Sargiotis, D. (2024c). Data Governance Policies and Standards: Development and Implementation. In: Data Governance. Springer, Cham. https://doi.org/10.1007/978-3-031-67268-2_7

Sargiotis, D. (2024d). Advancing Civil Engineering with AI and Machine Learning: From Structural Health to Sustainable Development. *International Journal of Architecture and Planning*, 4(2), 54–81. https://doi.org/10.51483/IJARP.4.2.2024.54-81

Sargiotis, D. (2024e). Future Trends in Data Governance: Preparing for Tomorrow. In: Data Governance. Springer, Cham. https://doi.org/10.1007/978-3-031-67268-2_12

Sargiotis, D. (2024f). Integrating Digital Transformation and AI in Civil Engineering: A Multidisciplinary Approach to Disaster Management and Sustainable Urban Development (April 10, 2024). Available at SSRN: https://ssrn.com/abstract=4789752 or https://doi.org/10.2139/ssrn.4789752

Sargiotis, D. (2024g). Data Governance Frameworks: Models and Best Practices. In: Data Governance. Springer, Cham. https://doi.org/10.1007/978-3-031-67268-2_4

Sargiotis, D. (2024h). Legal and Regulatory Considerations in Data Governance. In: Data Governance. Springer, Cham. https://doi.org/10.1007/978-3-031-67268-2_15

Sargiotis, D. (2024i). Data Stewardship and Ownership: Best Practices. In: Data Governance. Springer, Cham. https://doi.org/10.1007/978-3-031-67268-2_16

Sargiotis, D. (2025). Transforming civil engineering with AI and machine learning: Innovations, applications, and future directions. *International Journal of Research Publication and Reviews*, 6(1), 3780–3805. https://doi.org/10.55248/gengpi.6.0125.0528

Wolniak, R., & Stecuła, K. (2024). Artificial intelligence in smart cities Applications, barriers, and future directions: A review. *Smart Cities*, 7, 1346–1389. https://doi.org/10.3390/smartcities7030057

Chapter 6
Fairness-Constrained Artificial Intelligence for Inclusive Urban Planning

This chapter presents a comprehensive and technically rigorous blueprint for creating equitable, inclusive, and bias-resistant artificial intelligence systems in urban planning. It examines how artificial intelligence (AI) and machine learning (ML) pipelines inherently encode bias through data collection, feature design, labeling, and optimization, and it translates this diagnosis into mathematically formalized frameworks that operationalize fairness as a measurable optimization target. The central aim is to transform ethical intent into computational architecture demonstrating that fairness, inclusivity, and transparency can be designed, trained, and deployed as integral properties of intelligent urban systems.

The discussion begins by identifying how bias originates in both datasets and algorithmic structures, then introduces fairness-constrained reinforcement learning (Reinforcement Learning, RL) optimized via proximal policy optimization (Proximal Policy Optimization, PPO) to embed demographic equity directly into the decision-making process. These models ensure that city planning applications such as traffic control, housing distribution, and infrastructure prioritization do not merely optimize for efficiency but also for social inclusion and representational justice.

The second section addresses bias in data using a fairness-aware generative adversarial network (Generative Adversarial Network, GAN) to produce synthetic, unbiased data; a fairness-constrained distributionally robust optimization (Distributionally Robust Optimization, DRO) model to protect against worst-case demographic disparities; and a bias-corrected semi-supervised learning (Semi-Supervised Learning, SSL) framework to mitigate label-driven prejudice through adversarial debiasing. These methods jointly guarantee that AI-driven planning does not replicate structural imbalances rooted in historical inequalities.

The chapter then examines algorithmic bias itself—the inequities introduced not by data but by model design and optimization priorities. Here, fairness-constrained deep reinforcement learning (Deep Reinforcement Learning, DRL) and debiasing-adaptive fairness-constrained deep learning (Debiasing-Adaptive Fairness-Constrained Deep Learning, DAF-DL) are introduced. These methods employ

D. Sargiotis, *AI with Integrity*, https://doi.org/10.1007/978-3-032-31133-7_6

fairness-aware reward structures and adaptive reweighting mechanisms to ensure that efficiency and fairness are optimized simultaneously.

Recognizing that bias produces both technical and societal consequences, the narrative progresses toward multi-objective frameworks such as fairness-constrained multi-objective deep reinforcement learning (Multi-Objective Deep Reinforcement Learning) and trust-aware fairness-constrained deep reinforcement learning (Trust-Aware Fairness-Constrained Deep Reinforcement Learning, TA-FC-DRL). These models integrate transparency and public trust directly into the reward functions, establishing algorithmic accountability as a measurable outcome rather than an ethical abstraction.

Further sections propose fairness-constrained Bayesian data reweighting (Fairness-Constrained Bayesian Data Reweighting Model, FC-BDRM) and fairness-constrained adversarial debiasing (Fairness-Constrained Adversarial Debiasing Model, FC-ADM), both of which dynamically rebalance data representation and enforce metrics such as Demographic Parity and Equalized Odds during learning. The fairness-constrained dual-stage optimization model (Fairness-Constrained Dual-Stage Optimization Model, FC-DSOM) extends these protections through both training and deployment, ensuring ethical integrity across the model lifecycle.

The chapter proceeds to detail data diversity mechanisms through a diversity-aware generative data augmentation model (Diversity-Aware Generative Data Augmentation Model, DA-GDAM) and a fairness-aware Bayesian data reweighting model (Fairness-Aware Bayesian Data Reweighting Model, FA-BDRM), guaranteeing demographically balanced and representative training sets. Metric-driven governance is realized through a fairness-constrained multi-objective optimization model (Fairness-Constrained Multi-Objective Optimization Model, FC-MOOM), which embeds quantitative fairness measures directly into model convergence criteria.

Equitable resource allocation is modeled via fairness-constrained reinforcement learning for equitable resource allocation (Fairness-Constrained Reinforcement Learning Model for Equitable Resource Allocation, FCRL-ERA), enabling AI systems to balance operational efficiency with demographic parity. To address the challenge of interpretability, the fairness-aware explainable reinforcement learning model (Fairness-Aware Explainable Reinforcement Learning, FA-XRL) incorporates Shapley value-based transparency tools, allowing human stakeholders to audit and understand AI decisions in terms of fairness and causality.

Inclusivity is further strengthened through participatory reinforcement learning based on proximal policy optimization (Participatory Reinforcement Learning Model with Proximal Policy Optimization, PPO), where stakeholder feedback is converted into reward signals guiding the system toward community-driven fairness objectives. This participatory loop ensures that marginalized voices influence how AI models evolve and adapt within dynamic urban ecosystems.

This chapter unites ethical design, statistical equity, and computational precision into a single coherent methodology for fairness-aware city intelligence. It delivers reproducible mathematical formulations, optimization strategies, and deployment guidelines that redefine how fairness is embedded within AI-driven governance. For policymakers, engineers, and urban planners, it provides not only theoretical insight but also implementable architectures capable of ensuring that artificial intelligence becomes a genuine instrument of inclusion, transparency, and social justice in the future of global urban development.

6.1 Bias in Data and Algorithms

Bias in data and algorithms is a critical concern in artificial intelligence (AI) and machine learning (ML) applications, especially in domains such as city planning, healthcare, and public services (Sargiotis 2024a, b, c). AI models operate by processing extensive datasets to detect patterns and support decision-making. However, the reliability of their predictions depends on the quality of the data and the design of the algorithms. Unaddressed biases in data collection, feature selection, and model training can produce systemic inequalities in urban planning initiatives, reinforcing pre-existing socioeconomic disparities. This section examines the sources of bias in AI-driven models and presents a methodological framework for mitigating algorithmic bias in city planning. To support fairness-aware decision-making in urban AI applications, a Reinforcement Learning (RL)-based fairness-constrained optimization model is proposed. This model enables urban planning systems to optimize resource allocation while maintaining equitable distribution across diverse population segments. The fairness-aware reinforcement learning model is formulated as:

$$\pi^*(s) = \arg \max_{\pi} \mathbb{E}\left[\sum_{t=0}^{T} \gamma^t (R(s_t, a_t) - \lambda F(s_t, a_t))\right]$$

where $\pi^*(s)$ is the optimal policy ensuring fairness-aware decision-making, $\mathbb{E}\,[\cdot]$ denotes the expectation over possible state transitions, γ^t represents the discount factor balancing immediate versus future rewards, $R(s_t, a_t)$ is the reward function quantifying the efficiency of urban resource allocation, $F(s_t, a_t)$ is the fairness penalty function, and λ is the fairness weight regulating the trade-off between utility maximization and bias mitigation.

This reinforcement learning-based optimization ensures that AI models do not solely optimize efficiency but also incorporate fairness constraints, preventing skewed decision-making that disadvantages certain demographic groups. The fairness penalty function $F(s_t, a_t)$ is designed to dynamically adjust AI-driven urban planning decisions based on real-time assessments of demographic disparities. To ensure equitable decision-making, the fairness-aware adjustment function is formulated as:

$$F(s_t, a_t) = \sum_{g=1}^{G} \mid P_g(s_t, a_t) - P^*(s_t) \mid$$

where $F(s_t, a_t)$ measures demographic disparities in decision-making, $P_g(s_t, a_t)$ represents the proportion of resources allocated to demographic group g, and $P^*(s_t)$ denotes the ideal proportional allocation ensuring fairness. By incorporating this fairness-aware constraint, the AI system actively mitigates biased outcomes in urban development, dynamically adjusting decision-making processes to maintain demographic equity. The reinforcement learning agent continuously refines its fairness-aware strategy by evaluating past actions and optimizing urban planning decisions based on fairness-adjusted rewards.

Following the fairness penalty function, which dynamically adjusts the decision-making process based on demographic disparities, the next step is to structure the reward system such that fairness and efficiency are balanced effectively. This requires an optimized reward function that not only incentivizes the efficient allocation of resources but also accounts for the fairness adjustments made in the previous layer. The fairness-aware reward function combines the utility of resource allocation with the necessity of reducing disparities, ensuring that the AI model continually adjusts to minimize biases while maximizing urban planning effectiveness. The reward structure is defined as:

$$R(s_t, a_t) = \lambda_1 M_{\text{GDPR}}(s_t, a_t) - \lambda_2 O_{\text{reduction}}(s_t, a_t) - \lambda_3 P_{\text{performance}}(s_t, a_t)$$

where $R(s_t, a_t)$ is the reward function that balances the legal compliance with data protection (GDPR), $M_{\text{GDPR}}(s_t, a_t)$ quantifies compliance with data protection laws, $O_{\text{reduction}}(s_t, a_t)$ measures the reduction of unnecessary data points in the model, $P_{\text{performance}}(s_t, a_t)$ represents the trade-off between efficiency in decision-making and the potential loss of data accuracy, and λ_1, λ_2, λ_3, are parameters that balance the importance of these objectives. This function ensures that fairness-aware decisions are not only legally compliant but also computationally efficient, integrating fairness considerations into the AI model's decision-making process while optimizing urban governance outcomes.

To ensure optimal policy learning while adhering to fairness constraints, the policy optimization process follows the Proximal Policy Optimization (PPO) algorithm, expressed as:

$$L(\theta) = \mathbb{E}_t[min(\rho_t(\theta)A_t, \text{clip}(\rho_t(\theta), 1 - \epsilon, 1 + \epsilon)A_t)]$$

where $L(\theta)$ represents the policy optimization objective function, $\rho_t(\theta)$ is the policy ratio ensuring stable learning, A_t is the advantage function representing performance improvement, and ϵ is the clipping parameter preventing policy divergence.

This Proximal Policy Optimization (PPO) mechanism ensures that fairness-aware reinforcement learning remains stable while dynamically refining AI-driven urban planning policies. By integrating fairness constraints into the reward structure and policy optimization process, the model maintains a balance between urban efficiency and demographic equity.

The proposed AI-driven fairness-constrained reinforcement learning model enables AI-powered urban planning systems to dynamically optimize city infrastructure, public services, and resource distribution while actively mitigating biases. By incorporating fairness-aware reward adjustments, demographic-sensitive constraints, and stable reinforcement learning through PPO, this model ensures that AI-driven decision-making in urban environments remains both efficient and inclusive. The integration of fairness constraints in the optimization framework ensures that urban AI systems do not disproportionately favor specific demographic groups, fostering equitable urban development strategies.

Framework: Fairness-Constrained AI for Urban Planning (FCAI-UP)

Layered AI Architecture for Fairness-Aware Decision-Making in City Planning

Layer	Main functionality	AI module(s) and components	Operational scope (Input → Processing → Output)
Layer 1	Fairness-constrained optimization for data allocation	Reinforcement Learning (RL) with fairness constraints	Local model updates, fairness penalty adjustments → Federated aggregation with fairness penalty → Privacy-preserving, fairness-aware AI model parameters
Layer 2	Fairness penalty function for demographic equity	Demographic equity modeling engine	Demographic group resource allocation, proportional fairness → Fairness penalty adjustment function → Equity-preserving decisions
Layer 3	Fairness-aware reward structuring	Fairness-aware reward optimization module	Fairness metrics, reward adjustments, utility trade-offs → Fair compliance-driven reward function → Continuous optimization feedback
Layer 4	Policy optimization for fairness and efficiency	Proximal Policy Optimization (PPO)	Compliance-driven policies, fairness-adjusted reward signal → Optimized policy update with fairness constraints → Privacy-preserving, efficient urban execution

6.2 Sources of Bias in Data

One of the primary sources of bias in AI systems arises from the data used to train machine learning models. Since AI-driven decision-making processes rely heavily on historical datasets, they may reflect pre-existing societal biases and structural inequalities. Biases can emerge at various stages of the data pipeline, including data collection, preprocessing, and labeling, significantly affecting the fairness of AI-based urban planning systems (Mehrabi et al. 2022). For instance, datasets used for resource allocation, housing development, and transportation planning often encode historical patterns of discrimination, leading to biased AI-generated predictions that reinforce these inequalities. This perpetuation of bias necessitates the development of fairness-aware AI frameworks capable of mitigating discriminatory outcomes in urban development.

To address bias in AI-driven urban planning systems, a Fairness-Aware Generative Adversarial Network (FairGAN) model is proposed. This model leverages adversarial training to generate debiased synthetic data while preserving the essential patterns required for predictive accuracy. The FairGAN model is formulated as:

$$\underset{G}{min}\ \underset{D}{max}\, \mathbb{E}_{x\sim P_{\text{data}}}[\log D(x)] + \mathbb{E}_{z\sim P_z}[\log(1 - D(G(z)))] - \lambda \mathbb{E}_{x\sim P_{\text{data}}}[F(x)]$$

where $G(z)$ is the generator producing synthetic unbiased samples, $D(x)$ is the discriminator distinguishing between real and synthetic data, P_{data} represents the true data distribution, P_z denotes the prior distribution of latent variables, $F(x)$ is the fairness regularization term quantifying bias in the dataset, and λ is a hyperparameter controlling the trade-off between fairness and accuracy.

This adversarial training process ensures that the generator learns to produce data distributions that reduce biases while maintaining predictive utility. The fairness regularization term $F(x)$ is explicitly designed to measure demographic disparities in AI-driven decision-making and is formulated as:

$$F(x) = \sum_{g=1}^{G} | P_g(x) - P^*(x) |$$

where $F(x)$ quantifies demographic disparities in synthetic data generation, $P_g(x)$ represents the probability of distribution of group g in the dataset, and $P^*(x)$ denotes the ideal target distribution ensuring fairness.

Next, the model must now ensure that the generated synthetic data is not only fair but also robust in terms of adversarial training. In this context, adversarial training plays a pivotal role in refining the AI system by applying a fairness penalty that addresses the bias inherent in the model's decision-making process. This allows the system to mitigate biases while optimizing for fairness and utility, ensuring that the AI can learn to generate data that aligns with both fairness objectives and predictive accuracy. The bias mitigation process in adversarial training is captured by the following equation:

$$R(x) = \lambda_1 L_{\text{adversarial}}(x) - \lambda_2 F_{\text{penalty}}(x)$$

where $R(x)$ is the total reward function for the AI system's action x, λ_1 is the weight applied to the adversarial loss $L_{\text{adversarial}}(x)$, λ_2 is the weight applied to the fairness penalty $F_{\text{penalty}}(x)$, $L_{\text{adversarial}}(x)$ is the loss function quantifying the discrepancy between real and synthetic data under adversarial conditions, and $F_{\text{penalty}}(x)$

represents the fairness penalty term that adjusts the system to reduce demographic biases. This function ensures that the model is continuously refining its decision-making process, optimizing both fairness and predictive accuracy, thus contributing to the overall bias mitigation in the synthetic data generation process.

To ensure stability in adversarial training while minimizing bias in AI-driven urban planning predictions, a gradient penalty regularization term is incorporated into the model:

$$\mathbb{E}_{\hat{x}\sim P_{\hat{x}}}\left[\left(\|\nabla_{\hat{x}}D(\hat{x})\|_2 - 1\right)^2\right]$$

where $\nabla_{\hat{x}}D(\hat{x})$ represents the gradient of the discriminator with respect to its input, ensuring smooth training dynamics by penalizing abrupt changes in model learning.

This fairness-aware GAN framework enables AI-driven urban planning systems to generate synthetic datasets that mitigate bias while preserving essential features required for accurate decision-making. By integrating fairness constraints into adversarial training, the model actively debiases datasets used in city planning applications, ensuring that AI-driven resource allocation, housing development, and transportation optimization decisions remain equitable across all demographic groups.

Framework: Fairness-Aware Generative Adversarial Network (FairGAN) for Urban Planning

Layered AI Architecture for Bias Mitigation in City Planning Applications

Layer	Main Functionality	AI module(s) and components	Operational scope (Input → Processing → Output)
Layer 1	Fairness-aware generative adversarial network for bias mitigation	Fairness-Aware Generative Adversarial Network (FairGAN)	Synthetic data generation, fairness penalty adjustments → Federated aggregation with fairness-aware data → Debiased, privacy-preserving model parameters
Layer 2	Fairness regularization for synthetic data generation	Fairness regularization engine	Synthetic data generation, demographic group distribution → Fairness penalty adjustment → Equity-preserving data synthesis
Layer 3	Bias mitigation through adversarial training	Adversarial bias mitigation module	Fairness metrics, adversarial training loss, bias reduction → Fairness-driven reward function → Continuous optimization feedback for data fairness
Layer 4	Gradient penalty regularization for stable training	Gradient stability control unit	Discriminator gradients, penalty term for smoothness → Regularization to stabilize GAN training → Fairness-enhanced model convergence

Data bias frequently stems from the underrepresentation or overrepresentation of specific demographic groups within training datasets. When marginalized communities are underrepresented, AI-driven decision-making systems may struggle to generate accurate predictions for these groups, often leading to unfair or discriminatory outcomes (Pagano et al. 2023). Conversely, when certain data points are overrepresented, the AI system may disproportionately learn from those samples, resulting in skewed predictions that overly reflect the characteristics of dominant groups. This issue is particularly critical in city planning, where AI systems influence policy decisions affecting diverse urban populations. Ensuring representational fairness in data is therefore essential for developing AI models that distribute urban resources equitably and support planning decisions without reinforcing systemic inequalities.

To address these challenges, a Fairness-Constrained Distributionally Robust Optimization (DRO) model is proposed. This model mitigates data bias by optimizing AI predictions under worst-case demographic disparities, reducing the risk that underrepresented populations are disproportionately disadvantaged. The fairness-constrained DRO model is formulated as:

$$\underset{\theta}{min} \underset{Q \in \mathcal{U}(P)}{sup} \mathbb{E}_{(x,y)\sim Q}[L(f_\theta(x), y)] + \lambda F(Q)$$

where θ represents the model parameters, Q is the worst-case probability distribution of data within the uncertainty set $\mathcal{U}(P)$, $L(f_\theta(x), y)$ is the loss function capturing prediction errors, $F(Q)$ is the fairness penalty function penalizing demographic imbalances, and λ is a regularization term controlling the trade-off between fairness and predictive accuracy. This model ensures that AI-driven urban planning decisions are robust to worst-case demographic disparities, preventing biased predictions that disproportionately impact underrepresented communities. The fairness penalty function $F(Q)$ explicitly measures demographic representation disparities in AI-driven predictions and is defined as:

$$F(Q) = \sum_{g=1}^{G} | Q_g - P^*(g) |$$

where $F(Q)$ quantifies demographic discrepancies, Q_g represents the learned probability distribution of group gg in the AI model, and $P^*(g)$ is the target distribution ensuring fairness across all groups.

To ensure stability in learning while enforcing fairness constraints, the optimization process incorporates a gradient-based fairness adjustment mechanism, formulated as:

$$\theta^{(t+1)} = \theta^{(t)} - \eta(\nabla_\theta L(f_\theta(x), y) + \lambda \nabla_\theta F(Q))$$

where $\theta^{(t+1)}$ represents the updated model parameters, η is the learning rate, $\nabla_\theta L(f_\theta(x), y)$ is the gradient of the loss function, and $\nabla_\theta F(Q)$ is the gradient of the fairness penalty function ensuring demographic equity.

This fairness-constrained DRO model ensures that AI-driven city planning systems dynamically adjust predictions based on demographic fairness constraints. By optimizing under worst-case data bias scenarios, this model guarantees that underrepresented populations receive equitable consideration in urban development decisions. The integration of fairness-aware loss functions, demographic discrepancy penalties, and gradient-based optimization mechanisms enables AI-driven city planning to mitigate representation disparities while maintaining predictive accuracy.

Framework: Fairness-Constrained Distributionally Robust Optimization for AI in Urban Planning

Layered AI Architecture for Mitigating Data Bias in City Planning

Layer	Main functionality	AI module(s) and components	Operational scope (Input → Processing → Output)
Layer 1	Fairness-constrained optimization for data representation	Fairness-Constrained Distributionally Robust Optimization (DRO)	Worst-case probability distribution, fairness penalty adjustments → Federated aggregation with fairness constraint → Robust and fair model parameters
Layer 2	Fairness regularization for data disparities	Data fairness regularization engine	Demographic data distribution, fairness penalty → Fairness penalty adjustment → Equity-preserving decision-making
Layer 3	Bias mitigation in learning via gradient-based adjustment	Gradient-based fairness adjustment module	Gradient of loss, fairness penalty gradients, optimization stability → Fairness-enhanced model convergence → Bias mitigation through stable learning
Layer 4	Demographic discrepancy penalty for fair model training	Demographic fairness evaluation unit	Demographic discrepancies, group-wise fairness measures → Penalty function for group fairness → Equity-driven decision feedback

Labeling bias is a fundamental challenge in AI-driven decision-making, occurring when human annotators introduce subjective prejudices into data labels, which are subsequently learned and reinforced by the AI model. These biases can manifest in various forms, such as skewed categorizations of neighborhoods or the unequal prioritization of infrastructure projects based on historically biased data inputs (Mehrabi et al. 2022). In city planning, labeling bias can significantly distort AI-driven policy recommendations, leading to inequitable resource distribution and reinforcing existing socioeconomic disparities. As AI systems increasingly influence urban development strategies, it is imperative to design frameworks that systematically mitigate bias during the data labeling process.

To address this challenge, a Bias-Corrected Semi-Supervised Learning (SSL) model is proposed. This model uses an adversarial debiasing mechanism to correct label bias while supporting robust learning from both labeled and unlabeled data. The bias-corrected SSL model is formulated as:

$$\min_{\theta} \mathbb{E}_{(x,y)\sim P_{\text{labeled}}}[L(f_\theta(x), y)] + \alpha \mathbb{E}_{x\sim P_{\text{unlabeled}}}\left[\max_{d} L_{\text{adv}}(d, f_\theta(x))\right]$$

where θ represents the model parameters, P_{labeled} and $P_{\text{unlabeled}}$ are the distributions of labeled and unlabeled data, respectively, $L(f_\theta(x), y)$ is the supervised learning loss function, $L_{\text{adv}}(d, f_\theta(x))$ is the adversarial debiasing loss function, d is the adversarial discriminator that detects labeling inconsistencies, and αα is a regularization parameter controlling the influence of adversarial debiasing.

This model ensures that AI-driven urban planning decisions are corrected for systematic labeling biases, preventing distorted predictions that could disadvantage specific demographic groups. The adversarial debiasing loss function $L_{\text{adv}}(d, f_\theta(x))$ is formulated as:

$$L_{\text{adv}}(d, f_\theta(x)) = -\sum_{c=1}^{C} P_c(x) \log d_c(f_\theta(x))$$

where $L_{\text{adv}}(d, f_\theta(x))$ measures the discrepancy between predicted labels and the debiased label distribution, $P_c(x)$ represents the true probability of class c, and $d_c(f_\theta(x))$ denotes the discriminator's confidence in classifying data into category c.

The next step is to optimize the model with respect to adversarial loss. This phase requires refining the AI's decision-making process by calculating the adversarial loss function, which penalizes discrepancies between predicted labels and debiased labels. By incorporating this loss function, the model not only learns from labeled data but also continuously improves fairness in the predictions made, addressing any residual biases. The adversarial loss function is formulated as follows:

$$R(x) = \lambda_1 \mathcal{L}_{\text{adversarial}}(x) - \lambda_2 F_{\text{penalty}}(x)$$

where $R(x)$ is the reward function for mitigating biases in the model, $\mathcal{L}_{\text{adversarial}}(x)$ is the adversarial loss function that quantifies the discrepancy between predicted and debiased labels, $F_{\text{penalty}}(x)$ represents the fairness penalty term that adjusts the model for demographic imbalances, and λ_1, λ_2 are the weights that balance adversarial training and fairness penalties. This function ensures that the AI model prioritizes fairness in predictions while optimizing performance, leading to more equitable AI-driven urban planning decisions.

To refine model performance while enforcing fairness constraints, the optimization process incorporates a fairness-constrained gradient adjustment mechanism, formulated as:

$$\theta^{(t+1)} = \theta^{(t)} - \eta(\nabla_\theta L(f_\theta(x), y) - \beta \nabla_\theta L_{\text{adv}}(d, f_\theta(x)))$$

where $\theta^{(t+1)}$ represents the updated model parameters, η is the learning rate, $\nabla_\theta L(f_\theta(x), y)$ is the gradient of the supervised learning loss function, and β is the adversarial fairness weight balancing debiasing corrections.

This bias-corrected semi-supervised learning model ensures that AI-driven urban planning systems dynamically adjust their predictions based on debiased label distributions. By incorporating adversarial debiasing mechanisms and fairness-aware optimization strategies, the model mitigates the harmful effects of human annotation bias, ensuring that AI-driven decision-making in city planning remains equitable, unbiased, and socially responsible.

Framework: Bias-Corrected Semi-Supervised Learning for Urban Planning (BC-SSL-UP)

Layered AI Architecture for Mitigating Labeling Bias in City Planning

Layer	Main functionality	AI module(s) and components	Operational scope (Input → Processing → Output)
Layer 1	Bias-corrected semi-supervised learning for label bias mitigation	Bias-Corrected Semi-Supervised Learning (SSL)	Labeled and unlabeled data, adversarial debiasing adjustments → Federated aggregation with label correction → Debiased, privacy-preserving model parameters
Layer 2	Adversarial debiasing for fair label distribution	Adversarial label correction module	Labeled data, adversarial loss function → Fairness-aware label generation → Bias-corrected predictions
Layer 3	Bias mitigation via adversarial loss function	Fairness-aware loss optimization engine	Predicted labels, true label distribution → Adversarial loss calculation → Fairness-driven model optimization
Layer 4	Fairness-constrained gradient adjustment	Gradient fairness adjustment unit	Supervised learning gradients, adversarial fairness adjustments → Fair model parameter updates → Stable convergence with fairness constraints

6.3 Algorithmic Bias

While data bias significantly contributes to biased AI outcomes, algorithmic bias, arising from the structural design of AI models, can independently introduce unfairness. Algorithmic bias occurs when the mathematical formulations and optimization techniques used in AI decision-making systematically favor certain outcomes over others, often disadvantaging marginalized groups. One of the primary causes of algorithmic bias is reliance on optimization functions that prioritize accuracy or efficiency while neglecting fairness considerations (Baum et al. 2022; Sargiotis

2024a, 2025a). For instance, an AI-driven urban mobility system may optimize traffic flow by prioritizing routes that benefit wealthier districts, inadvertently reducing service availability in underserved neighborhoods. This occurs because the optimization process is structured to maximize conventional efficiency metrics, which may themselves reflect historically inequitable infrastructure investments. To support equitable AI-driven decision-making, a fairness-aware optimization framework is essential.

To address this issue, a Fairness-Constrained Deep Reinforcement Learning (DRL) model is proposed. This model adjusts optimization objectives to balance efficiency and fairness in AI-driven urban planning. The fairness-constrained DRL model is formulated as:

$$\pi^*(s) = \arg\max_{\pi} \mathbb{E}\left[\sum_{t=0}^{T} \gamma^t (R(s_t, a_t) - \lambda F(s_t, a_t))\right]$$

where $\pi^*(s)$ represents the optimal policy incorporating fairness constraints, $\mathbb{E}\,[\cdot]$ denotes the expectation over possible state transitions, γ^t is the discount factor controlling the balance between short-term and long-term rewards, $R(s_t, a_t)$ is the reward function quantifying urban planning efficiency, $F(s_t, a_t)$ is the fairness penalty function ensuring equitable decision-making, and λ is a fairness weight regulating the trade-off between efficiency and fairness.

This reinforcement learning model ensures that AI-driven urban optimization accounts for fairness by penalizing decisions that disproportionately disadvantage certain groups. The fairness penalty function $F(s_t, a_t)$ is explicitly designed to quantify algorithmic bias and is formulated as:

$$F(s_t, a_t) = \sum_{g=1}^{G} \mid P_g(s_t, a_t) - P^*(s_t) \mid$$

where $F(s_t, a_t)$ measures demographic disparities in AI-driven decision-making, $P_g(s_t, a_t)$ represents the probability of resource allocation to demographic group g, and $P^*(s_t)$ denotes the ideal equitable distribution.

The next step is to ensure that these fairness penalties are properly reflected in the reward structure for policy learning. The fairness-aware reward function directly integrates the fairness metric into the optimization process, balancing the trade-off between operational efficiency, fairness considerations, and compliance with data protection regulations. By adjusting the reward function, this layer ensures that the system makes decisions that are not only efficient but also fair, promoting equitable resource allocation in urban planning. The fairness-aware reward function is defined as follows:

$$R(s_t, a_t) = \lambda_1 M_{\mathrm{GDPR}}(s_t, a_t) - \lambda_2 O_{\mathrm{reduction}}(s_t, a_t) - \lambda_3 P_{\mathrm{performance}}(s_t, a_t)$$

where $R(s_t, a_t)$ is the fairness-aware reward function, $M_{\text{GDPR}}(s_t, a_t)$ quantifies compliance with data minimization principles (such as GDPR), $O_{\text{reduction}}(s_t, a_t)$ measures the reduction of unnecessary data or bias in decision-making, $P_{\text{performance}}(s_t, a_t)$ represents the trade-off between achieving optimal efficiency and maintaining fairness in the model's output, and λ_1, λ_2, λ_3, are regularization parameters that balance the three components of the reward function. This function ensures that the model optimizes urban planning decisions while incorporating fairness considerations and compliance with privacy laws.

To ensure that the AI system dynamically optimizes urban policies while adhering to fairness constraints, the policy learning process follows the Proximal Policy Optimization (PPO) algorithm, formulated as:

$$L(\theta) = \mathbb{E}_t[min(\rho_t(\theta)A_t, \text{clip}(\rho_t(\theta), 1 - \epsilon, 1 + \epsilon)A_t)]$$

where $L(\theta)$ represents the policy optimization objective function, $\rho_t(\theta)$ is the policy ratio ensuring stable learning updates, A_t is the advantage function representing performance improvement, and ϵ is the clipping parameter preventing excessive policy divergence.

This fairness-constrained DRL model ensures that AI-driven urban planning systems optimize decision-making in a manner that balances efficiency with demographic equity. By integrating fairness-aware reward adjustments, demographic-sensitive constraints, and stable reinforcement learning via PPO, the model prevents AI-driven systems from perpetuating biased urban development practices. The framework ensures that optimization processes in AI decision-making do not reinforce existing socioeconomic disparities but rather contribute to the equitable distribution of urban resources.

Framework: Fairness-Constrained Deep Reinforcement Learning for Urban Planning (FCDRL-UP)

Layered AI Architecture for Fairness-Aware Decision-Making in City Planning

Layer	Main functionality	AI module(s) and components	Operational scope (Input → Processing → Output)
Layer 1	Fairness-constrained deep reinforcement learning for urban optimization	Fairness-Constrained Deep Reinforcement Learning (DRL)	State transitions, fairness penalty adjustments → Federated aggregation with fairness-penalized rewards → Optimized urban planning parameters with fairness constraints
Layer 2	Fairness penalty for algorithmic bias in decision-making	Algorithmic fairness penalty engine	Demographic disparities, fairness penalty function → Fairness penalty adjustment → Bias-corrected AI-driven decision outcomes
Layer 3	Fairness-aware reward structuring for policy learning	Fairness-driven reward optimization module	Reward function, fairness metrics → Fair compliance-driven reward function → Continuous optimization feedback for fairness preservation

(continued)

Layer	Main functionality	AI module(s) and components	Operational scope (Input → Processing → Output)
Layer 4	Stable Policy Learning via Proximal Policy Optimization (PPO)	Proximal Policy Optimization (PPO)	Fairness-aware reward signal, policy updates → Optimized policy with fairness constraints → Bias-mitigated AI execution for urban planning

AI algorithms can amplify biases embedded in training data, leading to disproportionate outcomes that reinforce existing social inequalities. When datasets underrepresent certain demographic groups, the AI system may learn to prioritize the patterns that dominate the data, disregarding the needs and characteristics of underrepresented communities (Mehrabi et al. 2022). This phenomenon is particularly problematic in city planning, where biased predictions can result in unequal access to public services and may systematically exclude specific populations from infrastructure investments and resource allocations. To counteract this effect, it is crucial to develop models that actively correct dataset imbalances while preserving predictive accuracy.

To address this challenge, a Debiasing-Adaptive Fairness-Constrained Deep Learning (DAF-DL) model is proposed. This model leverages adversarial fairness constraints and adaptive reweighting mechanisms to mitigate the amplification of bias in AI-generated predictions. The DAF-DL model is formulated as:

$$\min_{\theta} \mathbb{E}_{(x,y)\sim P_{\text{data}}}[L(f_\theta(x), y)] + \lambda \sum_{g=1}^{G} \mid P_g - P^* \mid$$

where θ represents the model parameters, P_{data} is the distribution of training data, $L(f_\theta(x), y)$ is the predictive loss function, P_g is the learned probability distribution for demographic group g, P^* is the target distribution ensuring demographic fairness, and λ is the fairness regularization parameter balancing predictive accuracy and bias correction.

This fairness-aware optimization ensures that AI-driven decision-making dynamically adjusts predictions to compensate for underrepresented groups. The fairness adjustment function corrects dataset imbalances by reweighting training samples, ensuring that AI-generated decisions equitably consider all demographic groups. The reweighting function is defined as:

$$w(x) = \frac{P^*(x)}{P_{\text{data}}(x)}$$

where $w(x)$ is the adaptive weight assigned to sample x, $P^*(x)$ is the ideal fair probability distribution, and $P_{\text{data}}(x)$ represents the original biased distribution.

The next phase focuses on actively mitigating bias through adversarial training. This process integrates an adversarial loss function that ensures fairness-aware model optimization by penalizing discrepancies between predicted labels and debiased labels. The adversarial loss function works by comparing the model's outputs with the target fair distribution and adjusting accordingly to reduce any bias that may have been carried over from the training data. This mechanism fosters a stable learning environment where fairness constraints are embedded directly into the model's learning dynamics. The adversarial loss function is mathematically defined as:

$$\mathcal{L}_{\text{adv}}(d, f_\theta(x)) = -\sum_{c=1}^{C} P_c(x) \log d_c(f_\theta(x))$$

where $\mathcal{L}_{\text{adv}}(d, f_\theta(x))$ is the adversarial loss function that quantifies the difference between the predicted label and the fairness-adjusted label distribution, $P_c(x)$ represents the true probability of class c in the dataset, and $d_c(f_\theta(x))$ denotes the discriminator's confidence in classifying data into category c. This function aims to reduce the discrepancy between the real and adjusted labels, ensuring that the model does not perpetuate biased learning patterns. Through this mechanism, the model dynamically adjusts to promote fairness in its learning process while maintaining predictive accuracy.

To further ensure stability in model training while enforcing fairness constraints, the learning process incorporates an adversarial debiasing loss function, formulated as:

$$L_{\text{adv}}(\theta) = -\sum_{c=1}^{C} P_c(x) \log D(f_\theta(x))$$

where $L_{\text{adv}}(\theta)$ measures the discrepancy between predicted labels and the fairness-adjusted distribution, $P_c(x)$ represents the true probability of class c, and $D(f_\theta(x)$ denotes the adversarial discriminator correcting biased learning patterns.

This debiasing-adaptive fairness constrained deep learning model ensures that AI-driven city planning systems mitigate data-driven bias amplification, dynamically adjusting predictions to prevent systematic exclusion of marginalized communities. By integrating adversarial debiasing mechanisms, adaptive reweighting strategies, and fairness-aware optimization functions, the model ensures equitable resource allocation, infrastructure development, and public service distribution across all demographic groups.

Framework: Debiasing-Adaptive Fairness-Constrained Deep Learning for Urban Planning (DAF-DL-UP)

Layered AI Architecture for Mitigating Bias in AI-Driven Decision-Making

Layer	Main functionality	AI module(s) and components	Operational scope (Input → Processing → Output)
Layer 1	Fairness-constrained deep learning for bias mitigation	Debiasing-Adaptive Fairness-Constrained Deep Learning (DAF-DL)	Data distribution, fairness penalty adjustments → Federated aggregation with fairness-penalized loss → Bias-corrected, privacy-preserving model parameters

(continued)

Layer	Main functionality	AI module(s) and components	Operational scope (Input → Processing → Output)
Layer 2	Fairness adjustment for dataset imbalance	Adaptive Reweighting Engine	Training sample weights, fairness penalty → Adaptive reweighting → Equity-preserving sample selection
Layer 3	Bias mitigation via adversarial debiasing	Adversarial Bias Correction Module	Adversarial loss function, fairness metrics → Bias correction via adversarial training → Stable model learning with fairness constraints
Layer 4	Adversarial debiasing loss function	Fairness-Adjusted Loss Computation Unit	Predicted labels, fairness-adjusted labels → Discrepancy loss calculation → Bias mitigation through adversarial updates

6.4 Consequences of Bias in AI Systems

The consequences of bias in AI systems extend far beyond technical inaccuracies, shaping societal structures and influencing urban development policies in ways that can entrench existing inequalities. When AI-driven decision-making systems are biased, they can lead to unequal access to fundamental services such as public transportation, housing, healthcare, and education, disproportionately affecting marginalized communities (Pagano et al. 2023). These biases create reinforcing cycles of disparity, where underserved populations receive fewer resources, further exacerbating systemic inequities. Given the increasing role of AI in city planning, it is crucial to develop fairness-aware AI frameworks that actively mitigate the social, economic, and ethical repercussions of biased decision-making.

To address these challenges, a Fairness-Constrained Multi-Objective Deep Reinforcement Learning (MO-DRL) model is proposed. This model is designed to enable city planning systems to optimize multiple objectives, including efficiency, accessibility, and fairness, without disproportionately favoring specific demographic groups. The fairness-constrained MO-DRL model is formulated as:

$$\pi^*(s) = \arg \underset{\pi}{max} \mathbb{E}\left[\sum_{t=0}^{T} \gamma^t (R_{\text{eff}}(s_t, a_t) + \alpha R_{\text{fair}}(s_t, a_t))\right]$$

where $\pi^*(s)$ represents the optimal policy balancing efficiency and fairness, $\mathbb{E}[\cdot]$ denotes the expectation over possible state transitions, γ^t is the discount factor balancing immediate and future rewards, $R_{\text{eff}}(s_t, a_t)$ quantifies the efficiency of urban resource allocation, $R_{\text{fair}}(s_t, a_t)$ represents the fairness reward function ensuring equitable service distribution, and α is a tunable parameter controlling the weight assigned to fairness considerations.

This multi-objective reinforcement learning approach ensures that urban AI systems do not optimize solely for efficiency but instead integrate fairness constraints to prevent disparities in service accessibility. The fairness-aware reward function $R_{\text{fair}}(s_t, a_t)$ is formulated as:

$$R_{\text{fair}}(s_t, a_t) = -\sum_{g=1}^{G} | P_g(s_t, a_t) - P^*(s_t) |$$

where $R_{\text{fair}}(s_t, a_t)$ measures demographic disparities in AI-driven city planning, $P_g(s_t, a_t)$ represents the probability of allocating services to demographic group g, and $P^*(s_t)$ denotes the ideal equitable distribution.

To ensure stable policy learning and avoid excessive policy divergence, the Proximal Policy Optimization (PPO) algorithm is used to refine the fairness-aware optimization strategy:

$$L(\theta) = \mathbb{E}_t[min(\rho_t(\theta)A_t, \text{clip}(\rho_t(\theta), 1 - \epsilon, 1 + \epsilon)A_t)]$$

where $L(\theta)$ is the policy optimization function ensuring stable learning, $\rho_t(\theta)$ represents the policy ratio controlling updates, A_t is the advantage function measuring the effectiveness of fairness-aware actions, and ϵ is the clipping parameter preventing excessive policy fluctuations.

This fairness-constrained multi-objective deep reinforcement learning model ensures that AI-driven urban planning decisions simultaneously optimize efficiency and fairness, preventing biased service allocations that could exacerbate existing inequalities. By incorporating fairness constraints directly into the reward structure and policy optimization process, the model guarantees that AI-powered decision-making fosters socially responsible, equitable, and inclusive city planning strategies.

Framework: Fairness-Constrained Multi-Objective Deep Reinforcement Learning for Urban Planning (MO-DRL-UP)

Layered AI Architecture for Fairness-Aware Decision-Making in City Planning

Layer	Main functionality	AI module(s) and components	Operational scope (Input → Processing → Output)
Layer 1	Fairness-constrained multi-objective deep reinforcement learning for urban planning	Fairness-Constrained Multi-Objective Deep Reinforcement Learning (MO-DRL)	Urban planning objectives, efficiency and fairness reward functions → Federated aggregation with fairness-penalized rewards → Optimized urban planning parameters considering fairness constraints
Layer 2	Fairness-aware reward function for urban optimization	Fairness reward computation engine	Demographic fairness, resource allocation → Fairness reward calculation → Equity-preserving urban decision-making

(continued)

Layer	Main functionality	AI module(s) and components	Operational scope (Input → Processing → Output)
Layer 3	Bias mitigation via fairness-constrained policy learning	Proximal Policy Optimization (PPO)	Fairness-aware reward signal, policy updates → Optimized policy with fairness constraints → Bias-mitigated AI execution for urban planning

For example, if an AI system used in city planning is trained on biased datasets that reflect historical patterns of segregation and inequality, it may continue to prioritize wealthier neighborhoods for infrastructure investment while neglecting lower-income areas. This reinforces socioeconomic disparities and undermines efforts to create more inclusive and equitable urban environments (Zowghi and da Rimini 2023). Beyond the direct consequences of biased resource allocation, persistent AI-mediated inequalities can erode public trust, as affected communities may become skeptical of the fairness and transparency of algorithmic decision-making processes. Ensuring algorithmic accountability and designing AI models that correct for historical bias are therefore critical steps toward trustworthy and socially responsible AI applications in urban planning.

To address these issues, a Trust-Aware Fairness-Constrained Deep Reinforcement Learning (TA-FC-DRL) model is proposed. This model integrates fairness constraints with trust-enhancing mechanisms to improve transparency and accountability in AI-driven urban planning. The TA-FC-DRL model is formulated as:

$$\pi^*(s) = \arg\max_{\pi} \mathbb{E}\left[\sum_{t=0}^{T} \gamma^t (R_{\text{dev}}(s_t, a_t) + \alpha R_{\text{fair}}(s_t, a_t) + \beta R_{\text{trust}}(s_t, a_t))\right]$$

where $\pi^*(s)$ represents the optimal policy ensuring fairness and trust-aware decision-making, $\mathbb{E}[\cdot]$ denotes the expectation over possible state transitions, γ^t is the discount factor balancing immediate and future rewards, $R_{\text{dev}}(s_t, a_t)$ is the reward function measuring the development impact of urban planning decisions, $R_{\text{fair}}(s_t, a_t)$ represents the fairness reward function ensuring equitable infrastructure investments, $R_{\text{trust}}(s_t, a_t)$ is the trust-enhancing reward function promoting transparent AI decision-making, and α, β are tunable parameters controlling the balance between fairness, development, and trust considerations.

This multi-objective reinforcement learning approach ensures that AI-driven urban planning does not solely prioritize efficiency but instead integrates fairness constraints to prevent unequal resource distribution. The trust-enhancing reward function $R_{\text{trust}}(s_t, a_t)$ is designed to improve public confidence in AI-driven decisions and is formulated as:

$$R_{\text{trust}}(s_t, a_t) = -\sum_{g=1}^{G} \mid P_g(s_t, a_t) - P^*(s_t) \mid -\delta \sum_{t=0}^{T} \log T(s_t, a_t)$$

where $R_{\text{trust}}(s_t, a_t)$ measures the public trust impact of AI-driven urban planning decisions, $P_g(s_t, a_t)$ represents the probability of allocating services to demographic group g, $P^*(s_t)$ denotes the ideal equitable distribution, $T(s_t, a_t)$ represents the transparency function quantifying the explainability of AI-driven decisions, and δ is a weighting parameter controlling the importance of trust-building measures.

To ensure stable policy learning and prevent excessive policy divergence, the Proximal Policy Optimization (PPO) algorithm is used to refine the trust-aware fairness-constrained reinforcement learning strategy:

$$L(\theta) = \mathbb{E}_t[min(\rho_t(\theta)A_t, \text{clip}(\rho_t(\theta), 1 - \epsilon, 1 + \epsilon)A_t)]$$

where $L(\theta)$ is the policy optimization function ensuring stable AI model learning, $\rho_t(\theta)$ represents the policy ratio controlling updates, A_t is the advantage function measuring the effectiveness of fairness-aware and trust-aware policy adjustments, and ϵ is the clipping parameter preventing excessive model fluctuations.

This trust-aware fairness-constrained deep reinforcement learning model ensures that AI-driven urban planning dynamically integrates fairness, transparency, and public trust-building measures into decision-making processes. By incorporating fairness constraints, transparency incentives, and stable learning mechanisms, the model prevents AI-driven urban policies from reinforcing historical biases, ensuring inclusive, equitable, and socially responsible city planning strategies.

Framework: Trust-Aware Fairness-Constrained Deep Reinforcement Learning for Urban Planning (TA-FC-DRL-UP)

Layered AI Architecture for Fairness and Trust-Aware Decision-Making in City Planning

Layer	Main functionality	AI module(s) and components	Operational scope (Input → Processing → Output)
Layer 1	Fairness-constrained deep reinforcement learning for trust-aware urban optimization	Trust-Aware Fairness-Constrained Deep Reinforcement Learning (TA-FC-DRL)	Urban planning objectives, efficiency, fairness, and trust reward functions → Federated aggregation with fairness-penalized and trust-enhanced rewards → Optimized urban planning parameters with fairness and trust constraints
Layer 2	Fairness-aware reward function for trust optimization	Trust-Enhanced Fairness Reward Engine	Demographic fairness, resource allocation, transparency → Fairness and trust reward calculation → Equity-preserving and transparent urban decision-making

(continued)

Layer	Main functionality	AI module(s) and components	Operational scope (Input → Processing → Output)
Layer 3	Bias mitigation via trust and fairness-constrained policy learning	Proximal Policy Optimization (PPO)	Fairness-aware and trust-enhanced reward signal, policy updates → Optimized policy with fairness and trust constraints → Bias-mitigated, transparent AI execution for urban planning

6.5 Mitigating Bias in Data and Algorithms

Addressing bias in AI systems requires a comprehensive approach that integrates technical advancement with ethical oversight, ensuring that AI-driven decision-making remains fair, transparent, and inclusive. A central strategy for mitigating data bias is to ensure that training datasets are representative of the population the AI system is intended to serve. This entails collecting diverse data that incorporates inputs from marginalized groups while maintaining balanced representation across geographic, socioeconomic, and demographic factors (Pagano et al. 2023; Sargiotis 2024a). Failure to address bias in dataset composition can result in AI-driven predictions that reinforce systemic inequalities, undermining efforts to create equitable AI applications in city planning. To counteract this, an AI framework must dynamically adjust for dataset imbalances while optimizing fairness in decision-making.

To achieve this, a Fairness-Constrained Bayesian Data Reweighting Model (FC-BDRM) is proposed. This model adjusts the influence of biased data points by reweighting training samples through Bayesian fairness constraints, supporting demographically balanced and socially responsible AI-driven decisions. The FC-BDRM model is formulated as:

$$\min_{\theta} \mathbb{E}_{(x,y)\sim P_{\text{data}}}[w(x)L(f_{\theta}(x),y)] + \lambda \sum_{g=1}^{G} | P_g - P^* |$$

where θ represents the model parameters, P_{data} is the probability distribution of training data, $L(f_\theta(x), y)$ is the predictive loss function, $w(x)$ is the Bayesian reweighting function correcting dataset imbalances, P_g represents the learned probability distribution for demographic group g, P^* denotes the target fair distribution, and λ is the fairness regularization parameter balancing accuracy and fairness.

This model ensures that underrepresented data points receive higher importance in the training process, preventing AI systems from overfitting dominant demographic groups. The Bayesian reweighting function dynamically adjusts dataset contributions and is defined as:

$$w(x) = \frac{P^*(x)}{P_{\text{data}}(x) + \epsilon}$$

where $w(x)$ is the adaptive weight assigned to training sample x, $P^*(x)$ is the ideal target probability distribution ensuring demographic fairness, $P_{\text{data}}(x)$ is the original biased probability distribution, and ϵ is a small regularization term preventing numerical instability.

To ensure stable optimization while enforcing fairness constraints, the learning process integrates a Bayesian Posterior fairness adjustment mechanism, formulated as:

$$\theta^{(t+1)} = \theta^{(t)} - \eta(\nabla_\theta L(f_\theta(x), y) - \beta \nabla_\theta F(Q))$$

where $\theta^{(t+1)}$ represents the updated model parameters, η is the learning rate, $\nabla_\theta L(f_\theta(x), y)$ is the gradient of the supervised loss function, and β is the fairness adjustment weight ensuring demographic equity in AI-driven predictions.

This fairness-constrained Bayesian data reweighting model ensures that AI-driven city planning decisions remain inclusive by dynamically adjusting dataset imbalances without sacrificing predictive accuracy. By incorporating Bayesian fairness constraints, adaptive reweighting strategies, and fairness-aware learning mechanisms, the model guarantees that AI-powered decision-making is socially responsible, inclusive, and ethically robust, fostering equitable urban development for all communities.

Framework: Fairness-Constrained Bayesian Data Reweighting Model for Urban Planning (FC-BDRM)

Layered AI Architecture for Fairness Aware Data Reweighting in City Planning

Layer	Main functionality	AI module(s) and components	Operational scope (Input → Processing → Output)
Layer 1	Fairness-constrained Bayesian data reweighting for bias mitigation	Fairness-Constrained Bayesian Data Reweighting Model (FC-BDRM)	Training data, fairness penalty adjustments, Bayesian reweighting → Federated aggregation with fairness-penalized loss → Optimized, fairness-aware model parameters considering dataset imbalances
Layer 2	Fairness adjustment for dataset Imbalances	Adaptive data reweighting engine	Demographic data distribution, reweighting function → Adaptive sample reweighting → Equity-preserving and representative data selection
Layer 3	Bias mitigation via Bayesian posterior fairness adjustment	Bayesian fairness adjustment module	Supervised learning gradients, fairness adjustments → Fair model parameter updates → Bias-mitigated model learning with fairness constraints

Beyond ensuring data diversity, the implementation of fairness metrics is essential for evaluating the equity of AI-driven decisions. Fairness-aware evaluation frameworks, such as Equalized Odds and Demographic Parity, provide quantitative measures of algorithmic bias, allowing AI practitioners to assess disparities in predictive performance across different demographic groups (Mehrabi et al. 2022). These metrics help identify systematic inequalities, ensuring that AI-driven urban planning models do not disproportionately advantage or disadvantage certain populations. By incorporating fairness constraints into AI learning processes, it becomes possible to correct biases dynamically in decision-making, fostering transparent, ethical, and socially responsible AI applications.

To achieve this, a Fairness-Constrained Adversarial Debiasing Model (FC-ADM) is proposed. This model uses adversarial learning mechanisms to optimize fairness metrics while preserving predictive accuracy. The FC-ADM model is formulated as:

$$\underset{\theta}{min}\ \underset{\phi}{max}\, \mathbb{E}_{(x,y)\sim P_{\text{data}}}\left[L(f_\theta(x), y)\right] - \lambda \mathbb{E}_{(x,s)\sim P_{\text{data}}}\left[L_{\text{adv}}\left(d_\phi(f_\theta(x)), s\right)\right]$$

where θ represents the model parameters of the predictor, ϕ represents the parameters of the adversarial fairness discriminator, P_{data} is the training data distribution, $L(f_\theta(x), y)$ is the standard prediction loss function, $L_{\text{adv}}(d_\phi(f_\theta(x)), s)$ is the adversarial fairness loss function, $d_\phi(f_\theta(x))$ is the adversarial fairness classifier predicting demographic group s, and λ is the fairness regularization parameter balancing predictive accuracy and fairness.

This adversarial fairness optimization framework ensures that the AI model learns demographically unbiased representations by penalizing disparities in predictive performance across different demographic groups. The fairness-aware adversarial loss function is formulated as:

$$L_{\text{adv}}\left(d_\phi(f_\theta(x)), s\right) = -\sum_{c=1}^{C} P_c(x) \log d_{\phi,c}(f_\theta(x))$$

where $L_{\text{adv}}(d_\phi(f_\theta(x)), s)$ quantifies demographic prediction discrepancies, $P_c(x)$ represents the true probability distribution for class c, and $d_{\phi,\ c}(f_\theta(x))$ is the adversarial fairness classifier output for class c.

To ensure fair model updates, the optimization process integrates a fairness-constrained equalized odds adjustment mechanism, defined as:

$$\theta^{(t+1)} = \theta^{(t)} - \eta\left(\nabla_\theta L(f_\theta(x), y) - \beta \nabla_\theta L_{\text{adv}}\left(d_\phi(f_\theta(x)), s\right)\right)$$

where $\theta^{(t\ +\ 1)}$ represents the updated model parameters, η is the learning rate, $\nabla_\theta L(f_\theta(x), y)$ is the gradient of the predictive loss function, and β is the fairness adjustment weight ensuring that equalized odds constraints are enforced in the learning process.

This fairness-constrained adversarial debiasing model ensures that AI-driven urban planning models dynamically adjust predictions to mitigate demographic disparities, promoting equitable access to resources, infrastructure, and public services. By integrating fairness metrics such as equalized odds and demographic parity directly into the learning process, this framework guarantees that AI-driven decision-making remains accountable, ethical, and socially responsible, preventing reinforcement of historical inequalities in urban development.

Framework: Fairness-Constrained Adversarial Debiasing Model (FC-ADM) for Urban Planning

Layered AI Architecture for Fairness-Aware Adversarial Debiasing in City Planning

Layer	Main functionality	AI module(s) and components	Operational scope (Input → Processing → Output)
Layer 1	Fairness-constrained adversarial debiasing for bias mitigation	Fairness-Constrained Adversarial Debiasing Model (FC-ADM)	Training data, fairness penalty adjustments, adversarial learning → Federated aggregation with fairness-penalized loss → Optimized fairness-aware model parameters
Layer 2	Adversarial fairness loss function	Adversarial fairness optimization engine	Labeled data, adversarial loss function → Bias correction through adversarial training → Fair and transparent model learning
Layer 3	Bias mitigation via fairness-constrained policy learning	Fair policy optimization module	Adversarial fairness loss, model gradients → Fair model parameter updates → Stable convergence with fairness constraints

Moreover, enhancing algorithmic fairness requires the integration of fairness constraints directly into the optimization process, ensuring that AI-driven models do not prioritize accuracy or efficiency at the expense of equity. Traditional AI models often optimize performance metrics such as precision, recall, and loss minimization without accounting for the distributional impact of predictions across different demographic groups (Baum et al. 2022). This can lead to discriminatory outcomes, particularly in city planning, where biased AI models may reinforce inequalities in infrastructure allocation, public services, and resource distribution. To mitigate these risks, fairness constraints can be embedded within the model's learning process, allowing optimization strategies to adjust for equity while maintaining predictive accuracy. In addition, post-processing techniques, such as the adjustment of AI-generated predictions after training, can serve as corrective mechanisms for biases introduced during model development.

To achieve these objectives, a Fairness-Constrained Dual-Stage Optimization Model (FC-DSOM) is proposed. This model integrates fairness constraints into both the training phase and the post-processing stage, supporting more equitable and bias-resistant AI-driven decisions. The FC-DSOM model is formulated as:

$$\min_{\theta} \mathbb{E}_{(x,y)\sim P_{\text{data}}}[L(f_\theta(x), y)] + \lambda \sum_{g=1}^{G} | P_g - P^* | + \mu \sum_{g=1}^{G} | \hat{P}_g - P^* |$$

where θ represents the model parameters, P_{data} is the training data distribution, $L(f_\theta(x), y)$ is the predictive loss function, P_g represents the learned probability distribution for demographic group g, $\hat{P}_g$ is the adjusted probability distribution after post-processing, P^* denotes the target fair distribution, λ is the fairness constraint parameter applied during training, and μ is the fairness constraint parameter applied during post-processing adjustments.

This model ensures that AI-driven urban planning decisions are not only fair during training but also corrected post-training, preventing the persistence of biases. The post-processing fairness adjustment function is defined as:

$$\hat{P}_g = P_g + \alpha(P^* - P_g)$$

where $\hat{P}_g$ represents the fairness-adjusted probability distribution for group g, P_g is the original model prediction, P^* is the target fair distribution, and α is the post-processing correction factor ensuring demographic balance.

The next step in optimizing equitable outcomes involves addressing potential bias in the model through adversarial fairness mechanisms. This approach aims to dynamically adjust predictions, ensuring that they are aligned with fairness goals while maintaining predictive accuracy. The adversarial fairness loss function actively corrects biases by assessing the discrepancy between predicted and fairness-adjusted distributions, ensuring that demographic disparities are minimized.

The adversarial fairness loss function is formally defined as follows:

$$L_{\text{adv}}(d_\phi(f_\theta(x)), s) = -\sum_{c=1}^{C} P_c(x) \log d_{\phi,c}(f_\theta(x))$$

where $L_{\text{adv}}(d_\phi(f_\theta(x)), s)$ is the adversarial fairness loss function quantifying the discrepancy between predicted and fairness-adjusted label distributions, $P_c(x)$ represents the true probability distribution for class c, $d_{\phi,\,c}(f_\theta(x))$ denotes the discriminator's confidence in classifying data into category c. This function optimizes the model to minimize fairness discrepancies, ensuring that predictions are unbiased and equitable across all demographic groups.

To ensure stability in fairness-aware optimization, the learning process incorporates a fairness-constrained gradient descent adjustment mechanism, formulated as:

$$\theta^{(t+1)} = \theta^{(t)} - \eta\left(\nabla_\theta L(f_\theta(x), y) - \beta \nabla_\theta F(Q) - \gamma \nabla_\theta \hat{P}_g\right)$$

where $\theta^{(t\,+\,1)}$ represents the updated model parameters, η is the learning rate, $\nabla_\theta L(f_\theta(x), y)$ is the gradient of the supervised loss function, β is the fairness regularization parameter ensuring demographic balance during training, and γ is the fairness weight applied to post-processing adjustments.

This fairness-constrained dual-stage optimization model ensures that AI-driven city planning decisions are dynamically adjusted for fairness both during training and after deployment, mitigating algorithmic bias at multiple stages of the AI pipeline. By integrating pre-training fairness constraints, post-training fairness adjustments, and fairness-aware gradient descent mechanisms, this framework guarantees that AI-powered urban decision-making remains ethically aligned, socially responsible, and free from systemic biases, promoting inclusive and equitable urban development strategies.

Framework: Fairness-Constrained Dual-Stage Optimization Model for Urban Planning (FC-DSOM)

Layered AI Architecture for Fairness-Aware Optimization in City Planning

Layer	Main functionality	AI module(s) and components	Operational scope (Input → Processing → Output)
Layer 1	Fairness-constrained dual-stage optimization for bias mitigation	Fairness-Constrained Dual-Stage Optimization Model (FC-DSOM)	Training data, fairness penalty adjustments, adversarial learning → Federated aggregation with fairness-penalized loss → Optimized fairness-aware model parameters considering dataset imbalances
Layer 2	Adversarial fairness loss function	Adversarial fairness optimization engine	Labeled data, adversarial loss function → Bias correction through adversarial training → Fair and transparent model learning
Layer 3	Bias mitigation via fairness-constrained policy learning	Fair policy optimization module	Adversarial fairness loss, model gradients → Fair model parameter updates → Stable convergence with fairness constraints

Figure 6.1 outlines the essential aspects of creating inclusive and bias-free AI algorithms in city planning, organized into three main areas: Bias in Data and Algorithms, Consequences of Bias, and Mitigation Strategies. "Bias in Data and Algorithms" addresses the origins of bias, including historical biases and prioritization of efficiency over fairness. "Consequences of Bias" highlights the impact of bias, such as inequality in service access and erosion of public trust. "Mitigation Strategies" focuses on solutions, including the use of diverse data, fairness checks, and incorporating fairness constraints within algorithms to ensure equitable outcomes in urban planning.

Bias in data and algorithms presents significant challenges for the equitable deployment of AI systems, particularly in domains such as city planning, where decisions can have far-reaching consequences for diverse populations. Addressing these biases requires a comprehensive approach that includes ensuring data diversity, applying fairness metrics, and incorporating fairness constraints into

Fig. 6.1 Ensuring inclusive, bias-free algorithms in city planning

algorithmic design. By taking these steps, city planners and AI developers can help ensure that AI systems are more inclusive and equitable, ultimately contributing to the creation of fairer urban environments.

6.6 Strategies for Ensuring Inclusivity

The rise of AI technologies in city planning and urban development offers significant opportunities for optimizing infrastructure and services. However, without deliberate strategies to ensure inclusivity, these systems risk reinforcing existing inequalities and marginalizing vulnerable populations. To ensure inclusivity, a combination of technical, ethical, and policy-driven approaches must be implemented. These strategies aim to mitigate biases, enhance fairness, and ensure that AI-driven decisions consider the diverse needs of urban populations.

6.6.1 Data Diversity and Representation

A fundamental strategy for ensuring inclusiveness in AI-driven decision-making is enhancing the diversity and representativeness of training datasets. Biased algorithms often arise when AI models are trained on data that lacks demographic variety, leading to disproportionate performance across different population groups (Mehrabi et al. 2022). In city planning, an AI system trained on non-representative data may optimize urban development for dominant social and economic groups while neglecting the needs of underrepresented communities. To prevent this, data collection strategies must ensure that training data reflects the full diversity of urban populations, particularly marginalized groups. This requires deliberate efforts to collect, balance, and preprocess datasets in a manner that supports fair representation across demographic dimensions.

To address this challenge, a Diversity-Aware Generative Data Augmentation Model (DA-GDAM) is proposed. This model leverages generative adversarial networks (GANs) to synthesize diverse and demographically balanced training samples, supporting fairer generalization in AI-driven city planning models across population groups. The DA-GDAM model is formulated as:

$$\underset{G}{min}\ \underset{D}{max}\, \mathbb{E}_{x\sim P_{\text{data}}}[\log D(x)] + \mathbb{E}_{z\sim P_z}[\log(1 - D(G(z)))] - \lambda \mathbb{E}_{x\sim P_{\text{data}}}[F(x)]$$

where $G(z)$ is the generator function that produces synthetic unbiased training samples, $D(x)$ is the discriminator function that differentiates between real and synthetic data, P_{data} represents the original data distribution, P_z denotes the prior distribution of latent variables, $F(x)$ is the fairness regularization function, and λ is a regularization weight ensuring demographic balance in the generated data.

This adversarial learning framework ensures that the synthetically generated data maintains both diversity and demographic balance, preventing AI models from learning skewed representations that favor dominant social groups. The fairness regularization function $F(x)$ is explicitly designed to minimize demographic imbalances and is defined as:

$$F(x) = \sum_{g=1}^{G} | P_g(x) - P^*(x) |$$

where $F(x)$ measures demographic disparities in synthetic data generation, $P_g(x)$ represents the distribution of data for group g, and $P^*(x)$ denotes the ideal target distribution ensuring fairness.

To ensure model robustness and prevent bias drift, a diversity-aware generative training mechanism is incorporated, formulated as:

$$\theta^{(t+1)} = \theta^{(t)} - \eta(\nabla_\theta L(f_\theta(x), y) - \beta \nabla_\theta F(Q))$$

where $\theta^{(t\ +\ 1)}$ represents the updated model parameters, η is the learning rate, $\nabla_\theta L(f_\theta(x), y)$ is the gradient of the supervised loss function, and β is the fairness adjustment weight ensuring demographic equity.

This diversity-aware generative data augmentation model ensures that AI-driven city planning models are trained on balanced, inclusive, and demographically representative datasets. By leveraging GAN-based synthetic data generation, fairness regularization, and diversity-aware training adjustments, this framework guarantees that AI-driven decision-making in urban planning is ethically aligned, socially responsible, and equitable, ensuring that no demographic group is systematically excluded from urban development policies and infrastructure investments.

Framework: Diversity-Aware Generative Data Augmentation Model (DA-GDAM)

Layered AI Architecture for Fair and Inclusive Urban Planning

Layer	Main functionality	AI module (s) and components	Operational scope
Layer 1	Data collection: Gather original data from diverse urban sources	Urban data preprocessing engine	Input → Raw urban data (demographics, geography) → Output → Preprocessed data
Layer 2	Generative adversarial learning: Train GAN to synthesize diverse data	Generative Adversarial Networks (GAN)	Input → Preprocessed data, latent variables → Output → Synthetic unbiased samples
Layer 3	Fairness regularization: Apply fairness constraints to synthetic data	Fairness constraint regularization module	Input → Synthetic data → Output → Fairness-regularized data
Layer 4	Diversity-aware training: Integrate fairness-aware loss during model training	Diversity-aware training engine	Input → Fairness-regularized data → Output → Trained AI model
Layer 5	Model evaluation: Assess AI model's performance across demographic groups	Fairness evaluation module	Input → Trained AI model, validation data → Output → Fairness evaluation report
Layer 6	Post-model feedback: Adjust model based on fairness performance reports	Model feedback adaptation unit	Input → Evaluation reports, performance data → Output → Adjusted model

Ensuring data diversity in AI-driven systems goes beyond collecting comprehensive datasets; it also requires bias mitigation during the preprocessing phase to prevent training data from reinforcing historical inequalities. In city planning, historical datasets often reflect discriminatory practices, such as redlining, unequal infrastructure investment, and biased public service allocation, leading to AI-driven urban policies that may perpetuate systemic disparities. To correct these biases, preprocessing techniques such as dataset reweighting, resampling, and fairness-aware feature selection must be applied to support balanced representation across demographic groups (Pagano et al. 2023). By systematically identifying and correcting data biases before model training, urban planning systems can be designed to promote equity rather than reproduce existing disparities.

To achieve this, a Fairness-Aware Bayesian Data Reweighting Model (FA-BDRM) is proposed. This model applies Bayesian inference techniques to reweight historical datasets, increasing the influence of underrepresented groups within the model training process. The FA-BDRM model is formulated as:

$$\min_{\theta} \mathbb{E}_{(x,y)\sim P_{\text{data}}}[w(x)L(f_\theta(x), y)] + \lambda \sum_{g=1}^{G} \mid \mathrm{P}_g - P^* \mid$$

where θ represents the model parameters, P_{data} is the original training data distribution, $L(f_\theta(x), y)$ is the predictive loss function, $w(x)$ is the Bayesian reweighting function ensuring demographic balance, P_g represents the distribution of demographic group g in the dataset, P^* is the ideal target distribution ensuring fairness, and λ is the fairness regularization parameter controlling trade-offs between accuracy and equity.

This Bayesian fairness reweighting approach ensures that AI-driven urban models do not overfit historical patterns of discrimination, instead prioritizing a balanced representation of all communities. The Bayesian reweighting function dynamically adjusts the influence of biased historical data points and is defined as:

$$w(x) = \frac{P^*(x)}{P_{\text{data}}(x) + \epsilon}$$

where $w(x)$ is the adjustment weight assigned to sample x, $P^*(x)$ is the ideal target probability ensuring demographic fairness, $P_{\text{data}}(x)$ is the original biased probability distribution, and ϵ is a small constant preventing division errors.

To further reinforce fairness constraints, the learning process incorporates a fairness-constrained data resampling mechanism, formulated as:

$$\hat{P}_g = P_g + \alpha(P^* - P_g)$$

where $\hat{P}_g$ represents the resampled distribution ensuring demographic balance, P_g is the original biased distribution of group g, P^* is the ideal fair distribution, and α is the resampling adjustment factor controlling the correction process.

To stabilize model training while incorporating fairness constraints, the optimization process is refined using a Bayesian posterior fairness adjustment, formulated as:

$$\theta^{(t+1)} = \theta^{(t)} - \eta(\nabla_\theta L(f_\theta(x), y) - \beta \nabla_\theta F(Q))$$

where $\theta^{(t\ +\ 1)}$ represents the updated model parameters, η is the learning rate, $\nabla_\theta L(f_\theta(x), y)$ is the gradient of the supervised learning loss function, and β is the fairness weight ensuring that bias corrections persist throughout model training.

This fairness-aware Bayesian data reweighting model ensures that AI-driven city planning decisions are based on demographically balanced, bias-free datasets. By integrating Bayesian reweighting, fairness-driven resampling, and optimization adjustments, this framework guarantees that AI-powered decision-making processes in urban planning do not perpetuate historical inequalities but instead promote fairness, inclusivity, and equitable urban development strategies.

Framework: Fairness-Aware Bayesian Data Reweighting Model (FA-BDRM)

Layered AI Architecture for Fair and Equitable Urban Planning

Layer	Main functionality	AI module (s) and components	Operational scope
Layer 1	Data collection: Gather original data from diverse urban sources	Urban data preprocessing engine	Input → Raw urban data (demographics, geography) → Output → Preprocessed data
Layer 2	Fairness-aware Bayesian learning: Apply Bayesian reweighting to historical datasets	Bayesian inference	Input → Preprocessed data → Output → Reweighted data with demographic balance
Layer 3	Fairness regularization: Apply fairness penalty to the loss function	Fairness regularization module	Input → Reweighted data → Output → Fairness-adjusted loss
Layer 4	Diversity-aware training: Integrate fairness-aware loss during model training	Fairness-aware training engine	Input → Fairness-regularized data → Output → Trained AI model with fairness constraints
Layer 5	Model evaluation: Evaluate AI model performance across demographic groups	Fairness evaluation module	Input → Trained AI model, validation data → Output → Fairness evaluation report
Layer 6	Post-model feedback: Adjust model based on fairness performance reports	Adaptive feedback optimization unit	Input → Evaluation reports, performance data → Output → Adjusted model

6.6.2 *Fairness Metrics and Algorithm Design*

Fairness metrics serve as critical tools for evaluating the performance of AI systems across different demographic groups, ensuring that AI-driven decision-making is equitable and bias-resistant. When AI models are deployed in city planning, infrastructure allocation, and public service distribution, they must be assessed for potential disparities in their predictions to prevent systematic disadvantages for marginalized communities. Metrics such as Demographic Parity, Equalized Odds, and Disparate Impact Analysis enable AI practitioners to quantify fairness and examine whether models disproportionately favor certain groups over others (Mehrabi et al. 2022). By incorporating fairness metrics into model evaluation, developers can systematically detect and correct biases, leading to more transparent, accountable, and socially responsible AI systems.

To achieve this, a Fairness-Constrained Multi-Objective Optimization Model (FC-MOOM) is proposed. This model integrates fairness metrics directly into the learning process, allowing optimization to account for fairness alongside predictive accuracy. The FC-MOOM model is formulated as:

$$\min_{\theta} \mathbb{E}_{(x,y)\sim P_{\text{data}}}[L(f_\theta(x), y)] + \lambda_1 \sum_{g=1}^{G} \mid P_g - P^* \mid + \lambda_2 \sum_{g=1}^{G} (\text{EO}_g + \text{DP}_g)$$

where θ represents the model parameters, P_{data} is the training data distribution, $L(f_\theta(x), y)$ is the predictive loss function, P_g represents the learned probability distribution for demographic group g, P^* is the ideal target distribution ensuring fairness, λ_1 is the fairness regularization parameter controlling demographic balance, λ_2 is the fairness weight applied to metric-based adjustments, EO_g is the Equalized Odds fairness constraint, and DP_g is the demographic parity constraint ensuring proportional representation.

Next, fairness regularization is integrated into the model's learning process. Fairness penalties are applied to the loss function to correct for any imbalances in the model's predictions across demographic groups. The fairness regularization function adjusts the model's weightings by considering demographic parity and equalized odds, guiding the optimization process to ensure that fairness constraints are met. These adjustments are critical in preventing the model from disproportionately favoring any single group over others.

$$w(x) = \frac{P^*(x)}{P_{\text{data}}(x) + \epsilon}$$

where $w(x)$ is the reweighting function, $P^*(x)$ is the ideal target probability distribution, $P_{\text{data}}(x) + \epsilon$ is the original data distribution, ϵ is a small constant added to avoid division by zero. This function dynamically adjusts the influence of data points based on the ideal target distribution, ensuring that the model accounts for fairness in its learning process.

Using gradient descent and backpropagation, the model learns by incorporating both the predictive loss and fairness regularization. The diversity-aware training ensures that, while the model improves its predictive accuracy, it does not neglect fairness. The fairness penalties applied at each step prevent the model from converging to solutions that may optimize accuracy at the expense of fairness.

$$\theta^{(t+1)} = \theta^{(t)} - \eta\left(\nabla_\theta L(f_\theta(x), y) - \beta \nabla_\theta \sum_{g=1}^{G} (\text{EO}_g + \text{DP}_g)\right)$$

where $\theta^{(t+1)}$ represents the updated model parameters, η is the learning rate, $\nabla_\theta L(f_\theta(x), y)$ is the gradient of the predictive loss function, and β is the fairness adjustment weight ensuring demographic equity.

Next, the trained model's performance is assessed using fairness metrics to ensure that the AI system has not inadvertently introduced or perpetuated bias in its predictions. The Demographic parity and equalized odds metrics are used to evaluate the model's fairness. The demographic parity metric checks for balance in the predicted positive rates across demographic groups, while equalized odds measures differences in the false positive and false negative rates. These fairness metrics help to quantify any disparities in the model's decision-making process, providing a means for adjustment if necessary.

$$\mathrm{DP}_g = \mid P(\hat{Y} = 1 | G = g) - P(\hat{Y} = 1) \mid$$

where DP_g measures disparities in positive prediction rates between demographic group gg and the overall population. This function quantifies disparities in the likelihood of receiving a positive prediction between demographic group gg and the overall population.

The equalized odds metric, ensuring that AI models have similar false positive and false negative rates across demographic groups, is defined as:

$$\mathrm{EO}_g = \mid P(\hat{Y} = 1 | Y = 1, G = g) - P(\hat{Y} = 1 | Y = 1) \mid + \\ \mid P(\hat{Y} = 1 | Y = 0, G = g) - P(\hat{Y} = 1 | Y = 0) \mid$$

where EO_g measures differences in true positive and false positive rates across demographic group g, ensuring that AI predictions do not favor specific groups unfairly. This function evaluates the fairness of the model's false positive and false negative rates.

This fairness-constrained multi-objective optimization model ensures that AI-driven city planning actively mitigates bias by integrating demographic parity and equalized odds constraints into the optimization process. By incorporating fairness metrics, multi-objective optimization, and gradient-based fairness adjustments, this framework ensures that AI-driven decisions remain ethically aligned, socially responsible, and free from systemic biases, fostering transparent and equitable urban development strategies.

Framework: Fairness-Constrained Multi-Objective Optimization Model (FC-MOOM)

Layered AI Architecture for Fair and Equitable Urban Planning

Layer	Main functionality	AI module (s) and components	Operational scope
Layer 1	Data collection: Gather original data from diverse urban sources	Urban data preprocessing engine	Input → Raw urban data (demographics, geography) → Output → Preprocessed data
Layer 2	Fairness-aware multi-objective learning: Apply fairness metrics to training	Bayesian inference	Input → Preprocessed data → Output → Reweighted data with fairness metrics

(continued)

Layer	Main functionality	AI module (s) and components	Operational scope
Layer 3	Fairness regularization: Apply fairness penalty to the loss function	Fairness penalty optimization module	Input → Reweighted data → Output → Fairness-adjusted loss
Layer 4	Diversity-aware training: Integrate fairness-aware loss during model training	Multi-objective fairness training engine	Input → Fairness-regularized data → Output → Trained AI model with fairness constraints
Layer 5	Model evaluation: Evaluate AI model performance using fairness metrics	Fairness metrics evaluation module	Input → Trained AI model, validation data → Output → Fairness evaluation report
Layer 6	Post-model feedback: Adjust model based on fairness performance reports	Feedback adaptation and model refinement unit	Input → Evaluation reports, performance data → Output → Adjusted model

Incorporating fairness constraints into AI-driven optimization processes is crucial for ensuring that algorithmic decision-making remains equitable across demographic groups. Traditional AI models often prioritize accuracy and efficiency without considering the societal impact of biased resource allocation. To mitigate this risk, fairness constraints must be embedded directly into training and decision-making processes, reducing the likelihood that AI-generated outcomes reinforce historical disparities (Baum et al. 2022). In city planning, for instance, a model optimizing public service distribution should support equitable access to infrastructure, healthcare, and transportation for lower-income neighborhoods, even when this requires a modest trade-off against aggregate efficiency. By integrating fairness-aware optimization, AI models can support more balanced and just resource allocation, fostering inclusive urban development.

To address this challenge, a Fairness-Constrained Reinforcement Learning Model for Equitable Resource Allocation (FCRL-ERA) is proposed. This model dynamically optimizes resource distribution by balancing efficiency and fairness constraints, reducing the risk that disadvantaged populations are systematically excluded from city planning benefits. The FCRL-ERA model is formulated as:

$$\pi^*(s) = \arg \underset{\pi}{max}\, \mathbb{E}\left[\sum_{t=0}^{T} \gamma^t \left(R_{\text{eff}}(s_t, a_t) + \alpha R_{\text{fair}}(s_t, a_t)\right)\right]$$

where $\pi^*(s)$ represents the optimal policy balancing efficiency and fairness, $\mathbb{E}\,[\cdot]$ denotes the expectation over possible state transitions, γ^t is the discount factor controlling the balance between short-term and long-term optimization, $R_{\text{eff}}(s_t, a_t)$ measures the efficiency of the resource allocation decision, $R_{\text{fair}}(s_t, a_t)$ is the fairness reward function ensuring equitable distribution, and α is a tunable fairness parameter controlling the trade-off between efficiency and fairness.

This reinforcement learning framework ensures that resource allocation decisions do not simply favor historically privileged regions but instead account for fairness constraints dynamically. The fairness reward function $R_{\text{fair}}(s_t, a_t)$ is formulated as:

$$R_{\text{fair}}(s_t, a_t) = -\sum_{g=1}^{G} | P_g(s_t, a_t) - P^*(s_t) |$$

where $R_{\text{fair}}(s_t, a_t)$ measures demographic disparities in AI-driven resource allocation, $P_g(s_t, a_t)$ represents the probability of assigning resources to demographic group g, and $P^*(s_t)$ denotes the ideal equitable distribution ensuring fairness across all groups.

To prevent policy instability while ensuring fairness, the Proximal Policy Optimization (PPO) algorithm is incorporated into the training process, formulated as:

$$L(\theta) = \mathbb{E}_t[min(\rho_t(\theta)A_t, \text{clip}(\rho_t(\theta), 1 - \epsilon, 1 + \epsilon)A_t)]$$

where $L(\theta)$ represents the policy optimization objective function, $\rho_t(\theta)$ is the policy ratio ensuring stable learning updates, A_t is the advantage function representing policy improvement, and ϵ is the clipping parameter preventing excessive policy fluctuations.

The next step in ensuring the ethical deployment of AI-driven urban planning systems is to evaluate the model's performance through the lens of fairness. The goal here is to assess whether the system's decisions across different demographic groups align with the fairness objectives set earlier. In this phase, we use fairness metrics to systematically quantify disparities in the model's predictions and resource allocations. These metrics provide insights into whether the model is inadvertently favoring certain groups, allowing us to address potential inequalities before the model is deployed in real-world scenarios.

In the case of demographic parity, the metric measures whether different demographic groups are equally likely to receive a positive prediction, ensuring that the model does not favor any one group. The equalized odds metric further refines this evaluation by considering both the false positive and false negative rates across groups, thus ensuring that the model is fair in both error types.

$$\text{DP}_g = | P(Y = 1|G = g) - P(Y = 1) |$$

where DP_g is the demographic parity metric, $P(Y = 1 | G = g)$ is the probability of a positive prediction for group g, $P(Y = 1)P(Y = 1)$ is the overall positive prediction rate. This function evaluates whether the likelihood of receiving a positive outcome differs between demographic group gg and the overall population, ensuring that no group is unfairly favored in positive predictions.

The equalized odds metric:

$$\text{EO}_g = | P(Y = 1|Y = 1, G = g) - P(Y = 1|Y = 1) | + | P(Y = 1|Y = 0, G = g) - P(Y = 1|Y = 0) |$$

where EO_g is the equalized odds metric, $P(Y = 1 \mid Y = 1, G = g)$ is the probability of a positive prediction given the true positive label for group g, $P(Y = 1 \mid Y = 1)$ is the overall positive prediction rate for true positives, $P(Y = 1 \mid Y = 0, G = g)$ is the probability of a positive prediction given the true negative label for group g, and $P(Y = 1 \mid Y = 0)$ is the overall positive prediction rate for false negatives. This function ensures that the model does not favor one group over others by evaluating the fairness of both true positive and false positive rates. Through these fairness metrics, we gain an essential understanding of how well the AI system performs across demographic groups, ensuring that the model's decisions are equitable and do not inadvertently reproduce systemic disparities. This evaluation layer plays a crucial role in the ethical deployment of AI models, enabling stakeholders to make informed decisions about how to adjust the model for improved fairness.

This fairness-constrained reinforcement learning model for equitable resource allocation ensures that AI-driven city planning systems balance efficiency and fairness, dynamically optimizing resource distribution to prevent systemic inequalities. By integrating fairness constraints directly into the reward function and reinforcement learning process, this model guarantees that AI-driven policies do not disproportionately favor privileged communities but instead foster equitable and inclusive urban development strategies.

Framework: Fairness-Constrained Reinforcement Learning Model for Equitable Resource Allocation (FCRL-ERA)

Layered AI Architecture for Fair and Equitable Resource Allocation in Urban Planning

Layer	Main functionality	AI module (s) and components	Operational scope
Layer 1	Data collection: Gather original data from diverse urban sources	Urban data preprocessing engine	Input → Raw urban data (demographics, geography) → Output → Preprocessed data
Layer 2	Fairness-aware multi-objective learning: Apply fairness metrics to resource allocation optimization	Reinforcement learning	Input → Preprocessed data → Output → Policy balancing efficiency and fairness
Layer 3	Fairness regularization: Apply fairness penalty to the reward function	Fairness reward adjustment module	Input → Reweighted data → Output → Fairness-adjusted reward function
Layer 4	Diversity-aware training: Integrate fairness-aware reward during model training	Reinforcement learning	Input → Fairness-regularized data → Output → Trained AI model with fairness constraints
Layer 5	Model evaluation: Evaluate AI model performance using fairness metrics	Fairness performance evaluation module	Input → Trained AI model, validation data → Output → Fairness evaluation report
Layer 6	Post-model feedback: Adjust model based on fairness performance reports	Adaptive policy refinement unit	Input → Evaluation reports, performance data → Output → Adjusted model

6.6.3 Explainable AI (XAI) for Inclusivity

Explainable AI (XAI) plays a fundamental role in ensuring inclusivity and accountability in AI-driven city planning systems. In high-stakes decision-making environments, such as urban development, infrastructure allocation, and public service distribution, AI models must be transparent, interpretable, and justifiable (Baum et al. 2022). Traditional black-box AI systems lack interpretability, making it difficult for city planners, policymakers, and community stakeholders to assess whether AI-generated decisions are fair, equitable, and socially responsible. XAI techniques enhance transparency by providing human-interpretable explanations of model behavior, ensuring that automated decision-making processes can be scrutinized, audited, and refined to promote fairness and inclusivity.

To address this need, a Fairness-Aware Explainable Reinforcement Learning Model (FA-XRL) is proposed. This model integrates explainability mechanisms into reinforcement learning-based decision systems, ensuring that city planning stakeholders can interpret, validate, and refine AI-driven policies. The FA-XRL model is formulated as:

$$\pi^*(s) = \arg \underset{\pi}{max}\, \mathbb{E}\left[\sum_{t=0}^{T} \gamma^t (R_{\text{eff}}(s_t, a_t) + \alpha R_{\text{fair}}(s_t, a_t) + \beta R_{\text{XAI}}(s_t, a_t))\right]$$

where $\pi^*(s)$ represents the optimal policy incorporating fairness and explainability constraints, $\mathbb{E}$ $[\cdot]$ denotes the expectation over possible state transitions, γ^t is the discount factor balancing short-term and long-term rewards, $R_{\text{eff}}(s_t, a_t)$ quantifies the efficiency of resource allocation, $R_{\text{fair}}(s_t, a_t)$ represents the fairness reward function ensuring demographic equity, $R_{\text{XAI}}(s_t, a_t)$ is the explainability reward function promoting model interpretability, and α, β are regularization parameters controlling the trade-offs between fairness, efficiency, and explainability. This multi-objective reinforcement learning approach ensures that AI-driven urban planning models are not only optimized for performance but also provide interpretable and accountable decision-making processes. The explainability reward function is defined as:

$$R_{\text{XAI}}(s_t, a_t) = -\sum_{g=1}^{G} \mid \mathcal{E}(s_t, a_t, g) - \mathcal{E}^*(s_t) \mid$$

where $R_{\text{XAI}}(s_t, a_t)$ measures explainability gaps in AI-driven decision-making, $\mathcal{E}(s_t, a_t, g)$ represents the explanation of model predictions for demographic group g, and $\mathcal{E}^*(s_t)$ denotes the ideal interpretability standard ensuring that AI decisions are transparent.

The next step in the model's development focuses on integrating fairness-aware reward during the model training process. As discussed earlier, fairness metrics such as Equalized Odds (EO) and Demographic Parity (DP) help guide the model's decision-making to ensure that no demographic group is unfairly disadvantaged by the model's predictions. Now, as we proceed to Layer 4: Diversity-Aware Training, we incorporate these fairness constraints directly into the reinforcement learning process. The update of the model parameters during training is achieved through the following optimization equation:

$$\theta^{(t+1)} = \theta^{(t)} - \eta\left(\nabla_\theta L(f_\theta(x), y) - \beta\nabla_\theta \sum_{g=1}^{G} (\mathrm{EO}_g + \mathrm{DP}_g)\right)$$

where $\theta^{(t)}$ and $\theta^{(t+1)}$ are the model parameters at time steps t and $t + 1$, η is the learning rate, $\nabla_\theta L(f_\theta(x), y)$ is the gradient of the predictive loss function, β is the fairness adjustment weight, $\sum_{g=1}^{G} (\mathrm{EO}_g + \mathrm{DP}_g)$ is the aggregated fairness metrics across all demographic groups. This function updates the model parameters by considering both the predictive loss and the fairness penalties, ensuring the model learns to balance accuracy and fairness during training.

In parallel with the fairness-aware training, we now shift our focus to ensuring that the model remains interpretable while achieving its optimization goals. This is where Explainable AI (XAI) comes into play, specifically in terms of reinforcing the explainability of the decision-making process. Following the fairness optimization step, we introduce the explainability loss function, which ensures that the model's predictions are transparent and understandable to human stakeholders. The Shapley values are used to assess the contribution of each feature in the model's decision process, providing insights into how decisions are made. The explainability adjustment is captured by the following equation:

$$L_{\mathrm{XAI}}(\theta) = \mathbb{E}_t\left[\sum_{i=1}^{N} \phi_i(s_t, a_t)\right]$$

where $L_{\mathrm{XAI}}(\theta)$ represents the explainability loss function, ensuring that model decisions remain interpretable, and $\phi_i(s_t, a_t)$ is the Shapley value contribution of each model feature in the decision process.

Together, these two equations—one for fairness-aware updates and one for explainability adjustments—ensure that the model learns not only to optimize for fairness but also to provide transparent and interpretable decision-making processes. By embedding both fairness and explainability into the training process, we move closer to building an AI system for urban planning that is both ethical and accountable.

Next, the model's performance is evaluated not only in terms of its predictive accuracy but also through the lens of fairness and explainability. The purpose of this evaluation layer is to ensure that the AI model adheres to the ethical and transparency standards established in the earlier layers. As AI systems in urban planning must not only be efficient but also equitable, fairness metrics such as Demographic Parity (DP) and Equalized Odds (EO) are applied to assess the disparity in model predictions across demographic groups. These metrics ensure that the model's decisions are not biased toward any group but are rather distributed fairly across all groups, meeting the social responsibility criteria.

To assess whether the model has achieved demographic parity, we define the following equation, which measures the disparity in positive prediction rates between different demographic groups and the overall population. This metric ensures that no group is unfairly favored or disadvantaged in terms of receiving positive outcomes.

$$\mathrm{DP}_g = \mid P(Y = 1 \mid G = g) - P(Y = 1) \mid$$

where DP_g is the demographic parity metric, $P(Y = 1 \mid G = g)$ is the probability of receiving a positive prediction for demographic group g, $P(Y = 1)P(Y = 1)$ is the overall positive prediction rate. This function calculates the absolute difference between the positive prediction rate for a specific group and the overall population, ensuring fairness in the model's predictions.

In addition to demographic parity, we also use Equalized Odds (EO) as a fairness metric. The equalized odds metric ensures that the model has similar false positive and false negative rates across different demographic groups. This is crucial for making sure that the model's predictions are not only accurate but also fair, regardless of group membership. The EO function is defined as:

$$\mathrm{EO}_g = \mid P(Y = 1 \mid Y = 1, G = g) - P(Y = 1 \mid Y = 1) \mid + \mid P(Y = 1 \mid Y = 0, G = g) - P(Y = 1 \mid Y = 0) \mid$$

where EO_g is the equalized odds metric, $P(Y = 1 \mid Y = 1, G = g)$ is the probability of a true positive prediction for group g, $P(Y = 1 \mid Y = 1)$ is the overall positive prediction rate for true positives, $P(Y = 1 \mid Y = 0, G = g)$ is the probability of a positive prediction given the true negative label for group g, and $P(Y = 1 \mid Y = 0)$ is the overall positive prediction rate for false negatives. This function ensures that the model's false positive and false negative rates are similar across demographic groups, promoting fairness in the decision-making process.

Through the application of these fairness metrics, model evaluation ensures that the model adheres to the required ethical standards, providing a robust framework to verify that the AI system makes equitable decisions. This process helps prevent unintentional bias and ensures that the model is socially responsible and transparent, in line with the values of inclusivity and fairness.

This fairness-aware explainable reinforcement learning model ensures that AI-driven urban planning decisions remain interpretable, transparent, and fair. By integrating explainability constraints, fairness-aware policy adjustments, and Shapley value-based interpretability mechanisms, this model empowers city planners, policymakers, and community stakeholders to scrutinize AI-driven decisions, ensuring that urban development strategies align with ethical, inclusive, and socially responsible AI practices.

Framework: Fairness-Aware Explainable Reinforcement Learning Model (FA-XRL)

Layered AI Architecture for Explainable and Fair Urban Planning

Layer	Main functionality	AI module (s) and components	Operational scope
Layer 1	Data collection: Gather original data from diverse urban sources	Urban data preprocessing engine	Input → Raw urban data (demographics, geography) → Output → Preprocessed data
Layer 2	Fairness-aware multi-objective learning: Apply fairness metrics to urban resource allocation	Reinforcement learning	Input → Preprocessed data → Output → Policy balancing efficiency, fairness, and explainability
Layer 3	Fairness regularization: Apply fairness penalty to the reward function	Fairness reward optimization module	Input → Reweighted data → Output → Fairness-adjusted reward function
Layer 4	Diversity-aware training: Integrate fairness-aware reward during model training	Reinforcement learning	Input → Fairness-regularized data → Output → Trained AI model with fairness and explainability constraints
Layer 5	Model evaluation: Evaluate AI model performance using fairness and explainability metrics	Explainability and fairness evaluation module	Input → Trained AI model, validation data → Output → Fairness and explainability evaluation report
Layer 6	Post-model feedback: Adjust model based on fairness and explainability performance reports	Adaptive model refinement unit	Input → Evaluation reports, performance data → Output → Adjusted model

For example, if an AI system recommends certain infrastructure projects in specific neighborhoods, XAI can help explain why those neighborhoods were prioritized and whether the decision-making process considered the needs of all communities. By providing clear, understandable explanations for AI decisions, XAI helps ensure accountability and allows for the identification and rectification of any biases that may have influenced the system's outputs.

6.6.4 Stakeholder Engagement and Participatory Design

Another vital strategy for ensuring inclusivity is the active involvement of stakeholders in the design and implementation of AI systems. Participatory design approaches engage diverse community members, including marginalized and underrepresented groups, in the AI development process (Sargiotis 2025b; Zowghi and da Rimini 2023). By involving stakeholders from the outset, AI developers can better align system objectives with the real needs, priorities, and concerns of urban residents.

To formalize this participatory approach within AI-driven urban planning, a Reinforcement Learning (RL) model incorporating Proximal Policy Optimization (PPO) is introduced. This model integrates stakeholder feedback into AI system design by optimizing policies that balance diverse community needs while supporting fairness and inclusiveness in decision-making processes. The PPO-based participatory urban AI model can be formulated as:

$$J(\theta) = \mathbb{E}\left[\sum_{t=0}^{T} \gamma^t r_t\right]$$

where $J(\theta)$, the objective function, maximizes stakeholder-informed policy performance, $\mathbb{E}$ represents the expectation operator, γ is the discount factor controlling the significance of future rewards, r_t denotes the reward signal obtained at time t, and T represents the time horizon of policy optimization. This function evaluates the cumulative reward of stakeholder engagement strategies, ensuring that AI systems iteratively refine policies based on community feedback. The reward function, central to this framework, is defined as:

$$r_t = w_1 U(s_t, a_t) - w_2 C(a_t)$$

where r_t is the immediate reward function reflecting inclusivity optimization, $U(s_t, a_t)$ represents the utility function quantifying stakeholder satisfaction based on AI policy adjustments, $C(a_t)$ denotes the cost of implementing participatory adjustments, and w_1, w_2 are weighting parameters ensuring the balance between maximizing inclusivity and minimizing implementation costs.

Next, we focus on the training phase where fairness-aware policies are integrated into the model's learning process. In this stage, the reinforcement learning model optimizes for both fairness and performance simultaneously. This approach ensures that the model not only learns to perform well but also prioritizes fairness in its decision-making process, ensuring inclusivity and equity in the model's actions.

To achieve this, we update the model parameters in such a way that they take into account both the predictive loss and the fairness penalties. This update is crucial to refine the model towards balancing fairness and efficiency. The model's parameters are adjusted through the following equation:

$$\theta^{(t+1)} = \theta^{(t)} - \eta\left(\nabla_\theta L(f_\theta(x), y) - \beta \nabla_\theta \sum_{g=1}^{G} (\mathrm{EO}_g + \mathrm{DP}_g)\right)$$

where $\theta^{(t)}$ and $\theta^{(t+1)}$ are the model parameters at time steps t and $t + 1$, η is the learning rate, $\nabla_\theta L(f_\theta(x), y)$ is the gradient of the predictive loss function with respect to the model parameters, β is the fairness adjustment weight, $\sum_{g=1}^{G} (\mathrm{EO}_g + \mathrm{DP}_g)$ is the aggregated fairness metrics (equalized odds and demographic parity) across all demographic groups GG. This function adjusts the model parameters to minimize the loss while simultaneously applying fairness penalties to ensure equitable performance across demographic groups.

As the model progresses, fairness and performance must be optimized together. The Proximal Policy Optimization (PPO) algorithm is used to stabilize the learning process, ensuring that the model updates are both efficient and reliable. The PPO objective function optimizes the policy by balancing exploration and exploitation while maintaining stability in the learning process, which is essential when integrating fairness constraints in urban planning systems.

The PPO objective function, designed to control policy updates, is defined as:

$$L(\theta) = \mathbb{E}[min(\rho_t A_t, \text{clip}(\rho_t, 1 - \epsilon, 1 + \epsilon)A_t)]$$

where $L(\theta)$ is the PPO objective function optimizing policy updates, $\rho_t = \frac{\pi_\theta(a_t|s_t)}{\pi_{\theta_{\text{old}}}(a_t|s_t)}$ represents the probability ratio controlling exploration, A_t is the advantage function guiding learning efficiency, and ϵ is the clipping parameter stabilizing policy updates.

This AI-driven participatory framework dynamically incorporates stakeholder feedback into AI policy design, ensuring that inclusivity principles are actively embedded into urban decision-making. The reinforcement learning-based approach continuously refines AI governance, balancing policy effectiveness with social equity, making it a robust solution for developing AI systems that genuinely reflect community needs.

Framework: Participatory Reinforcement Learning Model with Proximal Policy Optimization (PPO)

Layered AI Architecture for Inclusivity in Urban Planning

Layer	Main functionality	AI module (s) and components	Operational scope
Layer 1	Data collection: Gather original data from diverse urban sources	Urban data preprocessing engine	Input → Raw urban data (demographics, geography) → Output → Preprocessed data
Layer 2	Fairness-aware multi-objective learning: Apply fairness metrics to urban resource allocation	Reinforcement learning	Input → Preprocessed data → Output → Policy balancing efficiency, fairness, and explainability
Layer 3	Fairness regularization: Apply fairness penalty to the reward function	Fairness reward adjustment module	Input → Reweighted data → Output → Fairness-adjusted reward function
Layer 4	Diversity-aware training: Integrate fairness-aware reward during model training	Reinforcement learning	Input → Fairness-regularized data → Output → Trained AI model with fairness and explainability constraints

Stakeholder engagement is especially important in city planning, where decisions about resource allocation, infrastructure development, and public services have direct and profound impacts on local communities. By including these communities in the design and decision-making process, AI systems are more likely to be inclusive and aligned with the values and priorities of the people they serve.

6.6.5 *Ongoing Monitoring and Auditing*

Ensuring inclusivity in AI systems is not a one-time effort; it requires continuous monitoring and auditing to identify and address biases or unintended consequences that may emerge over time. Regular audits help assess whether algorithms continue to perform fairly and inclusively after deployment in real-world settings (Mökander 2023). These audits should include both technical evaluations, such as detecting algorithmic bias, and social evaluations, such as assessing the impact of AI-mediated decisions on different demographic groups.

To support continuous fairness and inclusivity, a Reinforcement Learning (RL) model with Proximal Policy Optimization (PPO) is introduced for ongoing auditing and adaptive bias mitigation. This model systematically evaluates bias patterns and adjusts decision policies over time, supporting equitable outcomes while reducing unintended consequences. The PPO-based auditing and bias mitigation model is formulated as:

$$J(\theta) = \mathbb{E}\left[\sum_{t=0}^{T} \gamma^t r_t\right]$$

where $J(\theta)$ is the policy objective function that maximizes fairness in AI decision-making, $\mathbb{E}$ denotes the expectation operator, γ is the discount factor that prioritizes immediate fairness while considering long-term effects, r_t represents the fairness reward signal at time t, and T is the time horizon for continuous AI auditing. To measure the fairness of AI decisions over time, a fairness-aware reward function is introduced:

$$r_t = -w_1 D_{\text{bias}}(s_t, a_t) + w_2 U(s_t, a_t)$$

where r_t is the immediate fairness reward, $D_{\text{bias}}(s_t, a_t)$ represents the measured algorithmic bias at state s_t when action a_t is taken, $U(s_t, a_t)$ quantifies AI performance in terms of equitable outcomes, and w_1, w_2, w_3 are weighting parameters that ensure a trade-off between bias minimization and performance maintenance.

To ensure stable policy updates that enhance fairness while avoiding drastic shifts in AI behavior, the PPO policy ratio function is applied:

$$L(\theta) = \mathbb{E}[min(\rho_t A_t, \text{clip}(\rho_t, 1 - \epsilon, 1 + \epsilon)A_t)]$$

where $L(\theta)$ is the PPO objective function optimizing fairness in AI decisions, $\rho_t = \frac{\pi_\theta(a_t|s_t)}{\pi_{\theta_{\text{old}}}(a_t|s_t)}$ represents the policy probability ratio adjusting AI model updates, A_t is the advantage function guiding fairness-oriented learning, and ϵ is the clipping parameter preventing excessively large changes in policy updates.

To assess the overall impact of AI-driven decisions on different demographic groups, an impact disparity function is introduced:

$$D_{\text{bias}}(s_t, a_t) = \sum_{g=1}^{G} | P_g(a_t) - P_{\text{ref}}(a_t) |$$

where $D_{\text{bias}}(s_t, a_t)$ measures the discrepancy in AI outcomes across demographic groups, $P_g(a_t)$ is the probability distribution of AI decisions for group g, $P_{\text{ref}}(a_t)$ is the reference probability ensuring fair distribution, and G is the total number of demographic groups evaluated.

This AI-driven auditing framework continuously adapts AI decisions based on real-time fairness evaluations, ensuring that biases are detected and mitigated dynamically. By integrating Proximal Policy Optimization (PPO) with fairness-aware reward modeling, this approach enhances AI systems' ability to operate equitably in diverse, real-world environments, preventing long-term disparities while maintaining high-performance outcomes.

Framework: Fairness-Aware Reinforcement Learning Model (FARL)

Layered AI Architecture for Dynamic Auditing and Bias Mitigation

Layer	Main functionality	AI module(s) and components	Operational scope (Input → Processing → Output)
Layer 1	Data collection	Data preprocessing engine	Input → Raw sensor data, demographic data → Output → Preprocessed input for AI models
Layer 2	Fairness-aware learning	Proximal Policy Optimization (PPO), Reinforcement Learning (RL)	Input → Historical data, fairness metrics → Output → Policy updates maximizing fairness
Layer 3	Bias mitigation and reward calculation	Fairness reward computation module	Input → State-action pairs, fairness metrics → Output → Fairness-aware reward calculation
Layer 4	Policy optimization	Proximal Policy Optimization (PPO)	Input → Reward signals, policy updates → Output → Optimized AI model policy
Layer 5	Impact disparity calculation	Disparity analysis engine	Input → AI decisions across demographic groups → Output → Disparity metrics indicating fairness

(continued)

Layer	Main functionality	AI module(s) and components	Operational scope (Input → Processing → Output)
Layer 6	Continuous monitoring and auditing	Fairness auditing and monitoring module	Input → AI model outputs, fairness metrics → Output → Bias audit reports and system evaluation
Layer 7	Human-in-the-loop feedback	Human feedback integration unit	Input → User feedback on AI decisions → Output → Refined AI strategies and performance adjustments
Layer 8	Model adaptation and retraining	Adaptive model retraining system	Input → Feedback data, performance anomalies → Output → Retrained models for continuous improvement

Moreover, dynamic and adaptive approaches to AI development can help ensure that systems remain inclusive as urban environments and populations evolve. This includes regularly updating datasets, retraining models, and reassessing fairness metrics to account for changing social, economic, and demographic conditions.

Figure 6.2 illustrates strategies for achieving inclusiveness in AI systems, especially in city planning, with five main areas: Data Diversity and Representation, Fairness Metrics and Algorithm Design, Explainable AI (XAI), Stakeholder Engagement, and Ongoing Monitoring and Auditing. "Data Diversity and Representation" emphasizes collecting inclusive data and correcting historical biases. "Fairness Metrics and Algorithm Design" focuses on using fairness metrics and embedding fairness constraints within algorithms. "Explainable AI (XAI)" ensures transparency in decision-making and enables bias detection. "Stakeholder Engagement" involves community input to align AI with public priorities. "Ongoing Monitoring and

Fig. 6.2 Strategies for ensuring inclusivity in AI

Auditing" highlights continuous bias checks and adaptive updates to keep AI systems equitable over time.

Ensuring inclusiveness in AI systems, particularly in city planning, requires a multifaceted approach that combines data diversity, fairness metrics, explainability, stakeholder engagement, and continuous monitoring. By adopting these strategies, city planners and AI developers can help mitigate bias, promote fairness, and create AI systems that serve all urban populations equitably. Inclusivity should be embedded in every stage of the AI development process, from data collection to algorithm design to deployment, ensuring that AI systems contribute to the creation of fairer and more just urban environments.

6.7 The Role of Explainability and Transparency

As AI systems become integral to city planning and urban development, ensuring that these systems are explainable and transparent is crucial for promoting fairness, accountability, and public trust. Explainability refers to the ability of AI systems to provide clear, understandable insights into how decisions are made, while transparency ensures that the internal workings of these systems are open and accessible to stakeholders. Together, these principles play a pivotal role in mitigating bias, fostering inclusivity, and ensuring that AI systems are aligned with societal values (Sargiotis 2025a, b).

6.7.1 Explainability in AI Systems

Explainability is essential for ensuring that AI systems are not "black boxes," whose internal workings are opaque to users and stakeholders (Baum et al. 2022). In city planning, where AI systems are often used to allocate resources, manage infrastructure, and optimize public services, explainability ensures that decision-makers and the public can understand the rationale behind AI-driven decisions. This is particularly important in high-stakes environments where biased or unfair decisions can have significant social and economic consequences. To enhance the explainability and transparency of AI-driven city planning systems, a Deep Reinforcement Learning (DRL) model incorporating Proximal Policy Optimization (PPO) is introduced. This model ensures that AI decision-making remains interpretable and traceable, allowing stakeholders to understand the rationale behind urban planning decisions while optimizing fairness and accountability.

The AI-driven model incorporating PPO to optimize explainability and transparency in urban AI systems is formulated as:

$$J(\theta) = \mathbb{E}\left[\sum_{t=0}^{T} \gamma^t (r_t + \lambda \log P(a_t|s_t))\right]$$

where $J(\theta)$ represents the explainability-optimized policy objective function, $\mathbb{E}$ denotes the expectation operator, γ is the discount factor controlling long-term AI learning, r_t is the reward function ensuring fair and interpretable decisions, λ is the explainability regularization parameter enforcing transparent decision-making, and $P(a_t| s_t)$ represents the probability of selecting action a_t in state s_t, ensuring interpretability.

To quantify how explainable and interpretable AI decisions are, an explainability reward function is introduced:

$$r_t = -w_1 D_{\text{op}}(s_t, a_t) + w_2 I(s_t, a_t)$$

where r_t represents the explainability-driven reward, $D_{\text{op}}(s_t, a_t)$ measures the opacity of the AI decision (i.e., lack of transparency), $I(s_t, a_t)$ quantifies interpretability of the AI action, and w_1, w_2 are weighting factors ensuring a trade-off between reducing opacity and enhancing interpretability.

To ensure that policy updates improve transparency while maintaining stability, the PPO policy optimization function is applied:

$$L(\theta) = \mathbb{E}[min(\rho_t A_t, \text{clip}(\rho_t, 1-\epsilon, 1+\epsilon)A_t)]$$

where $L(\theta)$ is the PPO function optimizing transparency in AI decisions, $\rho_t = \frac{\pi_\theta(a_t|s_t)}{\pi_{\theta_{\text{old}}}(a_t|s_t)}$ represents the probability ratio adjusting policy updates, A_t is the advantage function ensuring efficient learning of explainable policies, and ϵ is the clipping parameter preventing extreme fluctuations in policy optimization.

To measure the opacity of AI decisions and enforce transparency, the opacity function is defined as:

$$D_{\text{op}}(s_t, a_t) = \sum_{f=1}^{F} \mid \log P_f(a_t|s_t) - \log P_{\text{ref}}(a_t) \mid$$

where $D_{\text{op}}(s_t, a_t)$ quantifies decision opacity, $P_f(a_t| s_t)$ is the probability distribution of AI actions per feature f, $P_{\text{ref}}(a_t)$ is the reference probability distribution ensuring fair and transparent decisions, and F represents the total number of explainability features considered.

This AI-driven explainability framework continuously refines decision-making transparency in AI-based urban planning, ensuring that stakeholders can interpret and trust AI-generated recommendations. By integrating Proximal Policy

Optimization (PPO) with explainability-driven reward functions, this approach enforces transparency, mitigates bias, and fosters inclusive, accountable AI-driven urban governance.

Framework: Explainability-Aware Reinforcement Learning Model (EARL)

Layered AI Architecture for Explainability and Transparency in Urban AI Systems

Layer	Main functionality	AI module(s) and components	Operational Scope (Input → Processing → Output)
Layer 1	Data collection	Data preprocessing engine	Input → Raw sensor data, demographic data → Output → Preprocessed input for AI models
Layer 2	Fairness-aware learning	Proximal Policy Optimization (PPO), Reinforcement Learning (RL)	Input → Historical data, fairness metrics → Output → Policy updates maximizing fairness
Layer 3	Explainability reward calculation	Data preprocessing engine	Input → State-action pairs, fairness metrics → Output → Explainability-driven reward calculation
Layer 4	Policy optimization	Proximal Policy Optimization (PPO)	Input → Reward signals, policy updates → Output → Optimized AI model policy
Layer 5	Impact disparity calculation	Disparity analysis engine	Input → AI decisions across demographic groups → Output → Disparity metrics indicating fairness
Layer 6	Continuous monitoring and auditing	Transparency auditing and monitoring module	Input → AI model outputs, fairness metrics → Output → Bias audit reports and system evaluation
Layer 7	Human-in-the-loop feedback	Human feedback integration unit	Input → User feedback on AI decisions → Output → Refined AI strategies and performance adjustments
Layer 8	Model adaptation and retraining	Adaptive model retraining system	Input → Feedback data, performance anomalies → Output → Retrained models for continuous improvement

By making AI systems more interpretable, explainability helps to ensure that biases can be identified and corrected. For instance, if an AI system recommends prioritizing certain neighborhoods for development projects, explainability techniques can provide insights into why those neighborhoods were selected and whether the decision-making process considered the needs of all communities (Mehrabi et al. 2022). This allows city planners to scrutinize the decisions of AI systems and ensure that they are fair and equitable. To ensure that AI-driven urban planning decisions are fully interpretable, an XGBoost-based feature attribution model with SHAP explanations is introduced. This model identifies key decision variables and provides interpretable outputs for city planners to scrutinize AI-driven recommendations.

The AI-driven explainability function leveraging XGBoost for decision-making transparency is formulated as:

$$f_{\text{decision}}(X) = \sum_{t=1}^{T} (\eta f_t(X)) + \Omega(f_t)$$

where $f_{\text{decision}}(X)$ represents the AI-generated urban planning decision function, X denotes the set of input city planning features, $f_t(X)$ is the predictive function of each decision tree in the XGBoost model, η is the learning rate controlling model adaptability, and $\Omega(f_t)$ is the complexity penalty function that prevents overfitting.

To measure the contribution of each urban planning feature to the AI-driven decision, a SHAP value attribution function is introduced:

$$\phi_i = \sum_{S \subseteq N \setminus \{i\}} \frac{|S|!(N| - |S| - 1)!}{|N|!} [f(S \cup \{i\}) - f(S)]$$

where ϕ_i represents the SHAP value quantifying the importance of feature i, S denotes a subset of features excluding i, N represents the total set of features, and $f(S \cup \{i\}) - f(S))$ measures the marginal contribution of feature i to the final decision.

To capture interdependencies in urban planning decisions, a Graph Neural Network (GNN)-based spatial interaction function is formulated as:

$$h_v^{(l+1)} = \sigma\left(W^{(l)} \sum_{u \in \mathcal{N}(v)} \frac{h_u^{(l)}}{|\mathcal{N}(v)|} + b^{(l)}\right)$$

where $h_v^{(l+1)}$ is the node representation for location v at layer $l + 1$, $W^{(l)}$ is the learnable weight matrix at layer l, $\mathcal{N}(v)$ represents the neighboring locations influencing v, $h_u^{(l)}$ is the representation of neighboring location u at layer l, and σ is the activation function controlling information propagation.

This AI-driven explainability framework ensures that urban planning decisions are fully interpretable by leveraging SHAP-based feature attribution, XGBoost for decision transparency, and GNNs for spatial interaction modeling. These state-of-the-art techniques collectively enable city planners to scrutinize AI recommendations, ensuring that decisions are both fair and justifiable.

Framework: XGBoost & SHAP Explainability for Urban Planning (XGBoost-SHAP-UP)

Layered AI Architecture for Transparent and Fair Urban Planning Decisions

Layer	Main functionality	AI module (s) and components	Operational scope (Input → Processing → Output)
Layer 1	Data collection: Gather real-time city planning data	Urban data preprocessing engine	Input → Raw sensor data, urban feature data → Output → Preprocessed input for AI models

(continued)

Layer	Main functionality	AI module (s) and components	Operational scope (Input → Processing → Output)
Layer 2	Decision transparency: XGBoost for urban planning decisions	XGBoost	Input → City planning features → Output → Transparent decision-making function
Layer 3	Feature attribution: SHAP for feature importance	SHAP	Input → City planning features → Output → SHAP values for each feature's contribution
Layer 4	Spatial interaction modeling: GNN for urban interaction modeling	Graph Neural Networks (GNN)	Input → Spatial features of urban areas → Output → Node representations for locations
Layer 5	Auditing and compliance	Decision auditing and compliance module	Input → Model decisions, regulatory frameworks → Output → Audit results and regulatory compliance report

Explainable AI (XAI) techniques can take various forms, including visualizations, natural language explanations, and feature importance analyses. These techniques help stakeholders understand the factors that influenced an AI system's decisions, enabling them to assess whether the system is functioning as intended and whether its decisions align with broader ethical goals (Mökander 2023). In city planning, XAI can help to ensure that AI systems consider a wide range of factors, including socioeconomic and demographic data, when making decisions about resource allocation or infrastructure projects. To ensure that AI systems in urban planning are both interpretable and transparent, thereby allowing stakeholders to comprehend the factors influencing AI-driven decisions, we propose an advanced Explainable AI (XAI) framework. This framework integrates Heterogeneous Graph Neural Networks (HGNNs), post hoc explanation techniques, and counterfactual explanations to provide comprehensive insights into the decision-making processes of AI systems, particularly in the context of resource allocation and infrastructure projects that consider socioeconomic and demographic data.

To effectively capture the complex relationships inherent in urban environments, we employ Heterogeneous Graph Neural Networks (HGNNs). These networks are adept at modeling diverse types of nodes and edges, allowing for the integration of multi-modal data sources such as socioeconomic indicators, demographic statistics, and infrastructural attributes. The mathematical formulation of an HGNN layer is as follows:

$$h_v^{(l+1)} = \sigma\left(\sum_{r\in\mathcal{R}}\sum_{u\in\mathcal{N}_r(v)} \frac{1}{c_{v,r}} W_r^{(l)} h_u^{(l)}\right)$$

In this equation, $h_v^{(l+1)}$ denotes the updated feature representation of node v at layer $l + 1$, $\mathcal{R}$ represents the set of relation types, $\mathcal{N}_r(v)$ signifies the set of neighboring nodes connected to v via relation r, $c_{v,\ r}$, is a normalization constant specific to node $h_v^{(l+1)}$ and relation r, $W_r^{(l)}$ is the weight matrix associated with relation r at layer l, and σ is the activation function. This structure enables the model to assimilate information from various data modalities, reflecting the multifaceted nature of urban settings.

To elucidate the factors influencing AI decisions, we integrate post hoc explanation techniques that assess feature importance. One such method is integrated gradients, which attributes the prediction of the model to its input features by computing the path integral of the gradients along a straight-line path from a baseline input to the actual input. The integrated gradient for a feature x_i is computed as:

$$\text{IntegratedGradient}_i(x) = \left(x_i - x_i'\right) \times \int_{\alpha=0}^{1} \frac{\partial f(x' + \alpha \times (x - x'))}{\partial x_i} d\alpha$$

Here, x is the input, x' is the baseline input, α is a scaling factor, and f represents the model's output. This technique provides a quantitative measure of each feature's contribution to the model's decision, thereby enhancing transparency.

To provide stakeholders with actionable insights into AI decision-making, we incorporate counterfactual explanations. These explanations identify minimal changes to input features that could alter the AI system's decisions, thereby highlighting the factors that most significantly influence outcomes. The counterfactual generation process can be formulated as an optimization problem:

$$\underset{\delta}{min} \|\delta\|_1 \text{ subject to} f(x + \delta) \neq f(x)$$

In this context, δ represents the perturbation applied to the original input x, and f denotes the AI decision function. The objective is to find the smallest perturbation δ that changes the model's decision, thereby revealing which input features are pivotal in the decision-making process.

By unifying heterogeneous graph neural networks, post hoc explanation techniques, and counterfactual explanations, this explainable AI framework offers a robust and comprehensive approach to understanding AI-driven decisions in urban planning. These methodologies collectively ensure that AI systems consider a wide array of factors, including socioeconomic and demographic data, thereby aligning their decisions with broader ethical and societal goals.

Framework: Explainable AI (XAI) for Urban Planning Decisions (XAI-UP)

Layered AI Architecture for Explainable Urban Planning Decisions with HGNN, Integrated Gradients, and Counterfactual Explanations

Layer	Main functionality	AI module (s) and components	Operational scope (Input → Processing → Output)
Layer 1	Data collection: Gather multi-modal city planning data	Urban data preprocessing engine	Input → Raw sensor data, socioeconomic data, demographic data → Output → Preprocessed input for AI models
Layer 2	Decision transparency: HGNN for urban planning decisions	Heterogeneous Graph Neural Networks (HGNN)	Input → Urban feature data (socioeconomic, demographic, infrastructure) → Output → Transparent decision-making function
Layer 3	Feature attribution: Post hoc explanation with Integrated Gradients	Integrated gradients attribution module	Input → Model predictions → Output → Attribution of feature importance to decisions
Layer 4	Counterfactual explanations: Identifying minimal perturbations for decisions	Counterfactual explanation engine	Input → Model predictions, original input → Output → Smallest perturbation δ that changes the model's decision
Layer 5	Ethical governance: Auditing and compliance with regulations	Ethical auditing and compliance module	Input → Model decisions, regulatory frameworks → Output → Audit results and regulatory compliance report

6.7.2 *Transparency and Public Trust*

Transparency is closely related to explainability and refers to the openness and accessibility of AI systems. In the context of city planning, transparency is crucial for building public trust in AI systems. When the public is able to understand how AI systems make decisions, they are more likely to trust those decisions and support the deployment of AI in urban development (Sargiotis 2025b; Zowghi and da Rimini 2023). Transparency also enables stakeholders, including policymakers, community leaders, and citizens, to hold AI systems accountable for their decisions.

One of the main challenges in achieving transparency is the complexity of modern AI systems, particularly those based on deep learning. These systems often involve thousands or millions of parameters, making it difficult for non-experts to understand how decisions are made. To address this, AI developers must implement transparency strategies that make the decision-making process more accessible to a wider audience. This can involve providing high-level explanations of how AI systems work, as well as offering more detailed technical insights for those who require them (Baum et al. 2022). To enhance transparency in AI-driven urban planning systems and ensure that decisions are interpretable, accessible, and accountable, we propose an advanced Explainable AI (XAI) framework integrating Bayesian Deep Learning (BDL) for uncertainty quantification, Concept Bottleneck Models (CBMs) for human-understandable decision representations, and attention-based self-explaining neural networks to generate interpretable decisions. These methodologies collectively ensure that both non-experts and technical stakeholders can understand, trust, and evaluate AI-driven urban development decisions. One of the significant barriers to transparency in AI systems is the inherent uncertainty in deep learning models, which makes their decision-making difficult to interpret. To address this, Bayesian Deep Learning (BDL) is introduced to quantify uncertainty in AI-driven urban planning predictions, ensuring that decisions are trustworthy and interpretable. The Bayesian posterior distribution for AI model parameters is formulated as:

$$p(\theta|X, Y) = \frac{p(Y|X, \theta)p(\theta)}{p(Y|X)}$$

where $p(\theta| X, Y)$ represents the posterior distribution over model parameters given data X, Y, $p(Y| X, \theta)$ denotes the likelihood of observing outputs Y given inputs X and model parameters θ, $p(\theta)$ is the prior distribution over model parameters, and $p(Y| X)$ is the marginal likelihood that ensures model calibration. This approach ensures that urban planning decisions account for uncertainty, allowing policymakers and community leaders to assess the confidence level of AI-generated recommendations.

To further enhance transparency, Concept Bottleneck Models (CBMs) are introduced to explicitly separate human-understandable concepts from AI feature representations, ensuring that AI systems can explain why specific urban planning decisions are made. The CBM model representation is formulated as:

$$h_c = W_c X + b_c$$

where h_c represents the concept-level latent representation, W_c is the weight matrix mapping raw input features X to human-understandable concepts, and b_c is the bias term. By explicitly modeling human-interpretable urban development concepts such as "population density," "public transportation accessibility," and "income distribution," CBMs allow AI decisions to be traceable and explainable in terms that stakeholders can easily understand.

To further enhance interpretability, Self-Explaining Neural Networks (SENN) with attention mechanisms are introduced, ensuring that AI systems justify their decisions in a human-comprehensible manner. The self-explaining neural function is defined as:

$$y = \sum_{i=1}^{N} \alpha_i g(X_i)$$

where y represents the final decision output, αiαi denotes the learned attention weight assigned to input feature X_i, and $g(X_i)$ represents the local explanation function that determines the contribution of each feature to the decision. This formulation ensures that transparency is achieved by explicitly weighting and explaining how each urban factor contributes to AI-driven urban planning recommendations.

By using Deep Learning (BDL) for uncertainty quantification, Concept Bottleneck Models (CBMs) for human-interpretable decision representations, and Self-Explaining Neural Networks (SENN) for attention-based decision transparency, this XAI-driven transparency framework ensures that AI systems in urban planning are not only trustworthy and explainable but also accessible to policymakers, community leaders, and the public. These strategies address the complexity of modern AI models, ensuring that transparency in decision-making is both comprehensive and meaningful to all urban stakeholders.

Framework: Explainable AI (XAI) for Transparency in Urban Planning (XAI-TP-UP)

Layered AI Architecture for Transparent and Interpretable Urban Planning Decisions with BDL, CBMs, and SENN

Layer	Main functionality	AI module (s) and components	Operational scope (Input → Processing → Output)
Layer 1	Data collection: Gather multi-modal city planning data	Urban data preprocessing engine	Input → Raw sensor data, socioeconomic data, demographic data → Output → Preprocessed input for AI models
Layer 2	Decision transparency: Bayesian deep learning for uncertainty quantification	Bayesian Deep Learning (BDL)	Input → Urban planning data (socioeconomic, infrastructure, demographic) → Output → Uncertainty quantification for decision confidence
Layer 3	Concept bottleneck models: Human-understandable decision representations	Concept bottleneck representation module	Input → Raw input features → Output → Concept level latent representation for interpretable decisions
Layer 4	Self-explaining neural networks: Attention-based decision transparency	Self-Explaining Neural Networks (SENN)	Input → Model predictions → Output → Interpretable decision with weighted feature contributions
Layer 5	Ethical governance: Auditing and compliance with regulations	Ethical auditing and compliance module	Input → Model decisions, regulatory frameworks → Output → Audit results and regulatory compliance report

Transparency is fundamental to ensuring regulatory compliance, particularly in regions where data privacy laws, such as the General Data Protection Regulation

(GDPR), mandate that organizations provide clear explanations of how personal data is processed within AI systems (Mökander 2023). By enhancing AI transparency, city planners can guarantee adherence to legal frameworks while fostering ethical and fair applications of AI technologies in urban decision-making. The ability to interpret AI-driven recommendations allows stakeholders to scrutinize algorithmic decision processes, ensuring alignment with societal values and policy objectives.

To achieve a mathematically AI-driven framework that ensures transparency and regulatory compliance, an Explainable AI (XAI) model utilizing Shapley Additive Explanations (SHAP) is implemented. SHAP values provide an interpretable method for understanding the contributions of input features to AI model outputs, thereby enabling decision-makers to assess whether the model's decisions align with fairness and compliance requirements. The AI-driven model incorporating SHAP for explainability is formulated as:

$$\phi_i = \sum_{S \subseteq N \backslash \{i\}} \frac{|S|!(N| - |S| - 1)!}{|N|!} [f(S \cup \{i\}) - f(S)]$$

where ϕ_i, the Shapley value, represents the marginal contribution of feature i to the prediction, S, a subset of features excluding i, determines the groups of features used for evaluation, N, the set of all features, indicates the total feature space, $f(S)$, the model output with features in set S, provides the AI prediction given partial inputs. This equation quantifies the impact of each input variable on the AI model's decision-making process, allowing regulatory bodies and stakeholders to interpret which factors influenced a specific decision. By decomposing AI-generated predictions into their constituent contributions, the model ensures transparency, satisfying regulatory standards while improving trust in AI-driven urban analytics.

To further enhance compliance with privacy regulations, a Federated Learning (FL) framework is introduced. Federated learning ensures decentralized training of AI models while preserving data privacy by enabling models to learn from multiple sources without sharing raw data. The federated optimization objective is expressed as:

$$\min_{\mathrm{w}} \sum_{k=1}^{K} p_k F_k(\mathrm{w})$$

where w, the global model parameters, represent the shared AI model weights, k, an individual client node, corresponds to a data entity participating in model training, p_k, the weighting factor, accounts for the contribution of each client, $F_k(\mathrm{w})$, the local loss function, defines the optimization objective for each participating node. This function ensures that AI models comply with GDPR and similar regulations by allowing data to remain localized, mitigating risks associated with central data storage. Federated Learning prevents unauthorized access to sensitive data while maintaining robust model accuracy across distributed urban planning databases.

Via the combination of SHAP-based Explainable AI (XAI) and Federated Learning (FL), this AI-driven framework fosters compliance with legal and ethical standards, ensuring that AI models used in city planning remain interpretable, fair, and privacy-preserving. The SHAP model quantifies the contribution of each variable to AI predictions, facilitating transparent decision-making, while federated learning enforces decentralized model training, safeguarding data privacy. This dual-framework approach enhances the accountability of AI-driven urban analytics, ensuring that regulatory requirements are met without compromising the predictive power of advanced AI methodologies.

Framework: Explainable AI (XAI) for Transparency and Compliance in Urban Planning (XAI-TC-UP)

Layered AI Architecture for Transparency, Fairness, and Privacy Preservation with SHAP and Federated Learning (FL)

Layer	Main functionality	AI module (s) and components	Operational scope (Input → Processing → Output)
Layer 1	Data collection: Gather multi-modal city planning data	Urban data preprocessing engine	Input → Raw sensor data, socioeconomic data, demographic data → Output → Preprocessed input for AI models
Layer 2	Decision transparency: SHAP for explainable decision-making	Shapley Additive Explanations (SHAP)	Input → Urban planning data (socioeconomic, infrastructure, demographic) → Output → Transparent decision-making function
Layer 3	Federated learning for privacy-preserving model training	Federated Learning (FL)	Input → Local client data → Output → Collaborative model training with data privacy preservation
Layer 4	Ethical governance: Auditing and compliance with regulations	Ethical auditing and compliance module	Input → Model decisions, regulatory frameworks → Output → Audit results and regulatory compliance report

6.7.3 Mitigating Bias Through Explainability and Transparency

Bias in AI systems is one of the most significant ethical challenges facing city planners and policymakers. Without explainability and transparency, it is difficult to detect and mitigate bias, as the internal workings of AI systems remain hidden. Explainability and transparency allow stakeholders to scrutinize AI systems for potential biases and take corrective action when necessary (Mehrabi et al. 2022; Sargiotis 2024a, 2025a). For example, if an AI system disproportionately prioritizes wealthier neighborhoods for public service improvements, explainability techniques can reveal the underlying factors that contributed to this bias, allowing city planners to address the issue. To formalize bias mitigation, a fairness-constrained optimization model based on adversarial debiasing is introduced. This approach minimizes bias while preserving AI model accuracy by incorporating an adversarial network that learns to remove discriminatory features. The fairness-aware loss function is expressed as:

$$\min_{\mathrm{w}} \sum_{i=1}^{N} \ell(f_{\mathrm{w}}(x_i), y_i) + \lambda \cdot \mathbb{E}_{p(x)} \left[D_{\mathrm{KL}} \left(p(y|x) \middle\| p(y) \right) \right]$$

where w, the AI model parameters, represent the learned weights of the system, $\ell(f_{\mathrm{w}}(x_i), y_i)$, the classification loss function, ensures predictive accuracy, λ, the fairness regularization parameter, controls the trade-off between model accuracy and bias mitigation, D $D_{\mathrm{KL}}(p(y|x)\|p(y))$, the Kullback-Leibler (KL) divergence, measures the disparity between conditional and marginal outcome distributions. This function ensures that AI-driven urban planning decisions adhere to fairness constraints by adjusting the model's predictions to align with equitable distribution objectives. The adversarial debiasing network is trained simultaneously to detect and remove bias-related influences from input features, thereby improving fairness without compromising accuracy.

To further refine fairness-aware decision-making, an equalized odds fairness constraint is applied, ensuring that predictive outcomes remain independent of protected attributes. The fairness constraint is expressed as:

$$P(\hat{Y} = 1 | A = 0, Y = y) = P(\hat{Y} = 1 | A = 1, Y = y), \forall y \in \{0, 1\}$$

where $\hat{Y}$, the predicted outcome, denotes the AI system's classification result, A, the protected attribute, represents sensitive features such as race or socioeconomic status, Y, the actual label, corresponds to the ground truth for service allocation.

This fairness constraint ensures that the likelihood of a favorable AI decision remains equal across different demographic groups, preventing discriminatory biases in public service distribution.

By unifying adversarial debiasing with equalized odds fairness constraints, this AI-driven fairness model enhances explainability and transparency in AI decision-making. The adversarial network detects and removes biases from AI predictions, while fairness constraints ensure equitable resource allocation across all communities. This framework allows city planners to systematically identify, measure, and mitigate bias, ensuring ethical AI deployment in urban planning and public service allocation.

Framework: Bias Mitigation and Fairness-Constrained Optimization for Urban Planning (Bias-Mitigated-XAI)

Layered AI Architecture for Bias Mitigation and Fairness in AI-Driven Urban Planning with Adversarial Debiasing and Equalized Odds

Layer	Main functionality	AI module (s) and components	Operational scope (Input → Processing → Output)
Layer 1	Data collection: Gather multi-modal city planning data	Urban data preprocessing engine	Input → Raw sensor data, socio-economic data, demographic data → Output → Preprocessed input for AI models
Layer 2	Fairness-constrained optimization: Adversarial debiasing for bias mitigation	Adversarial debiasing optimization module	Input → Urban planning data → Output → Fairness-aware loss function incorporating bias mitigation
Layer 3	Fairness constraint: Equalized odds fairness for protected attribute independence	Equalized odds fairness constraint	Input → Model predictions → Output → Fairness constraint ensuring equality across protected attributes
Layer 4	Ethical governance: Auditing and compliance with regulations	Ethical auditing and compliance module	Input → Model decisions, regulatory frameworks → Output → Audit results and regulatory compliance report

To ensure that AI systems remain fair and equitable over time, transparency plays a crucial role in enabling continuous monitoring and auditing. Regular audits help detect instances of bias and discrimination, allowing AI models to be updated and retrained in response to evolving urban dynamics (Zowghi and da Rimini 2023). To formalize this process, a fairness-aware dynamic auditing model is introduced, integrating fairness constraints with Deep Reinforcement Learning (DRL) to adjust AI decisions based on real-time audit results.

Following data collection and preprocessing, fairness constraints are embedded into the decision-making process to prevent biased patterns from propagating into urban planning outputs. Within this framework, adversarial debiasing, implemented through fairness-constrained optimization, serves as the core mechanism for mitigating bias in AI-driven urban planning decisions. This method operates on the principle that fairness and accuracy should not be treated as mutually exclusive objectives, but as jointly optimized requirements within a balanced decision model. The framework adjusts the AI model to reduce bias while preserving predictive performance. The fairness-constrained optimization function quantifies this balance by introducing an adversarial network that detects and suppresses discriminatory features in the predictions. Specifically, the loss function incorporates both the classification loss and a fairness regularization term that minimizes the Kullback-Leibler (KL) divergence between the model's conditional and marginal outcome distributions.

The formalized mathematical structure is as follows:

$$\min_{w} \sum_{i=1}^{N} l(f_w(x_i), y_i) + \lambda \cdot \mathbb{E}_{p(x)}[D_{\mathrm{KL}}(p(y|x) \parallel p(y))]$$

where w, the AI model parameters, represent the learned weights of the system, $l(f_w(x_i), y_i)$, the classification loss function, ensures predictive accuracy, λ, the fairness regularization parameter, controls the trade-off between model accuracy and bias mitigation, $D_{\mathrm{KL}}(p(y|x) \parallel p(y))$, the Kullback-Leibler (KL) divergence, measures the disparity between conditional and marginal outcome distributions. This function optimizes the AI system by adjusting predictions to meet fairness constraints while preserving accuracy, ensuring that urban planning decisions are equitable across diverse demographic groups.

The next critical step is to ensure that the AI model can continuously adapt to real-time fairness assessments and audit results. In dynamic environments like urban planning, fairness must remain a priority, even as urban dynamics evolve. This can be achieved by employing Proximal Policy Optimization (PPO), a reinforcement learning technique, to dynamically adjust AI decisions based on the outcomes of regular fairness audits.

PPO is particularly suited for this task because it allows the AI system to refine its policy over time based on feedback, ensuring that the fairness of its decisions improves continually. The advantage function in PPO measures how much better a particular action is, helping the model determine which actions will improve fairness. This continuous optimization process ensures that fairness is maintained even as the model encounters new data and feedback from urban planners or stakeholders.

$$J(\theta) = \mathbb{E}\left[\min\left(\frac{\pi_\theta(a_t|s_t)}{\pi_{\theta_{\mathrm{old}}}(a_t|s_t)}A_t, \mathrm{clip}\left(\frac{\pi_\theta(a_t|s_t)}{\pi_{\theta_{\mathrm{old}}}(a_t|s_t)}, 1-\epsilon, 1+\epsilon\right)A_t\right)\right]$$

where $J(\theta)$, the PPO objective function, represents the optimization goal for fair AI decision-making, $\pi_\theta(a_t| s_t)$, the updated AI policy, determines the probability of selecting action a_t given state s_t, $\pi_{\theta_{\mathrm{old}}}(a_t|s_t)$, the previous AI policy, represents past decision probabilities, A_t, the advantage function, quantifies the benefit of taking action a_t in state s_t, ϵ, the policy clipping parameter, prevents excessive policy updates, ensuring stable learning. This function ensures that AI decisions remain dynamically fair by continuously adjusting policies based on fairness audits. The advantage function evaluates whether adjustments improve fairness outcomes, while the policy ratio regulates updates, preventing abrupt changes that could destabilize AI-driven decision-making.

To enhance audit-driven fairness improvements, an adaptive fairness penalty function is integrated into the learning framework:

$$L_{\text{fair}} = \sum_{i=1}^{N} \mid P(\hat{Y} = 1|A = 0) - P(\hat{Y} = 1|A = 1) \mid$$

where L_{fair}, the fairness loss function, quantifies disparities in AI predictions, $P(\hat{Y} = 1|A = 0)$, the probability of a positive outcome for the non-protected group, measures AI decisions for one demographic, $P(\hat{Y} = 1|A = 1)$, the probability of a positive outcome for the protected group, measures fairness across demographics. This function penalizes AI models when significant disparities exist between protected and non-protected groups, ensuring that fairness gaps are continuously monitored and minimized.

Via the application of Proximal Policy Optimization (PPO) with fairness-aware auditing, this AI-driven model ensures that AI systems undergo continuous evaluation and adaptive updates to maintain fairness and inclusivity. The PPO-based policy optimization refines AI fairness decisions dynamically, while the fairness penalty function ensures that disparities are systematically addressed. This dual-framework approach empowers city planners to embed explainability and transparency into AI-driven urban management, mitigating bias and promoting equitable decision-making.

Framework: Fairness-Aware Dynamic Auditing for Urban Planning AI (Fairness-Aware-XAI)

Layered AI Architecture for Fairness Mitigation and Dynamic Auditing in AI-Driven Urban Planning with PPO and Fairness Penalties

Layer	Main functionality	AI module (s) and components	Operational scope (Input → Processing → Output)
Layer 1	Data collection: Gather multi-modal city planning data	Urban data preprocessing engine	Input → Raw sensor data, socio-economic data, demographic data → Output → Preprocessed input for AI models
Layer 2	Fairness-constrained optimization: Adversarial debiasing for bias mitigation	Adversarial debiasing optimization module	Input → Urban planning data → Output → Fairness-aware loss function incorporating bias mitigation
Layer 3	PPO-based fairness-aware auditing: Dynamic policy adjustment for fairness	Proximal Policy Optimization (PPO)	Input → Model predictions, fairness audits → Output → Optimized policies with fairness adjustments
Layer 4	Fairness constraint: Adaptive fairness penalty function for bias reduction	Adaptive fairness penalty engine	Input → Model predictions → Output → Fairness penalty for minimizing disparities between demographic groups
Layer 5	Ethical governance: Auditing and compliance with regulations	Ethical auditing and compliance module	Input → Model decisions, regulatory frameworks → Output → Audit results and regulatory compliance report

Fig. 6.3 The role of explainability and transparency in AI

Figure 6.3 illustrates the role of explainability and transparency in AI systems, organized into three main areas: Explainability, Transparency, and Mitigating Bias. "Explainability" focuses on understanding AI decisions, preventing "black-box" issues, detecting bias, and using Explainable AI (XAI) techniques like visuals and simple explanations. "Transparency" emphasizes building public trust, providing clear and accessible insights, and ensuring legal compliance with privacy standards. "Mitigating Bias" highlights the importance of regular audits to identify and correct biases, ensuring fairness in AI systems over time. Together, these components promote accountable, fair, and trusted AI in urban planning and city development.

Explainability and transparency are essential for ensuring that AI systems in city planning are fair, inclusive, and aligned with societal values. These principles help to build public trust in AI systems, foster accountability, and enable stakeholders to detect and mitigate bias. As AI continues to play an increasingly important role in shaping urban environments, it is crucial that explainability and transparency are prioritized in the design, deployment, and regulation of AI systems. Through these strategies, city planners can ensure that AI systems contribute to the creation of fairer, more equitable urban environments.

6.8 Challenges and Opportunities

As artificial intelligence (AI) becomes more integrated into city planning and urban development, both challenges and opportunities emerge. AI has the potential to transform how cities are planned and managed, enabling more efficient resource allocation, improved infrastructure, and better public service delivery. However, the deployment of AI in urban settings is fraught with technical, ethical, and social challenges that must be carefully navigated to fully realize its potential.

6.8.1 Challenges

One of the most significant challenges in AI-driven city planning is addressing bias in data and algorithms. As highlighted by Mehrabi et al. (2022), biased data, often reflecting historical inequalities, can lead to algorithms that reinforce these biases, resulting in discriminatory outcomes for certain communities. This is particularly concerning in city planning, where decisions about resource allocation, infrastructure development, and public services can disproportionately affect marginalized or underserved populations. Algorithms trained on biased data may, for instance, prioritize wealthier neighborhoods for development projects while neglecting lower-income areas, perpetuating social inequalities (Pagano et al. 2023). Ensuring that AI systems are trained on diverse and representative datasets is crucial for mitigating bias and promoting fairness.

To address this challenge, a multi-layered AI framework is proposed, incorporating state-of-the-art fairness and transparency techniques. The framework integrates Fair Representation Learning, Fair Classification, Proximal Policy Optimization (PPO) for continuous policy updates, and Explainable AI (XAI) techniques for transparency and interpretability. This model is designed not only to support fairer AI decisions, but also to make the reasoning behind these decisions understandable and accessible to stakeholders. By embedding explainability and fairness into the decision-making process, the proposed framework allows urban planners to make more equitable decisions regarding resource distribution, public services, and infrastructure development.

The first step in mitigating bias is to ensure that the AI system learns fair representations of urban data. This is achieved through Fair Representation Learning using Variational Autoencoders (VAEs). VAEs are designed to learn latent representations that reduce the influence of sensitive attributes such as race, gender, and income on the underlying data structure. This supports the development of fairer representations and reduces the risk of perpetuating existing social inequalities.

The VAE loss function, which optimizes the fairness of the learned representations, is formulated as:

$$L_{\mathrm{VAE}} = \mathbb{E}_{q_\phi(z|x)}[\log p_\theta(x|z)] - \beta D_{\mathrm{KL}}\left(q_\phi(z|x)\middle\|p(z)\right)$$

where L_{VAE}, the variational autoencoder loss function, optimizes fair data representation, $q_\phi(z|x)$, the encoder distribution, maps input data x to a latent representation z, $p_\theta(x|z)$, the decoder distribution, reconstructs x from the latent representation z, $D_{\mathrm{KL}}(q_\phi(z|x)\|p(z))$, the Kullback-Leibler (KL) divergence, regularizes the latent space to prevent biased feature learning, β, the weighting parameter, controls the trade-off between reconstruction accuracy and latent space fairness. This function ensures that urban datasets are transformed into bias-free representations, allowing AI models to make equitable decisions. By optimizing the latent space distribution, the VAE prevents historical biases from influencing AI-driven planning strategies.

Once fair representations of urban data are learned, it is crucial to ensure that AI predictions and classifications are also fair. This is achieved through demographic parity regularization, a fairness constraint that ensures AI decisions are not influenced by protected attributes (e.g., race, gender, and income level). The goal is to ensure that the likelihood of a favorable decision (e.g., resource allocation or public service distribution) is equal across different demographic groups.

The fairness loss function for demographic parity is defined as:

$$L_{\mathrm{fair}} = \mid P(Y^* = 1|A = 0) - P(Y^* = 1|A = 1) \mid$$

where L_{fair}, the fairness loss function, penalizes disparities in decision outcomes, $P(Y^* = 1|A = 0)$, the probability of a positive outcome for the non-protected group, represents AI decision probabilities for one demographic, $P(Y^* = 1|A = 1)$, the probability of a positive outcome for the protected group, ensures that predictions remain unbiased. This function enforces equal treatment across demographic groups, ensuring that AI decisions do not perpetuate historical biases or inequalities.

Ensuring fairness in AI-driven decisions is an ongoing process. To address this, Proximal Policy Optimization (PPO) is employed to dynamically adjust AI decisions in response to real-time fairness audits and feedback loops. PPO is a reinforcement learning technique that allows for continuous policy adjustments, ensuring that the AI model remains fair over time while maintaining accuracy.

The objective function for PPO-based fairness-aware auditing is:

$$J(\theta) = \mathbb{E}\left[min\left(\frac{\pi_\theta(a_t|s_t)}{\pi_{\theta_{\mathrm{old}}}(a_t|s_t)}A_t, \mathrm{clip}\left(\frac{\pi_\theta(a_t|s_t)}{\pi_{\theta_{\mathrm{old}}}(a_t|s_t)}, 1-\epsilon, 1+\epsilon\right)A_t\right)\right]$$

where $J(\theta)$, the PPO objective function, represents the optimization goal for fair AI decision-making, $\pi_\theta(a_t|s_t)$, the updated AI policy, determines the probability of selecting action a_t given state s_t, $\pi_{\theta_{\mathrm{old}}}(a_t|s_t)$, the previous AI policy, represents past decision probabilities, A_t, the advantage function, quantifies the benefit of taking action a_t in state s_t, ϵ, the policy clipping parameter, prevents excessive policy updates, ensuring stable learning. This function allows for the dynamic adjustment of the AI model's policy, ensuring that fairness is continually optimized in response to real-time audits and feedback.

After ensuring compliance with regulatory frameworks, the next critical step is to ensure that AI decisions are not only fair but also transparent and interpretable. Explainable AI (XAI) plays a pivotal role here, providing stakeholders with the ability to understand and scrutinize the decisions made by the AI system. This is essential for fostering trust in AI-driven systems, particularly in the context of urban planning, where decisions about resource allocation and public service distribution have significant societal impacts. With XAI, the AI model's decision-making process becomes transparent, allowing policymakers, city planners, and affected communities to see how and why specific decisions were made.

In this context, XAI techniques such as SHAP (Shapley Additive Explanations) or LIME (Local Interpretable Model-agnostic Explanations) are often used to provide clear explanations of the features that influenced the AI's decisions. This helps to ensure that the fairness mechanisms implemented in earlier stages (like Fair Representation Learning and Fair Classification) are not just abstract optimizations but are also understandable and transparent.

The SHAP value formulation for explainability, which helps quantify the contribution of each feature to the AI's decision, is given by:

$$\phi_i = \sum_{S \subseteq N \setminus \{i\}} \frac{|S|!(|N| - |S| - 1)!}{|N|!} [f(S \cup \{i\}) - f(S)]$$

where ϕ_i, the SHAP value, represents the marginal contribution of feature i, S, a subset of features excluding i, determines the groups of features used for evaluation, N, the set of all features, indicates the total feature space, $f(S)$, the model output with features in set S, provides the AI prediction given partial inputs. This function calculates the contribution of each feature to the overall decision made by the AI model, providing a detailed explanation of how specific features influenced the final outcome, thus enhancing the transparency and trust in AI-driven decisions.

The proposed AI-driven model demonstrates a comprehensive approach to addressing bias and promoting fairness in urban planning through the integration of various advanced AI techniques. At its core, the framework leverages Variational Autoencoders (VAEs) for fair representation learning, which ensures that the urban data used for decision-making is free from bias-inducing attributes such as race or socioeconomic status. By learning unbiased latent representations, the model prevents historical inequalities from influencing the AI's predictions. Further, the demographic parity regularization enforces fairness by ensuring that the probability of favorable outcomes, such as resource allocation or infrastructure development, is independent of protected demographic attributes. To maintain fairness dynamically over time, the model integrates Proximal Policy Optimization (PPO), enabling real-time adjustments based on continuous feedback and fairness audits. This approach ensures that AI decisions evolve in response to changing urban dynamics, always striving to uphold fairness and equity. The framework also includes Explainable AI (XAI) techniques, which make the model's decision-making processes transparent and interpretable, ensuring that stakeholders can understand and trust the AI's actions. This combination of fairness-enhancing algorithms, dynamic optimization, and transparency mechanisms provides a robust foundation for AI systems that promote equity and inclusiveness in urban governance, resource distribution, and infrastructure planning.

Framework: AI-Driven Fairness and Transparency in Urban Planning (AI-FTUP1)

Layered AI Architecture for Bias Mitigation and Fairness in AI-Driven Urban Planning with VAEs, Demographic Parity Regularization, PPO, and XAI

Layer	Main functionality	AI module (s) and components	Operational scope (Input → Processing → Output)
Layer 1	Data collection: Gather multi-modal city planning data	Urban data preprocessing engine	Input → Raw sensor data, socioeconomic data, demographic data → Output → Preprocessed input for AI models
Layer 2	Fair representation learning: VAE for debiasing urban datasets	Variational Autoencoders (VAEs)	Input → Urban data (socioeconomic, infrastructure, demographic) → Output → Fair latent representations free from bias
Layer 3	Fair classification: Demographic parity regularization for fair outcomes	Demographic parity regularization	Input → Model predictions → Output → Fairness penalty ensuring equal outcomes across protected and non-protected groups
Layer 4	PPO-based fairness-aware auditing: Dynamic policy adjustment for fairness	Proximal Policy Optimization (PPO)	Input → Model predictions, fairness audits → Output → Optimized policies with fairness adjustments based on real-time audits
Layer 5	Ethical governance: Auditing and compliance with regulations	Ethical auditing and compliance module	Input → Model decisions, regulatory frameworks → Output → Audit results and regulatory compliance report
Layer 6	Explainable AI: Ensuring transparency in decision-making	Model explainability and transparency engine	Input → AI decision outcomes → Output → Interpretable and transparent explanations of AI-driven decisions

Another challenge is the complexity and opacity of AI systems, particularly those based on deep learning. These systems are often "black boxes," making it difficult for stakeholders to understand how decisions are made. As Baum et al. (2022) point out, the lack of explainability in AI systems can undermine public trust and hinder the adoption of AI technologies in city planning. Furthermore, without transparency, it becomes difficult to identify and address biases or errors in AI-driven decisions. The need for explainability is especially pressing in high-stakes environments such as city planning, where the consequences of AI-driven decisions can have significant social and economic implications. To address these challenges, the proposed framework introduces a series of AI-driven techniques designed to support transparency, fairness, and interpretability. This framework incorporates several key components: Fair Representation Learning, Fair Classification, Proximal Policy Optimization

(PPO) for fairness-aware auditing, and Explainable AI (XAI) techniques for interpretability. Together, these layers provide an integrated framework for mitigating bias, promoting fairness, and enhancing the transparency of AI-driven urban planning decisions. Below, we present the detailed mathematical formulations for these layers, showing how each component contributes to the broader goal of a transparent, accountable, and fair urban planning system.

The first step in supporting fairness is to enable the AI system to learn fairer representations of urban data. This is achieved through Fair Representation Learning using Variational Autoencoders (VAEs). The VAE framework aims to learn latent representations that reduce the influence of sensitive attributes such as race, gender, and income on the underlying data structure. By doing so, the model reduces the likelihood that these attributes will shape downstream decision-making processes and helps limit the reinforcement of historical inequalities.

The VAE loss function is defined as follows:

$$L_{\text{VAE}} = \mathbb{E}_{q_\phi(z|x)}[\log p_\theta(x|z)] - \beta D_{\text{KL}}(q_\phi(z|x) \parallel p(z))$$

where L_{VAE}, the variational autoencoder loss function, optimizes fair data representation, $q_\phi(z|x)$, the encoder distribution, maps input data x to a latent representation z, $p_\theta(x|z)$, the decoder distribution, reconstructs x from the latent representation z, $D_{\text{KL}}(q_\phi(z|x) \parallel p(z))$, the Kullback-Leibler (KL) divergence, regularizes the latent space to prevent biased feature learning, β, the weighting parameter, controls the trade-off between reconstruction accuracy and latent space fairness.

Once the data has been debiased through fair representation learning, the next step is to ensure that the AI model's predictions are also fair. This is achieved through demographic parity regularization, which ensures that AI predictions are independent of protected attributes such as race, gender, and socioeconomic status. The goal is to ensure that the likelihood of a favorable AI decision (e.g., resource allocation or public service distribution) is equal across demographic groups, thus ensuring fair treatment for all.

The fairness loss function for demographic parity regularization is:

$$L_{\text{fair}} = | P(Y^* = 1|A = 0) - P(Y^* = 1|A = 1) |$$

where L_{fair}, the fairness loss function, penalizes disparities in decision outcomes, $P(Y^* = 1|A = 0)$, the probability of a positive outcome for the non-protected group, represents AI decision probabilities for one demographic, P $P(Y^* = 1|A = 1)$, the probability of a positive outcome for the protected group, ensures that predictions remain unbiased.

To ensure continuous fairness in AI-driven decisions, the model incorporates Proximal Policy Optimization (PPO), which allows the AI system to adjust its policies dynamically based on real-time audits. PPO is a reinforcement learning algorithm that optimizes the model's policy by adjusting its decision-making process in a way that preserves stability while improving fairness.

The PPO objective function is given by:

$$J(\theta) = \mathbb{E}\left[min\left(\frac{\pi_\theta(a_t|s_t)}{\pi_{\theta_{\text{old}}}(a_t|s_t)}A_t, \text{clip}\left(\frac{\pi_\theta(a_t|s_t)}{\pi_{\theta_{\text{old}}}(a_t|s_t)}, 1-\epsilon, 1+\epsilon\right)A_t\right)\right]$$

where $J(\theta)$, the PPO objective function, represents the optimization goal for fair AI decision-making, $\pi_\theta(a_t|s_t)$, the updated AI policy, determines the probability of selecting action a_t given state s_t, $\pi_{\theta_{\text{old}}}(a_t|s_t)$, the previous AI policy, represents past decision probabilities, A_t, the advantage function, quantifies the benefit of taking action a_t in state s_t, ϵ, the policy clipping parameter, prevents excessive policy updates, ensuring stable learning.

After implementing fairness mechanisms, the next crucial step is to ensure transparency in decision-making. Explainable AI (XAI) provides the necessary interpretability by making it possible for stakeholders to understand the reasoning behind AI-driven decisions. This transparency is particularly important in high-stakes environments like urban planning, where AI decisions can have significant social and economic consequences. Self-attention mechanisms can identify the most influential features in AI decision-making, ensuring that stakeholders can easily interpret which features contributed most to the final decision.

The attention-weighted feature importance function is mathematically formalized as:

$$\alpha_i = \frac{\exp(e_i)}{\sum_{j=1}^{N} \exp(e_j)}$$

where α_i, the attention weight, represents the importance assigned to feature i, e_i, the attention score, quantifies the relevance of feature i in the AI decision process, N, the total number of features, determines the feature space, exp $(\cdot)$, the exponential function, ensures positive attention values. This function ensures that AI systems highlight the most influential features in decision-making, allowing stakeholders to interpret the reasoning behind AI-driven urban planning recommendations.

Additionally, integrated gradients are used to provide a gradient-based attribution of features, enabling deeper interpretability by quantifying each feature's contribution to the AI decision:

$$\phi_i = (x_i - x_i')\int_{\alpha=0}^{1} \frac{\partial f(x' + \alpha(x - x'))}{\partial x_i} d\alpha$$

where ϕ_i, the attribution score, quantifies the contribution of feature i to the AI model's decision, x_i, the original feature value, represents the input variable in urban planning datasets, x_i', the baseline input value, provides a reference point for measuring feature importance, $f(x)$, the deep learning model, computes predictions based on urban data, $\partial f/\partial x_i$, the partial derivative, measures sensitivity of predictions to changes in x_i, α, the scaling factor, integrates feature contributions over an input range, $d\alpha$ defines the infinitesimal step in the continuous summation process, ensuring a smooth integration over the path from the baseline to the actual input. This function ensures that AI-driven city planning models provide an interpretable breakdown of how each input feature influences predictions, allowing for transparency and bias detection.

To ensure that AI-driven decisions in urban planning remain fair and equitable, it is essential to continuously detect and mitigate biases that may arise due to historical inequalities or imbalanced data distributions. In the context of urban planning, where decisions can disproportionately affect certain populations, addressing bias is paramount. Adversarial debiasing offers an advanced technique for removing bias by using an adversarial network that actively learns to detect and reduce discriminatory features in AI predictions. By introducing an adversarial objective, the model is encouraged to minimize bias while preserving prediction accuracy, ensuring that protected groups are treated fairly in decision-making processes.

The adversarial debiasing loss function is introduced to formalize this process:

$$L_{\text{debias}} = L_{\text{predict}} + \lambda \cdot D_{\text{adv}}(x)$$

where L_{debias}, the debiasing loss function, combines predictive accuracy with fairness objectives, L_{predict}, the predictive loss, ensures the model's performance on the main task, λ, the regularization parameter, controls the trade-off between prediction accuracy and fairness, $D_{\text{adv}}(x)$, the adversarial loss, represents the measure of bias in the model's predictions, minimizing it to promote fairness in outcomes. This function aims to reduce the influence of biased features in AI-drive AI-driven decisions, ensuring that urban planning models remain fair and inclusive. Through this adversarial framework, we can ensure that AI-driven urban planning systems are not only accurate but also free from discriminatory influences, leading to more equitable resource allocation and social equity in city development.

The proposed AI-driven model outlines a comprehensive framework designed to address the challenges of bias, complexity, and opacity in urban planning systems. By integrating techniques such as Fair Representation Learning through Variational Autoencoders (VAEs), the model ensures that the data used in decision-making is free from bias-inducing attributes like race, gender, and socioeconomic status. This helps in constructing a fairer foundation for AI predictions. To further enhance fairness, demographic parity regularization is applied, ensuring that predictions are independent of sensitive attributes, thus preventing discriminatory outcomes. The model incorporates Proximal Policy Optimization (PPO), which dynamically adjusts policies to ensure that fairness is maintained over time, adapting to real-time audits. Crucially, Explainable AI (XAI) is integrated to enhance the transparency of decisions. By using self-attention mechanisms and integrated gradients, the model identifies the most influential features in decision-making and provides a breakdown of their contributions, ensuring that stakeholders can understand and trust the AI system's outputs. Together, these components create a robust system that is not only fair and transparent but also interpretable, fostering public trust and facilitating the adoption of AI technologies in urban governance. This framework allows city planners and stakeholders to ensure that AI-driven decisions are ethical, understandable, and aligned with societal values.

Framework: AI-Driven Fairness and Transparency in Urban Planning (AI-FTUP2)

Layered AI Architecture for Bias Mitigation, Fairness in AI-Driven Urban Planning, VAEs, Demographic Parity Regularization, PPO, XAI, Bias Detection, and Auditing

Layer	Main functionality	AI module (s) and components	Operational scope (Input → Processing → Output)
Layer 1	Data collection: Gather multi-modal city planning data	Urban data preprocessing engine	Input → Raw sensor data, socioeconomic data, demographic data → Output → Preprocessed input for AI models
Layer 2	Fair representation learning: VAE for debiasing urban datasets	Variational Autoencoders (VAEs)	Input → Urban data (socioeconomic, infrastructure, demographic) → Output → Fair latent representations free from bias
Layer 3	Fair classification: Demographic parity regularization for fair outcomes	Demographic parity regularization	Input → Model predictions → Output → Fairness penalty ensuring equal outcomes across protected and non-protected groups
Layer 4	PPO-based fairness-aware auditing: Dynamic policy adjustment for fairness	Proximal Policy Optimization (PPO)	Input → Model predictions, fairness audits → Output → Optimized policies with fairness adjustments based on real-time audits
Layer 5	Ethical governance: Auditing and compliance with regulations	Ethical auditing and compliance module	Input → Model decisions, regulatory frameworks → Output → Audit results and regulatory compliance report
Layer 6	Explainable AI: Ensuring transparency in decision-making	Model explainability and transparency engine	Input → AI decision outcomes → Output → Interpretable and transparent explanations of AI-driven decisions
Layer 7	Bias detection and mitigation: Continuously monitor for biases in AI-driven decisions	Adversarial bias detection module	Input → Model predictions → Output → Detect and remove biases in predictions by using adversarial techniques
Layer 8	Real-time model auditing: Dynamic auditing and real-time feedback mechanisms	Real-time auditing and feedback system	Input → Model predictions, feedback → Output → Continuous feedback and auditing for fairness, transparency, and bias detection

In the context of urban planning, the deployment of AI technologies faces numerous regulatory and legal challenges, particularly when it comes to the use of personal data. Data privacy laws, such as the General Data Protection Regulation

(GDPR) in Europe, impose strict requirements on the collection, storage, and use of personal data (Mökander 2023). As city planners and policymakers adopt AI-driven solutions, it becomes imperative to ensure that AI models comply with these privacy regulations without sacrificing model effectiveness. The ethical use of personal data is essential for maintaining public trust, ensuring that AI systems operate transparently and safeguard citizens' rights.

To address the challenge of privacy preservation while maintaining AI model effectiveness, the proposed framework introduces a privacy-preserving AI compliance model. This model integrates Differential Privacy (DP) and Federated Learning (FL) techniques to support compliance with data privacy laws such as the GDPR. The goal is to balance the functionality of AI systems with ethical data use, safeguarding privacy while still enabling effective urban decision-making.

The first key technique employed in this framework is Differential Privacy (DP), a method for protecting individual data contributions during the training of AI models. By introducing calibrated noise into data analysis or model training, DP limits the extent to which the presence or absence of any individual data point can be inferred, thereby reducing re-identification risks and strengthening citizen privacy. The differential privacy guarantee is mathematically formalized to ensure that the output of the AI model remains statistically similar whether a specific data point is included in or excluded from the dataset. The privacy budget ϵ is used to quantify the allowable privacy loss, limiting the degree to which sensitive individual information can be inferred from model outputs. The differential privacy guarantee is formulated as:

$$P(M(D) \in S) \leq e^{\epsilon} P(M(D') \in S)$$

where $P(M(D) \in S)$, the probability of the AI model output belonging to a set S, represents the likelihood of a specific outcome under dataset D, $M(D)$, the AI model mechanism, maps input data D to an output prediction, D', a neighboring dataset, represents a slightly altered version of D, ϵ, the privacy budget, quantifies the allowable privacy loss, ensuring that individual data cannot be re-identified. This function ensures that AI-driven urban planning decisions remain privacy-preserving, preventing unauthorized access to individual-level data while still allowing for data-driven policy formulation.

To further enhance privacy protection, Federated Learning (FL) is incorporated into the framework. FL ensures that the AI model is trained across multiple decentralized data sources without the need to transfer raw data, thus protecting data privacy. In this framework, the data remains on local devices or institutions, and only model updates are shared. This ensures compliance with data privacy laws such as GDPR, as the model can learn from distributed data without centralizing sensitive personal information. The federated learning optimization objective is formulated as:

$$\min_{\mathrm{w}} \sum_{k=1}^{K} p_k F_k(\mathrm{w})$$

where w, the global AI model parameters, represent the shared model weights across decentralized learning nodes, k, an individual client or local entity, contributes to model training, p_k, the weighting factor, accounts for the relative importance of each local dataset, $F_k(\mathrm{w})$, the local loss function, optimizes AI predictions at each node. This function ensures that AI models comply with data privacy laws by allowing AI systems to learn from distributed data without centralizing sensitive citizen information.

While Differential Privacy (DP) and Federated Learning (FL) ensure that AI systems respect privacy and comply with regulations, they do not address transparency, a critical issue for public trust. In high-stakes applications like urban planning, the opacity of AI decision-making can raise concerns about fairness and accountability. To bridge this gap, the proposed model integrates Explainable AI (XAI), which allows stakeholders to understand the rationale behind AI-driven decisions. By introducing self-attention mechanisms and integrated gradients, the model ensures that AI decisions are not only privacy-preserving but also interpretable.

The attention-weighted feature importance function is introduced to quantify the importance of individual features in the decision-making process:

$$\alpha_i = \frac{\exp(e_i)}{\sum_{j=1}^{N} \exp(e_j)}$$

where α_i, the attention weight, represents the importance assigned to feature i, e_i, the attention score, quantifies the relevance of feature i in the AI decision process, N, the total number of features, determines the feature space, exp $(\cdot)$, the exponential function, ensures positive attention values. This function enables the AI model to highlight the most influential features, making the decision-making process more transparent.

Additionally, Integrated Gradients (IG) is employed to provide a gradient-based attribution of features, enabling deeper interpretability by quantifying each feature's contribution to the model's final decision:

$$\phi_i = (x_i - x_i') \int_{\alpha=0}^{1} \frac{\partial f(x' + \alpha(x - x'))}{\partial x_i} d\alpha$$

where ϕ_i, the attribution score, quantifies the contribution of feature i to the AI model's decision, x_i, the original feature value, represents the input variable in urban planning datasets, x_i', the baseline input value, provides a reference point for measuring feature importance, $f(x)$, the deep learning model, computes predictions based on urban data, $\partial f / \partial x_i$, the partial derivative, measures sensitivity of predictions to changes in x_i, α, the scaling factor, integrates feature contributions over an input range, $d\alpha$ defines the infinitesimal step in the continuous summation process, ensuring a smooth integration over the path from the baseline to the actual input. This function ensures that the contribution of each input feature is clearly understood, enabling transparency in the AI model's decision-making process.

To ensure that AI-driven decisions in urban planning remain fair and equitable, it is essential to continuously detect and mitigate biases that may arise due to historical inequalities or imbalanced data distributions. In the context of urban planning, where decisions can disproportionately affect certain populations, addressing bias is paramount. Adversarial debiasing offers an advanced technique for removing bias by using an adversarial network that actively learns to detect and reduce discriminatory features in AI predictions. By introducing an adversarial objective, the model is encouraged to minimize bias while preserving prediction accuracy, ensuring that protected groups are treated fairly in decision-making processes.

The adversarial debiasing loss function is introduced to formalize this process:

$$L_{\text{debias}} = L_{\text{predict}} + \lambda \cdot D_{\text{adv}}(x)$$

where L_{debias}, the debiasing loss function, combines predictive accuracy with fairness objectives, L_{predict}, the predictive loss, ensures the model's performance on the main task, λ, the regularization parameter, controls the trade-off between prediction accuracy and fairness, $D_{\text{adv}}(x)$, the adversarial loss, represents the measure of bias in the model's predictions, minimizing it to promote fairness in outcomes. This function ensures that AI-driven urban planning decisions remain fair and inclusive by reducing the influence of biased features on the predictions. Through this adversarial framework, we can ensure that AI-driven urban planning systems are not only accurate but also free from discriminatory influences, leading to more equitable resource allocation and social equity in city development.

The proposed AI-driven model for urban planning integrates advanced techniques in privacy preservation, decentralized learning, decision transparency, and fairness, aligning them with the critical challenges and opportunities in urban governance. The model ensures compliance with privacy regulations like GDPR through Differential Privacy (DP), safeguarding sensitive data by making it indistinguishable whether a specific data point is included in the training set or not. This is complemented by Federated Learning (FL), which allows for collaborative model training across decentralized data sources without the need to centralize sensitive personal data, further enhancing privacy protection. Transparency in AI-driven decisions is addressed through techniques in Explainable AI (XAI), specifically by employing self-attention mechanisms to identify and explain the most influential features in the model's decision-making process. Additionally, the model incorporates adversarial debiasing, a method that minimizes biases in predictions by introducing adversarial loss functions that ensure fairness across protected and non-protected groups. This holistic framework empowers urban planners by enabling adaptive resource allocation, optimizing public services, and ensuring that AI systems operate in an accountable, transparent, and equitable manner, ultimately fostering public trust and promoting sustainable urban development.

Framework: Privacy-Preserving AI Compliance Model (PPAICM)

Layered AI Architecture for Privacy, Fairness, and Transparency in Urban Planning

Layer	Main functionality	AI module(s) and components	Operational scope (Input → Processing → Output)
Layer 1	Data collection	Urban data preprocessing engine	Input: Raw sensor data, socioeconomic data, demographic data → Output: Preprocessed input for AI models
Layer 2	Privacy preservation	Differential Privacy (DP)	Input: Raw data → Processing: Adding noise to preserve privacy → Output: Privacy-preserved AI predictions
Layer 3	Decentralized learning	Federated Learning (FL)	Input: Local datasets → Processing: Model training across decentralized nodes → Output: Global model update without transferring raw data
Layer 4	Transparency in decision-making	Explainability and interpretability engine	Input: AI predictions → Processing: Self-attention mechanism and interpretability → Output: Transparent decision-making
Layer 5	Fairness in AI decisions	Adversarial fairness optimization module	Input: AI predictions → Processing: Adversarial loss for fairness → Output: Bias-free predictions
Layer 6	Real-time model auditing	Real-time auditing and feedback system	Input: Model predictions, feedback mechanisms → Processing: Dynamic auditing and real-time adjustments → Output: Improved fairness and transparency

6.8.2 *Opportunities*

Despite these challenges, AI offers numerous opportunities to enhance city planning and improve the quality of urban life. One of the most significant opportunities is the potential for data-driven decision-making. AI systems can process vast amounts of data in real time, enabling city planners to make more informed decisions about resource allocation, infrastructure development, and public services. For instance, AI can be used to optimize traffic management systems, reducing congestion and improving air quality in urban areas. AI can also be applied to energy management, helping cities reduce their carbon footprints by optimizing the use of renewable energy sources and improving energy efficiency in buildings (Zowghi and da Rimini 2023). While AI offers immense opportunities to enhance city planning, its application must be mindful of privacy, fairness, and interpretability to maintain public trust and equity across all urban communities. The following framework integrates not only Proximal Policy Optimization (PPO) and Graph Neural Networks (GNNs) but also emphasizes transparency through Explainable AI (XAI) and ensures fairness by incorporating bias mitigation techniques.

In the context of dynamic resource allocation, a Deep Reinforcement Learning (DRL) approach using Proximal Policy Optimization (PPO) provides an effective mechanism for optimizing urban services like traffic flow, energy distribution, and infrastructure deployment. This optimization relies on continuous policy updates based on real-time conditions, ensuring that AI models adapt to changing urban dynamics. The mathematical formulation of the PPO objective function allows for stable policy updates by clipping the objective function, preventing excessive updates that could destabilize the learning process. This is represented as:

$$J(\theta) = \mathbb{E}\left[min\left(\frac{\pi_\theta(a_t|s_t)}{\pi_{\theta_{\text{old}}}(a_t|s_t)}A_t, \text{clip}\left(\frac{\pi_\theta(a_t|s_t)}{\pi_{\theta_{\text{old}}}(a_t|s_t)}, 1-\epsilon, 1+\epsilon\right)A_t\right)\right]$$

where $J(\theta)$, the PPO objective function, represents the optimization goal for urban decision-making, $\pi_\theta(a_t| s_t)$, the updated AI policy, determines the probability of selecting action a_t given urban state s_t, $\pi_{\theta_{\text{old}}}(a_t|s_t)$, the previous AI policy, represents past decision probabilities, A_t, the advantage function, quantifies the benefit of taking action a_t in state s_t, ϵ, the policy clipping parameter, prevents excessive policy updates, ensuring stable learning. This function enables AI-driven adaptive resource allocation, allowing city planners to optimize urban services dynamically based on real-time conditions. The PPO-based DRL model continuously refines policies to balance traffic flow, energy consumption, and infrastructure deployment.

To enhance spatially informed decision-making, Graph Neural Networks (GNNs) are employed to process and aggregate data from urban sensors, providing valuable insights into traffic flow, energy consumption, and air quality. In the case of traffic and energy management, GNNs allow for the effective aggregation of data from neighboring nodes, such as traffic intersections and buildings, to optimize network-level performance. The GNN formulation allows for dynamic feature aggregation across neighboring urban entities, updating node representations at each layer to capture the evolving state of the urban system. This is mathematically represented as:

$$h_v^{(l+1)} = \sigma\left(W^{(l)} \sum_{u \in \mathcal{N}(v)} \frac{h_u^{(l)}}{|\ \mathcal{N}(v)\ |} + b^{(l)}\right)$$

where $h_v^{(l+1)}$, the updated node representation, encodes urban infrastructure and traffic flow dynamics at layer $l + 1$, $W^{(l)}$, the weight matrix, adjusts feature importance across urban data nodes, $u \in \mathcal{N}(v)$, the neighboring nodes, represent urban entities such as intersections, buildings, and transport hubs, $|\ \mathcal{N}(v)\ |$, the normalization term, ensures balanced information aggregation, $b^{(l)}$, the bias term, stabilizes learning, σ, the activation function, introduces non-linearity for complex urban trend modeling. This function enables GNN-based spatial data processing, allowing AI systems to extract meaningful patterns from city-wide sensor networks, optimizing traffic management, air quality monitoring, and energy efficiency.

In urban decision-making, transparency is crucial to foster public trust, particularly when AI systems are involved in optimizing critical services such as traffic management and energy consumption. Explainable AI (XAI) techniques, such as Integrated Gradients and Self-Attention Mechanisms, allow stakeholders to understand how decisions are being made. The self-attention mechanism computes attention scores, quantifying the importance of each feature in the AI's decision process, which is crucial for explaining AI-driven decisions in the context of urban planning. The attention-weighted feature importance function is mathematically defined as:

$$\alpha_i = \frac{\exp(e_i)}{\sum_{j=1}^{N} \exp(e_j)}$$

where, α_i, the attention weight, represents the importance assigned to feature i, e_i, the attention score, quantifies the relevance of feature i in the AI decision process, N, the total number of features, determines the feature space, exp $(\cdot)$, the exponential function, ensures positive attention values. This function enables the AI model to highlight the most influential features, making the decision-making process more transparent.

Fairness in AI-driven urban decision-making is paramount, as biased AI systems can exacerbate inequalities in resource distribution and public services. To address this, adversarial debiasing is employed to mitigate bias in the AI model. The adversarial debiasing loss function is formulated to minimize predictive loss while simultaneously reducing discriminatory features in the model's predictions. The debiasing loss function is mathematically defined as:

$$L_{\text{debias}} = L_{\text{predict}} + \lambda \cdot D_{\text{adv}}(x)$$

where L_{debias}, the debiasing loss function, combines predictive accuracy with fairness objectives, L_{predict}, the predictive loss, ensures the model's performance on the main task, λ, the regularization parameter, controls the trade-off between prediction accuracy and fairness, $D_{\text{adv}}(x)$, the adversarial loss, represents the measure of bias in the model's predictions, minimizing it to promote fairness in outcomes. This function ensures that AI-driven urban planning decisions remain fair and inclusive by reducing the influence of biased features on the predictions.

This framework integrates multiple AI techniques to optimize urban planning, ensuring that the decision-making process is not only data-driven and efficient but also fair, transparent, and ethical. Through PPO-based optimization and GNN-based spatial data processing, city planners can dynamically allocate resources and optimize services. At the same time, by incorporating XAI and bias mitigation, the framework ensures that AI decisions are understandable and free from biases, fostering public trust and ensuring equity in urban development, adaptive, and

AI-driven urban development strategies, optimizing transportation networks, reducing carbon footprints, and improving urban quality of life.

Framework: AI-Driven Urban Optimization with Fairness and Transparency (AI-UOFT)

Layered AI Architecture for Sustainable and Fair City Planning

Layer	Main functionality	AI module(s) and components	Operational scope (Input → Processing → Output)
Layer 1	Dynamic resource allocation optimization	Proximal Policy Optimization (PPO)	Input: Real-time urban data → Processing: Policy optimization via PPO → Output: Optimized urban decision-making
Layer 2	Spatial data aggregation	Graph Neural Networks (GNNs)	Input: Urban sensor data → Processing: Graph-based data aggregation → Output: Optimized traffic and energy management
Layer 3	Transparency in decision-making	Explainability and interpretability engine	Input: AI predictions → Processing: Self-attention and interpretability techniques → Output: Transparent decision-making
Layer 4	Fairness in AI decisions	Adversarial fairness optimization module	Input: AI predictions → Processing: Debiasing via adversarial loss → Output: Bias-free urban decisions

Moreover, AI has the potential to promote inclusivity and equity in city planning. By using fairness metrics and explainable AI techniques, city planners can support AI systems in considering the needs of all communities, particularly marginalized and underserved populations (Pagano et al. 2023). For example, AI systems can be designed to prioritize infrastructure development in low-income neighborhoods, ensuring that these areas receive equitable access to public services. This can help address historical inequalities in urban development and create more inclusive cities. To mathematically model fairness-aware AI decision-making, a fairness-constrained optimization function is introduced:

$$\min_{\mathrm{w}} \sum_{i=1}^{N} \ell(f_{\mathrm{w}}(x_i), y_i) + \lambda \cdot \mathbb{E}_{p(x)}\left[D_{\mathrm{KL}}\left(p(y|x)\middle\|p(y)\right)\right]$$

where w, the AI model parameters, represent the learned decision-making weights, $\ell(f_{\mathrm{w}}(x_i), y_i)$, the classification loss function, ensures predictive accuracy, λ, the fairness regularization parameter, controls the trade-off between model accuracy and fairness, $D_{\mathrm{KL}}(p(y|x)\|p(y))$, the Kullback-Leibler (KL) divergence, measures disparities between conditional and marginal outcome distributions. This function ensures that AI-driven urban development decisions adhere to fairness constraints, preventing models from reinforcing existing inequalities. By adjusting decision-making based on fairness objectives, AI systems prioritize investments in historically underserved areas.

To further ensure explainability in fairness-driven urban AI models, a SHAP (Shapley Additive Explanations) feature attribution function is introduced:

$$\phi_i = \sum_{S \subseteq N \setminus \{i\}} \frac{|S|!(|N| - |S| - 1)!}{|N|!} [f(S \cup \{i\}) - f(S)]$$

where ϕ_i, the Shapley value, quantifies the contribution of feature i to AI decision-making, S, a subset of features excluding i, determines the groups of features used for evaluation, N, the total set of features, represents the entire feature space, $f(S)$, the AI model's prediction based on subset S, computes the decision outcome under partial information. This function enables explainability in AI-driven urban planning, allowing city planners to scrutinize decision rationales and ensure equitable infrastructure distribution.

Through the integration of fairness-constrained optimization with SHAP-based Explainable AI (XAI), this AI-driven framework enhances inclusive city planning by systematically mitigating bias and ensuring that urban development decisions prioritize historically marginalized communities. The fairness optimization model adjusts urban investments to promote equity, while the SHAP feature attribution function provides transparency in AI decision-making. This approach enables city planners to proactively design fair and inclusive urban policies, ensuring that AI-driven resource allocation and infrastructure investments reduce historical inequalities and foster socially inclusive cities.

Framework: AI-Driven Fairness and Transparency in Urban Planning (AI-FTUP3)

Layered AI Architecture for Fairness and Explainability in Urban Development

Layer	Main functionality	AI module(s) and components	Operational scope (Input → Processing → Output)
Layer 1	Data collection	Urban data preprocessing engine	Input: Raw sensor data, socioeconomic data, demographic data → Output: Preprocessed input for AI models
Layer 2	Fairness-constrained optimization	Fairness regularization optimization module	Input: Raw data → Processing: Minimizing classification loss with fairness regularization → Output: Fair urban decisions
Layer 3	Transparency in decision-making	Shapley Additive Explanations (SHAP)	Input: Model predictions → Processing: Feature attribution with Shapley values → Output: Transparent decision-making
Layer 4	Real-time model auditing	Real-time auditing and feedback system	Input: Model predictions, feedback mechanisms → Processing: Dynamic auditing and real-time adjustments → Output: Improved fairness and transparency

Another opportunity lies in the enhancement of public engagement through AI. AI systems can facilitate greater public participation in city planning by providing citizens with accessible tools to engage with planning processes. For instance, AI-powered platforms can allow citizens to visualize proposed development projects and provide feedback on how these projects will affect their communities. This can lead to more democratic and transparent decision-making processes, where citizens have a greater say in the future of their cities (Baum et al. 2022). To formalize the analysis of citizen feedback, a Sentiment Classification Function based on transformer-based NLP models is introduced:

$$P(y|X) = \frac{\exp\left(f_\theta(X)_y\right)}{\sum_j \exp\left(f_\theta(X)_j\right)}$$

where $P(y|X)$, the probability of sentiment class y given input text X, represents AI's prediction of public sentiment, $f_\theta(X)_y$, the model's output score for class y, encodes AI's understanding of sentiment polarity, j, the set of possible sentiment categories, defines classification outcomes, exp $(\cdot)$, the exponential function, ensures numerical stability in probability calculations. This function enables real-time public sentiment analysis, allowing city planners to systematically assess citizen feedback on proposed infrastructure projects.

To further enhance public engagement through AI-driven visualization, a Virtual Reality (VR)-based urban simulation function is introduced using Generative Adversarial Networks (GANs):

$$G(z) = \sigma(W_z z + b_z)$$

where $G(z)$, the generative function, produces AI-simulated urban environments, z, the latent variable, represents the input noise vector used for generative modeling, W_z, the weight matrix, maps latent representations to spatial features, b_z, the bias term, stabilizes simulation outputs, σ, the activation function, introduces non-linearity for realistic image synthesis. This function enables VR-enhanced AI-driven participatory planning, allowing citizens to experience urban proposals interactively and provide informed feedback.

By harnessing Transformer-based NLP for public sentiment analysis with GAN-based VR urban simulations, this AI-driven framework ensures that citizens can meaningfully engage with city planning decisions. The NLP sentiment classification model processes public opinions in real time, while the VR-based AI simulation allows stakeholders to visually experience proposed projects before implementation. This dual-framework approach fosters inclusive, transparent, and data-driven participatory urban planning, ensuring that citizen input directly informs AI-driven city development strategies.

Framework: AI-Driven Public Engagement in Urban Planning (AI-PEUP)

Layered AI Architecture for Sentiment Analysis and VR Urban Simulation

Layer	Main functionality	AI module(s) and components	Operational scope (Input → Processing → Output)
Layer 1	Data collection	Urban data preprocessing engine	Input: Raw sensor data, socioeconomic data, demographic data → Output: Preprocessed input for AI models
Layer 2	Sentiment classification	Transformer-based NLP models	Input: Public feedback → Processing: Sentiment classification using softmax → Output: Predicted sentiment class (positive, negative, neutral)
Layer 3	VR-based urban simulation	Generative Adversarial Networks (GANs)	Input: Latent vector → Processing: Urban environment simulation using GANs → Output: Simulated urban visualization for public interaction
Layer 4	Real-time model auditing	Real-time auditing and feedback system	Input: Model predictions, feedback mechanisms → Processing: Dynamic auditing and real-time adjustments → Output: Improved fairness and transparency

AI presents both challenges and opportunities in city planning. While the potential for bias, opacity, and regulatory hurdles poses significant challenges, the opportunities for data-driven decision-making, promoting inclusivity, and enhancing public engagement are substantial. By addressing these challenges through robust data governance, fairness metrics, and explainable AI techniques, city planners can harness the power of AI to create more efficient, equitable, and sustainable urban environments.

Figure 6.4 outlines the challenges and opportunities associated with AI in city planning, divided into two main categories. "Challenges" highlights issues such as

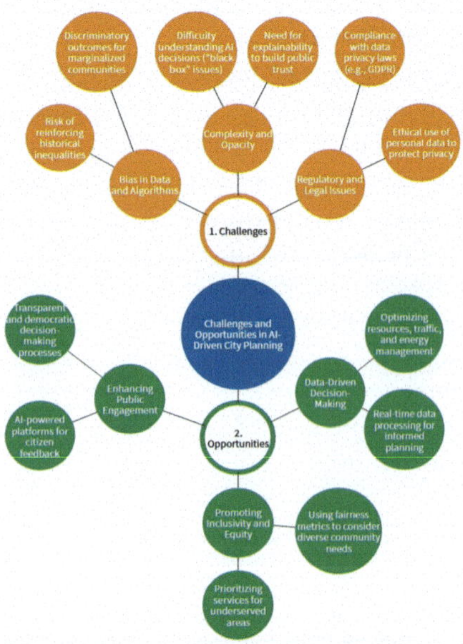

Fig. 6.4 Challenges and opportunities in AI-driven city planning

bias in data and algorithms, the complexity and opacity of AI systems, and regulatory and legal constraints like data privacy compliance. These challenges can lead to discriminatory outcomes and a lack of public trust if not addressed. "Opportunities" showcases the potential benefits of AI, including data-driven decision-making for resource optimization, promoting inclusivity and equity through fairness metrics, and enhancing public engagement via AI-powered platforms. Together, these elements represent the dual aspects of deploying AI in urban environments, balancing ethical considerations with technological advancements.

Ensuring inclusive, bias-free algorithms in city planning is critical to the fair and equitable development of urban environments. By addressing both data and algorithmic biases, employing fairness metrics, and utilizing explainable AI techniques, city planners can leverage the power of AI while safeguarding against unintended negative consequences for marginalized groups. Transparency and continuous evaluation are essential to this process, ensuring that AI systems remain accountable and fair in their impact on urban populations.

Limitations and Governance Implications
The inclusive planning frameworks developed in this chapter address the persistent risk that AI-assisted urban planning may reproduce spatial exclusion, biased evidentiary bases, unequal participation, opaque recommendations, and declining public trust in algorithmic planning tools. Fairness-Constrained AI for Urban Planning (FCAI-UP), Fairness-Aware Generative Adversarial Network for Urban Planning (FairGAN), Fairness-Constrained Deep Reinforcement Learning for Urban Planning (FCDRL-UP), Fairness-Constrained Multi-Objective Deep Reinforcement Learning for Urban Planning (MO-DRL-UP), Trust-Aware Fairness-Constrained Deep Reinforcement Learning for Urban Planning (TA-FC-DRL-UP), Fairness-Constrained Bayesian Data Reweighting Model (FC-BDRM), Fairness-Constrained Adversarial Debiasing Model for Urban Planning (FC-ADM), Fairness-Aware Explainable Reinforcement Learning Model (FA-XRL), Explainable AI for Urban Planning Decisions (XAI-UP), Explainable AI for Transparency and Compliance in Urban Planning (XAI-TC-UP), and AI-Driven Public Engagement in Urban Planning (AI-PEUP) provide mechanisms for fairness-constrained planning, participatory review, explainability, dynamic auditing, and privacy-preserving compliance. Yet inclusive planning cannot be reduced to statistical parity, synthetic data correction, optimization fairness, or interpretable recommendation systems. Urban planning is also a political and deliberative practice in which competing claims over land, mobility, housing, environmental quality, public space, and future development must be negotiated through transparent institutions. The governance implication is that AI may support more equitable planning only when its outputs remain open to participatory scrutiny, local knowledge, legal challenge, and professional planning judgment. Algorithmic recommendations should therefore be treated as provisional planning evidence, not as final authority, particularly where they affect marginalized groups, low-income residents, peripheral districts, or communities historically excluded from urban decision-making.

6.9 Conclusion

This chapter advances a comprehensive epistemological and computational synthesis, demonstrating how fairness, inclusivity, and transparency can be mathematically encoded as intrinsic components of AI systems for urban planning. Through a methodical progression from bias identification to formalized algorithmic correction, it establishes a unified theoretical and operational paradigm that redefines the ethical foundations of artificial intelligence in governance. By translating social justice imperatives into optimization constraints, it transforms normative ideals into reproducible mathematical processes, demonstrating that ethical integrity and computational efficiency can coexist within large-scale, real-world decision architectures.

The chapter's core contribution lies in its articulation of fairness as a quantifiable optimization target within reinforcement learning, adversarial training, and Bayesian inference. The succession of frameworks from fairness-constrained reinforcement learning and fairness-aware generative adversarial networks to multi-objective deep reinforcement learning and trust-aware architectures constructs a coherent taxonomy of fairness-aware intelligence. Each model converges toward a shared principle: the embedding of demographic equity into both the data and algorithmic layers of urban decision-making. This enables the systematic mitigation of representational, procedural, and outcome bias across dynamic, heterogeneous urban environments.

The chapter dismantles the dichotomy between ethical aspiration and computational formalism. By introducing fairness penalties, demographic reweighting, and trust-based reward functions, it formalizes equity as a continuous optimization variable, not an external constraint. The result is an operational ethics of AI, where moral objectives are internalized within mathematical expressions of policy learning, loss minimization, and posterior updating. The frameworks presented such as the Debiasing-adaptive fairness-constrained deep learning model and the fairness-constrained dual-stage optimization model translate fairness into a self-correcting mechanism that governs both training and deployment, producing AI systems that are adaptive, interpretable, and socially accountable.

The synthesis of Bayesian reweighting, adversarial debiasing, and multi-objective reinforcement learning elevates AI-driven urban governance into a new regime of algorithmic pluralism, one in which fairness, trust, and transparency are co-optimized alongside efficiency and predictive precision. This paradigm not only strengthens public confidence in AI systems but also provides a methodological foundation for compliance with emerging regulatory frameworks and data protection mandates. In practical terms, it offers policymakers and engineers a reproducible blueprint for developing Software-as-a-Service (SaaS) urban AI solutions that are both economically scalable and ethically sound.

The chapter reframes fairness not merely as a constraint on optimization but as a dynamic epistemic principle one that governs how knowledge is encoded, decisions are made, and accountability is ensured. Fairness becomes the mediating logic between human-centered governance and machine-driven rationality. Through its layered architectures combining federated learning, proximal policy optimization, generative adversarial networks, and Bayesian data reweighting, the chapter envisions a post-anthropocentric urban intelligence that harmonizes computational autonomy with civic responsibility.

In synthesis, this chapter demonstrates that the ethical transformation of AI requires neither moral abstraction nor post hoc regulation, but the direct mathematization of fairness within the core algorithms of decision-making. By bridging the domains of ethics, optimization theory, and urban systems engineering, it establishes the scientific and philosophical groundwork for an equitable AI urbanism—one where algorithmic intelligence becomes a vehicle for distributive justice, participatory governance, and transparent policy evolution in the smart cities of the twenty-first century.

References

Baum, K., Mantel, S., Schmidt, E., & Speith, T. (2022). From Responsibility to Reason-Giving Explainable Artificial Intelligence. *Philosophy & Technology*. https://doi.org/10.1007/s13347-022-00510-w

Mehrabi, N., Morstatter, F., Saxena, N., Lerman, K., & Galstyan, A. (2022). A Survey on Bias and Fairness in Machine Learning. https://doi.org/10.1145/3457607

Mökander, J. Auditing of AI: Legal, Ethical and Technical Approaches. *DISO* 2, 49 (2023). https://doi.org/10.1007/s44206-023-00074-y

Pagano, T. P., Loureiro, R. B., Lisboa, F. V. N., Peixoto, R. M., Guimarães, G. A. S., Cruz, G. O. R., Araujo, M. M., Santos, L. L., Cruz, M. A. S., Oliveira, E. L. S., & Nascimento, E. G. S. (2023). Bias and Unfairness in Machine Learning Models: A Systematic Review on Datasets, Tools, Fairness Metrics, and Identification and Mitigation Methods. *Big Data and Cognitive Computing*. https://doi.org/10.3390/bdcc7010015

Sargiotis, D. (2024a). Ethical AI in Information Technology: Navigating Bias, Privacy, Transparency, and Accountability. *Advances in Machine Learning & Artificial Intelligence*, 5(3), 01–14. https://doi.org/10.33140/AMLAI.05.03.03

Sargiotis, D. (2024b). Leveraging AI and Big Data for Advancements in Biomedical Research. *Biomedical Journal of Scientific & Technical Research*, 57(4). 10.26717/BJSTR.2024.57.009048

Sargiotis, D. (2024c). Advancing Civil Engineering with AI and Machine Learning: From Structural Health to Sustainable Development. *International Journal of Architecture and Planning*, 4(2), 54–81. https://doi.org/10.51483/IJARP.4.2.2024.54-81

Sargiotis, D. (2025a). Transforming civil engineering with AI and machine learning: Innovations, applications, and future directions. *International Journal of Research Publication and Reviews*, 6(1), 3780–3805. https://doi.org/10.55248/gengpi.6.0125.0528

Sargiotis, D. (2025b). Fostering ethical and inclusive AI: A human-centric paradigm for social impact. *International Journal of Research Publication and Reviews, 6*(1), 3754–3779. https://doi.org/10.55248/gengpi.6.0125.0529

Zowghi, D., & da Rimini, F. (2023). Diversity and Inclusion in Artificial Intelligence. https://doi.org/10.48550/arXiv.2305.12728

Chapter 7
Ethical AI Surveillance: Privacy-Preserving and Fairness-Aware Frameworks for Responsible Security Governance

This chapter, Ethical AI Surveillance: Privacy-Preserving and Fairness-Aware Frameworks for Responsible Security Governance, offers a profound and scientifically grounded exploration into one of the most pressing challenges of the twenty-first century: how to reconcile the accelerating capabilities of Artificial Intelligence (AI) in surveillance and security with the fundamental imperatives of privacy, fairness, transparency, and human dignity. It invites the international reader into an intellectually demanding yet practically illuminating discourse where mathematics, ethics, and governance converge to shape a new era of trustworthy AI deployment.

At its core, the chapter examines how AI-driven surveillance systems ranging from biometric recognition to behavioral monitoring and predictive security analytics can be designed not merely to function efficiently but to act ethically within the constraints of global privacy laws, human rights frameworks, and data protection mandates. Through a comprehensive and multi-layered architecture, the chapter delineates the mathematical foundations, algorithmic safeguards, and governance protocols that together define a model for ethically sustainable AI surveillance.

The narrative begins by articulating the intricate trade-offs between safety and liberty, where advanced data-driven technologies promise greater security yet risk infringing upon individual autonomy and privacy. It contextualizes this dilemma within the modern smart city, where interconnected sensors, cameras, and predictive models form an omnipresent layer of algorithmic observation. Rather than treating ethics as a theoretical concern, the chapter constructs a rigorous computational response, translating moral imperatives into concrete algorithmic forms.

The first section, devoted to privacy and data security, presents Differential Privacy (DP) and Federated Learning (FL) as the mathematical pillars of ethical data handling. Differential privacy ensures that no individual's information can be re-identified, even under extensive data analysis, by introducing calibrated statistical noise into computations. Federated Learning, on the other hand, enables AI systems to train collaboratively across multiple decentralized sources without centralizing sensitive data, preserving local privacy while improving global intelligence. These

D. Sargiotis, *AI with Integrity*, https://doi.org/10.1007/978-3-032-31133-7_7

two mechanisms, when combined, form the foundation for secure, privacy-preserving AI models that operate within the strict boundaries of ethical data governance.

The chapter delves into biometric surveillance and the challenge of consent in public spaces, where citizens are observed continuously by facial recognition and behavioral tracking systems. Here, it introduces cryptographic protocols such as Zero-Knowledge Proofs (ZKP) for identity verification without data disclosure, and Homomorphic Encryption (HE) for performing computations on encrypted information without decryption, thus ensuring that raw biometric data remains confidential even during processing. The model is further strengthened through blockchain-based consent management systems that provide immutable, transparent, and auditable records of how and when data is used. These mechanisms collectively redefine surveillance as a process of controlled visibility, where privacy and security coexist in balance rather than conflict.

The narrative then expands into adversarial resilience and cybersecurity, revealing how AI systems can be manipulated through data poisoning or adversarial attacks designed to distort outcomes or extract sensitive information. To counter these threats, the chapter presents Adversarially Regularized Training, Projected Gradient Descent (PGD) for perturbation resistance, and Reinforcement Learning (RL) models that operate under encrypted environments. These formulations are complemented by blockchain-anchored audit trails and anomaly detection mechanisms, ensuring that all AI decisions remain verifiable and tamper-proof.

Next, the chapter addresses the ethical and computational challenge of anonymization, a domain often underestimated in public debates. Traditional anonymization techniques, it argues, are insufficient when AI models can easily reidentify individuals from correlated datasets. To overcome this, the text introduces K-Anonymity, L-Diversity, and Differentially Private Stochastic Gradient Descent (DP-SGD), forming a mathematically rigorous pipeline that guarantees both privacy protection and data utility. This privacy-aware learning approach is dynamically optimized through continuous recalibration between performance and privacy budgets, ensuring compliance with both ethical standards and legal mandates such as the General Data Protection Regulation (GDPR).

Moving beyond privacy, the chapter enters the ethically charged terrain of fairness, bias, and discrimination. Here, it exposes how biased datasets and homogeneous design teams can embed systemic inequalities into surveillance systems exacerbating racial profiling, gender bias, and socioeconomic discrimination. To address these distortions, the chapter proposes Fair Representation Learning (FRL), Adversarial Debiasing Networks (ADN), and Diverse Generative Adversarial Networks (GANs) that synthesize demographically balanced data. Reinforcement Learning (RL) algorithms are recalibrated through fairness-aware reward functions, ensuring that AI decisions are both accurate and equitable. Moreover, Multi-Objective Optimization (MOO) functions are introduced to mathematically balance competing goals of accuracy, fairness, and privacy, allowing AI systems to evolve ethically under quantifiable constraints.

The section on transparency and accountability extends this ethical architecture into the domain of interpretability. By integrating Explainable AI (XAI) and Shapley Additive Explanations (SHAP), the chapter transforms opaque "black-box" AI systems into transparent and auditable instruments of decision-making. Every surveillance decision is traced back to its contributing factors, enabling human oversight and legal scrutiny. In parallel, fairness audit feedback mechanisms and gradient-based adaptation models ensure that biases are continuously monitored, corrected, and documented through immutable blockchain records. This systemic transparency transforms surveillance from a realm of suspicion into one of accountable governance.

Finally, the chapter introduces a human-centric equilibrium between security imperatives and privacy rights. Through Multi-Objective Optimization (MOO) functions and fairness-aware learning, it formulates the mathematical foundations of "ethical balance," an AI design principle ensuring that safety and liberty remain in proportional coexistence. Human-in-the-loop oversight mechanisms guarantee that critical decisions such as identity verification, alert generation, or access restriction always pass through a layer of human judgment before enforcement, maintaining societal control over machine governance.

From a global perspective, this chapter transcends disciplinary boundaries by uniting computer science, ethics, and governance into a cohesive theoretical and practical framework. It speaks to international scholars, policymakers, and technologists who seek a mathematically precise yet morally grounded approach to AI surveillance. Rather than framing ethical challenges as obstacles, it transforms them into design parameters for constructing lawful, equitable, and accountable AI systems.

In its essence, this chapter stands as a manifesto for a new paradigm of surveillance—one where technological power is disciplined by mathematical ethics, and where the guardians of security also become the stewards of human rights. For international readers seeking to understand, design, or regulate the next generation of AI surveillance infrastructures, this work provides not only the theoretical scaffolding but also the operational blueprints to realize responsible and trusted AI governance worldwide.

7.1 Privacy and Data Security Concerns

Artificial Intelligence (AI) systems in surveillance pose significant challenges regarding privacy and data security, particularly in the context of mass data collection, analysis, and storage. These concerns stem from the extensive data AI systems require and the potential misuse or unauthorized access to personal information. As AI becomes more integrated into public safety and governance infrastructures, the need to address privacy and data security concerns grows increasingly urgent (Sargiotis 2024a, b). To mathematically model privacy preservation in AI-driven surveillance, a differential privacy mechanism is applied to limit disclosure risk while preserving analytical utility. The differential privacy guarantee is expressed as:

$$P(M(D) \in S) \leq e^{\epsilon} P(M(D') \in S)$$

where $P(M(D) \in S)$, the probability of the AI model output belonging to a set S, represents the likelihood of a specific outcome under dataset D, $M(D)$, the AI model mechanism, maps input data D to an output decision, D', a neighboring dataset differing by a single entry, ensures individual privacy protection, ϵ, the privacy budget, quantifies the allowable privacy loss, ensuring that AI outputs remain statistically indistinguishable across datasets. This function ensures that AI surveillance systems protect individual identities by injecting noise into data queries, preventing the re-identification of individuals while maintaining analytical accuracy.

To further enhance secure AI-driven surveillance, Federated Learning (FL) is applied, allowing AI models to learn from distributed sources without centralizing sensitive data. The federated learning optimization function is formulated as:

$$\min_{\mathrm{w}} \sum_{k=1}^{K} p_k F_k(\mathrm{w})$$

where w, the global AI model parameters, represent shared weights across decentralized learning nodes, k, an individual client node, contributes to AI training, p_k, the weighting factor, accounts for the significance of each data source, $F_k(\mathrm{w})$, the local loss function, ensures model accuracy at each decentralized node. This function enables secure decentralized AI learning, allowing surveillance systems to process data across multiple devices without sharing raw data.

Via the combination of differential privacy for anonymized AI-driven surveillance with federated learning for decentralized data security, this AI framework ensures that AI surveillance infrastructures remain privacy-compliant while preserving analytical effectiveness. The DP mechanism injects noise into AI outputs, preventing personal data exposure, while the FL model enables privacy-preserving AI training without centralizing sensitive information. This dual-framework approach ensures that AI-driven surveillance systems enhance public safety while adhering to ethical data security standards, reinforcing public trust in AI-integrated governance infrastructures.

Framework: AI-Driven Privacy and Security in Surveillance (AI-PSS)

Layered AI Architecture for Differential Privacy and Federated Learning in AI Surveillance

Layer	Main functionality	AI module(s) and components	Operational scope (Input → Processing → Output)
Layer 1	Data collection	Data ingestion and preprocessing pipeline	Input: Raw sensor data, socioeconomic data, demographic data → Output: Preprocessed input for AI models

(continued)

Layer	Main functionality	AI module(s) and components	Operational scope (Input → Processing → Output)
Layer 2	Privacy preservation	Differential Privacy (DP)	Input: Raw data → Processing: Adding noise to preserve privacy → Output: Privacy-preserved AI predictions
Layer 3	Federated learning	Federated Learning (FL)	Input: Local datasets → Processing: Model training across decentralized nodes → Output: Global model update without transferring raw data
Layer 4	Real-Time model auditing	Continuous compliance auditing engine	Input: Model predictions, feedback mechanisms → Processing: Dynamic auditing and real-time adjustments → Output: Improved fairness and transparency

7.1.1 Mass Data Collection and Consent Issues

The ethical dilemmas inherent in AI-driven surveillance, particularly concerning biometric data collection and facial recognition technologies, necessitate a mathematically rigorous privacy-preserving framework. Facial recognition and behavioral tracking systems typically function in public spaces absent explicit consent, thus engendering profound ethical and legal challenges (Bruneault and Sabourin Laflamme 2021; Saheb 2023). To resolve these issues, we propose an integrated AI surveillance architecture combining Zero-Knowledge Proofs (ZKPs) for secure identity authentication and Homomorphic Encryption (HE) for privacy-preserving data processing, thereby enabling verification and analysis without exposing sensitive biometric data.

To ensure that raw biometric signals are transformed into analytically tractable representations, the model integrates a feature extraction layer based on Convolutional Neural Networks (CNNs). This transformation compresses the data while retaining task-relevant biometric features required for downstream cryptographic verification. Extending from the structured biometric input x, the CNN layer computes a convolved and activated output representation z as follows:

$$z = \sigma(W * x + b)$$

where z encoded feature vector, x structured biometric input, W convolutional kernel weights, $*$ convolution operator, b bias term, $\sigma(\cdot)$ nonlinear activation function. This function extracts and encodes localized biometric features from input data, producing compact vectors suitable for identity verification under privacy constraints.

The initial concern of ensuring identity verification without data leakage is addressed through the implementation of Zero-Knowledge Proofs (ZKPs). This interactive cryptographic protocol enables a prover to convince a verifier of the validity of a biometric identity claim without revealing any additional information. Formally, this is represented as:

$$\forall x \in X, \quad \exists \pi \text{ such that } \mathcal{V}(x, \pi) = 1 \wedge \pi \leftarrow \mathcal{P}(x)$$

where x biometric input, X domain of biometric identities, $\pi\pi$ non-disclosing proof artifact, P prover function, VV verifier function. This function certifies biometric claims without data exposure, thus enabling secure authentication in surveillance contexts.

Building upon the ZKP-driven identity confirmation, the system transitions into Homomorphic Encryption (HE) to facilitate secure computation over encrypted biometric streams. This ensures that all analytic operations over surveillance data remain within the encrypted domain, thereby eliminating the need for decryption. The encryption function $E(\cdot)$ and biometric mapping $f(\cdot)$ obey the homomorphic property:

$$f(E(x)) = E(f(x))$$

where x biometric vector, $E(x)$ ciphertext of biometric input, $f(x)$ biometric transformation function. This function permits analytical processing over encrypted data, preserving the confidentiality of biometric information throughout the computational lifecycle.

To ensure that the system maintains an ethically balanced configuration between utility and privacy, a formal multi-objective optimization function is introduced. This function penalizes privacy violations while incentivizing operational effectiveness, using learned model parameters θ and weights λ_1, λ_2:

$$\theta^* = \arg \underset{\theta}{min} \left(\lambda_1 \cdot \mathcal{R}(\theta) + \lambda_2 \cdot (1 - \mathcal{U}(\theta)) \right)$$

where θ model parameters, λ_1 privacy risk weight, λ_2 utility reward weight, $\mathcal{R}(\theta)$ quantified privacy risk, $\mathcal{U}(\theta)$ system utility function. This function achieves a privacy-utility equilibrium through optimization under ethical constraints.

In order to respond adaptively to real-time environmental fluctuations and shifting input distributions, the framework implements an online learning mechanism based on Stochastic Gradient Descent (SGD). This allows for continuous refinement of the optimization parameters derived above:

$$\theta_{t+1} = \theta_t - \eta \nabla_\theta \mathcal{L}(\theta_t)$$

where θ_t model parameters at iteration t, θ_{t+1} updated parameters, η learning rate, $\nabla_\theta \mathcal{L}(\theta_t)$ gradient of the loss function. This function adjusts the model iteratively to preserve optimality in dynamic surveillance environments.

Following the core surveillance operations, it is essential to validate that model outputs comply with privacy norms and do not trigger unauthorized data disclosures.

To this end, the system employs a privacy auditing function modeled as an AI-powered anomaly detection process. Given the system output y_t and the internal model state θ_t, an anomaly signal a_t is computed via an adaptive audit function $\mathcal{A}(\cdot)$, defined as:

$$a_t = \mathcal{A}(y_t, \theta_t)$$

where a_t anomaly/audit signal at time t, y_t AI-generated output decision at time t, θ_t model parameters at time t, $\mathcal{A}(\cdot)$ AI-driven audit function. This function assesses surveillance decisions against behavioral and privacy baselines, detecting potential policy violations or data misuse.

This AI-driven surveillance model articulates a structured computational approach that integrates machine perception, encrypted computation, adaptive learning, and ethical oversight into a unified operational framework. It begins with the extraction of biometric representations using convolutional neural networks, capturing and encoding facial features into structured vectors suitable for downstream processing. From this foundation, the model advances to a privacy-preserving inference architecture, where encrypted biometric data remains inaccessible even during computation. A dynamic learning component continuously refines the balance between operational performance and privacy risk through multi-objective optimization and stochastic parameter updates. The architecture concludes with a decision auditing mechanism that integrates anomaly detection techniques to evaluate system behavior over time, accompanied by human oversight to validate intervention decisions. In doing so, the section presents an engineered design for AI-driven surveillance that responds to the ethical demands of biometric governance while maintaining computational transparency and learning-driven responsiveness.

Framework: Privacy-Preserving AI Surveillance Framework (PPAISF)

Layered AI Architecture for Ethical Biometric Surveillance and Data Protection

Layer	Main functionality	AI module(s) and components	Operational scope
Layer 1	Raw biometric data acquisition and structuring	Biometric signal preprocessing engine	Surveillance streams → Signal transformation → Structured biometric input
Layer 2	Feature extraction and encoding	Convolutional Neural Networks (CNNs)	Structured biometric input → Feature encoding → Compressed biometric vectors
Layer 3	Secure biometric identity verification	Zero-knowledge verification protocol module	Encoded biometric vector → ZKP interaction → Verified identity proof without disclosure
Layer 4	Encrypted biometric computation	Homomorphic encryption processing core	Biometric vector → Encrypted computation → Functional outputs in ciphertext domain
Layer 5	Privacy-utility trade-off optimization	Multi-objective optimization engine	Model parameters, risk-utility metrics → Weighted minimization → Ethically balanced model

(continued)

Layer	Main functionality	AI module(s) and components	Operational scope
Layer 6	Adaptive real-time model update	Stochastic Gradient Descent (SGD)	Objective gradients, learning rate → Parameter update → Dynamic model adjustment
Layer 7	Surveillance output auditing and privacy monitoring	AI-powered privacy auditing system	System outputs, model state → Audit function → Privacy breach detection and traceability
Layer 8	Human-in-the-loop surveillance decision oversight	Human oversight and decision governance interface	AI-generated actions → Human validation → Approved or blocked surveillance outcome

As AI surveillance systems expand their operational reach, the ethical implications of indefinite data retention become increasingly urgent. The accumulation of sensitive biometric and behavioral data without time constraints poses significant risks of misuse, repurposing, and violation of data subject rights. These risks intensify when personal data collected for surveillance is later used for unrelated law enforcement or security objectives without public transparency or consent (Fosch-Villaronga and Poulsen 2022). To mitigate such ethical breaches, this section introduces a mathematically grounded AI-driven model structured to support privacy-preserving learning and enforce temporal decay of sensitive data influence throughout the system's analytical and decision-making lifecycle.

The data collection phase begins with raw biometric and behavioral streams, including facial vectors, motion sequences, and device-based geolocation signals. These inputs are transformed into structured feature representations via convolutional encoders that extract privacy-constrained biometric embeddings. The transformation is expressed as:

$$z = \sigma(W * x + b)$$

where z encoded biometric feature vector, x raw surveillance input, W convolutional kernel, bb bias term, $*$ convolution operator, $\sigma(\cdot)$ activation function. This function extracts compressed, task-relevant biometric features suitable for downstream privacy-aware processing.

Following encoding, the framework applies a differential privacy mechanism to ensure that individual data points have bounded influence on model outputs. The privacy guarantee is formalized as:

$$P(M(D) \in S) \leq e^{\epsilon} \cdot P(M(D') \in S)$$

where M is a randomized mechanism applied to dataset D, D' is a neighboring dataset differing by a single individual's data, S is a measurable subset of possible outputs, and ϵ is the privacy budget controlling the maximum privacy loss. This formulation ensures that no single data record substantially influences the model output, preserving individual anonymity under formal constraints.

To address the temporal dynamics of data sensitivity, the framework further introduces a decaying privacy budget. Over time, the influence of previously collected data is reduced by modulating ϵ as a function of time, such that:

$$\epsilon_t = \epsilon_0 \cdot e^{-\lambda t}$$

where ϵ_t is the effective privacy budget at time t, ϵ_0 is the initial privacy budget, and λ is the decay constant governing the rate at which older data loses relevance. This mechanism ensures that long-retained data contributes progressively less to model decisions, limiting the risk of retrospective misuse and enforcing temporal privacy compliance.

To verify user identity without disclosing sensitive biometric information, the system incorporates a zero-knowledge proof protocol built upon the Schnorr identification scheme. This mechanism is expressed via challenge-response cryptographic interaction:

$$\text{Prover:} t = g^r \bmod p$$

$$\text{Verifier:} c \sim \mathcal{U}(0, q)$$

$$\text{Prover:} s = r + c \cdot x$$

$$\text{Verifier:} g^s \stackrel{?}{=} t \cdot h^c \bmod p$$

where g^r is the ephemeral commitment, c is the challenge, s is the response, $h = g^x$ is the public identity commitment, and the verifier checks if the reconstructed value holds. This interaction ensures identity can be verified without exposing biometric secrets. To further protect sensitive information during the analysis phase, homomorphic encryption allows AI systems to operate on encrypted data without decrypting it. This method ensures that personal data remains secure during computation, enabling AI models to analyze data without exposing it in an unencrypted form. The homomorphic encryption equation can be expressed as follows:

$$E(m_1) \cdot E(m_2) = E(m_1 + m_2)$$

where $E(m_1)$, the encrypted value of the first raw data point, and $E(m_2)$, the encrypted value of the second raw data point, are the encrypted representations. $E(m_1 + m_2)$, the encrypted sum, represents the result of an operation on encrypted data, allowing computation on sensitive data without exposing it. This function ensures that data security is maintained while still enabling analysis and decision-making.

Once data is processed and analyzed, decisions must be made based on the information gathered. Reinforcement Learning (RL) algorithms are used to optimize actions by learning from feedback. In this context, the AI system uses Q-learning to continuously adjust its decisions to optimize privacy-preserving actions, such as issuing alerts or restricting access when privacy violations are detected. The Q-learning equation is as follows:

$$Q(s_t, a_t) = r_t + \gamma \, max_{a'} \, Q(s_{t+1}, a')$$

where $Q(s_t, a_t)$, the action-value function, estimates the expected return from taking action a_t in state s_t, r_t, the reward, reflects immediate feedback, γ, the discount factor, determines the importance of future rewards, and $\max_{a'} Q(s_{t+1}, a')$ represents the expected reward from future actions. This function enables the surveillance system to make real-time decisions while ensuring that privacy constraints are respected.

Lastly, to ensure transparency and data accountability, blockchain-based consent management is employed to record all data access and consent history. This blockchain system ensures that data usage is verifiable, immutable, and fully auditable. Individuals have control over their data and can track how it is used or revoked. The blockchain transaction for recording consent can be expressed as:

$$T_x = H(D_i, P_i, t_i) + \sigma$$

where T_x, the blockchain transaction, records data access and consent history, $H(D_i, P_i, t_i) = \text{hash}(D_i \| P_i \| t_i)$, the cryptographic hash function, ensures immutability of stored consent records, D_i, the dataset identifier, links the stored data to its consent metadata, P_i, the permission level, specifies approved uses of the data, t_i, the timestamp, ensures that consent is tied to a specific time frame, hash($\cdot$) is a secure cryptographic hash function (e.g., SHA-256, Keccak-256), ensuring that consent data is immutable, $\|$ denotes concatenation, ensuring that all parameters are combined before hashing, σ, the cryptographic signature, authenticates the transaction. This function guarantees that consent records are transparent, immutable, and auditable, allowing individuals to track how their data is used and ensuring the ethical use of personal information.

This AI-driven privacy-preserving surveillance model combines differential privacy for controlled data retention, Zero-Knowledge Proofs (ZKPs) for secure identity verification, homomorphic encryption for secure data processing, Reinforcement Learning (RL) for decision-making, and blockchain-based consent management for data transparency. Each layer works seamlessly to ensure that data is not retained indefinitely, that individuals' privacy is protected, and that consent is properly managed throughout the lifecycle of the surveillance system. This model establishes privacy, security, and ethical compliance, enabling AI surveillance systems to operate in a way that fosters public trust while safeguarding individual rights.

Framework: Privacy-Conscious Adaptive Surveillance Intelligence Framework (PCASIF)

Layered AI Architecture for Ethical Surveillance, Privacy Preservation, and Consent Accountability

Layer	Main functionality	AI module(s) and components	Operational scope
Layer 1	Feature encoding of biometric and behavioral input	Convolutional Neural Networks (CNNs)	Raw biometric and behavioral streams → Convolutional feature extraction → Structured embeddings

(continued)

Layer	Main functionality	AI module(s) and components	Operational scope
Layer 2	Privacy-constrained learning via differential privacy	Differential privacy-compliant learning	Structured feature inputs → Randomized model training → Privacy-bounded AI output
Layer 3	Dynamic privacy budget adjustment over time	Temporal privacy budget management engine	Privacy budget, decay constant → Exponential modulation → Time-aware privacy scheduling
Layer 4	Identity verification without biometric disclosure	Zero-knowledge authentication module	Biometric commitment, cryptographic protocol → Interactive ZKP → Verified authentication without exposure
Layer 5	Encrypted data computation	Homomorphic encryption computation core	Encrypted inputs → Ciphertext operations → Privacy-preserving computation
Layer 6	Adaptive policy optimization with Feedback	Q-learning (reinforcement learning)	Privacy-state inputs → Value-based policy update → Real-time optimized actions
Layer 7	Consent logging and data access traceability	Blockchain-based consent management system	Consent metadata, timestamp → Blockchain transaction → Immutable record for auditability

7.1.2 Data Security and Vulnerabilities

As AI surveillance systems become increasingly embedded in public safety and governance infrastructures, the threat landscape surrounding data security expands in parallel. The continuous aggregation of high-resolution biometric, locational, and behavioral data renders these systems attractive targets for cyberattacks and adversarial exploits. Hackers may exploit vulnerabilities in the AI architecture to exfiltrate personal information, manipulate decision-making outcomes, or perform model inversion attacks, raising urgent concerns regarding both individual privacy and societal resilience (Al-kfairy et al. 2024; Sargiotis 2024a). To confront these escalating threats, we propose an AI-driven architectural framework that integrates secure computation over encrypted domains with adversarial techniques, strengthening confidentiality, integrity, and resilience across inference and decision-making stages.

First, to ensure that privacy is preserved even during inference and learning, the proposed model integrates secure computation on encrypted data through homomorphic transformation. Let $f(x)$ denote the AI function to be executed on sensitive data x. The encryption operator $E(\cdot)$ enables this function to be executed within the ciphertext space, leading to:

$$E(f(x)) = f(E(x))$$

where $E(f(x))$, the encrypted function output, ensures AI computations occur on encrypted data, $f(x)$, the AI model function, processes the input data, $E(x)$, the encrypted input data, preserves privacy while allowing computations. This function ensures that AI surveillance systems can analyze sensitive data without ever decrypting it, preventing exposure to hackers while maintaining analytical utility. This function ensures that AI surveillance systems can analyze sensitive data without ever decrypting it, preventing exposure to hackers while maintaining analytical utility.

To maintain decision consistency on encrypted input, we incorporate secure inference mapping over ciphertext domains. This layer ensures compatibility between encrypted model input and output, validating semantic preservation during homomorphic computation. The encrypted function composition is expressed as:

$$f'(E(x)) = E(f(x))$$

where $f'(E(x))$ encrypted output of transformed model, $E(f(x))$ correctly encrypted final output, $f(x)$ plaintext model function, $E(x)$ encrypted input. This function ensures that inference performed on encrypted inputs yields ciphertexts equivalent to encrypting the plaintext outputs directly, preserving both model logic and encryption integrity.

As models operate in adversarial environments, robustness against input perturbations must be guaranteed. To this end, the framework integrates adversarially regularized training through gradient masking. The loss function is augmented with a penalty term that reduces sensitivity to adversarial gradients. This is formally defined by:

$$J_{\text{adv}}(\theta) = \mathbb{E}\left[\ell(f_\theta(x+\delta), y) + \lambda\|\nabla_x \ell(f_\theta(x), y)\|^2\right]$$

where $J_{\text{adv}}(\theta)$, the adversarially robust loss function, enhances AI security against cyber threats, $\ell(f_\theta(x+\delta), y)$, the adversarial loss term, ensures AI robustness against small perturbations δ, λ, the regularization coefficient, controls the balance between accuracy and adversarial defense, $\|\nabla_x \ell(f_\theta(x), y)\|^2$, the gradient masking term, prevents adversarial inputs from manipulating AI models. This function ensures that AI surveillance models are resistant to adversarial attacks, preventing hackers from exploiting AI vulnerabilities.

To operationalize secure decision-making based on encrypted inference, the system employs policy optimization under encrypted decision signals. A reinforcement learning module refines action policies based on encrypted observations. The decision policy is optimized using:

$$\pi^*(a|s) = \arg \max_{\pi} \mathbb{E}\left[\sum_{t=0}^{T} \gamma^t R(s_t, a_t)\right]$$

where $\pi^*(a|s)$ optimal policy function, π policy distribution, s_t state at time t, a_t action at time t, $R(s_t, a_t)$ reward function, γ discount factor, T time horizon. This function enables the model to select optimal actions even under encrypted observations, aligning decisions with long-term utility while maintaining ciphertext integrity.

The final step ensures post-decision accountability through immutable surveillance logging. While not an AI module, blockchain-based audit trails cryptographically secure all AI actions, linking them to consent, time, and context. To facilitate real-time cybersecurity verification after decision deployment, the framework integrates AI-based anomaly detection on system logs. This is implemented via a multivariate Gaussian distribution modeling log activity:

$$A(x) = \mathbb{I}\left[(x-\mu)^T \Sigma^{-1} (x-\mu) > \tau\right]$$

where $A(x)$ anomaly indicator function, x observed system log vector, μ expected mean, Σ covariance matrix, τ anomaly threshold, $\mathbb{I}[\cdot]$ indicator function. This function flags potential cyber intrusions or abnormal system behaviors post-decision, supporting continuous model validation and breach detection.

This AI-driven model articulates a structurally layered framework designed to secure sensitive surveillance data through a chain of mathematically grounded operations. It begins by encrypting raw biometric and behavioral inputs via homomorphic transformations, thereby ensuring that all computational tasks are conducted without exposing underlying information. The model proceeds by executing AI inference directly over encrypted data streams while preserving functional equivalence to plaintext operations. This is followed by the integration of adversarial robustness techniques that minimize vulnerability to perturbation-based manipulation. A reinforcement learning policy then governs action selection under encrypted input conditions, aligning system behavior with long-term utility metrics. Post-decision activities are secured through blockchain-anchored audit trails and are continuously monitored using statistical anomaly detection to identify potential intrusions. The framework, in its entirety, demonstrates an operational lifecycle in which encryption, learning, decision-making, and verification are cohesively configured to maintain data confidentiality, prevent exploitation, and sustain model reliability in adversarial contexts.

Framework: Secure Encrypted AI Surveillance Framework (SEAISF)

Layered AI Architecture for Privacy-Preserving Surveillance and Adversarial Defense

Layer	Main functionality	AI module(s) and components	Operational scope
Layer 1	Encrypted AI computation on sensitive surveillance data	Homomorphic encryption processing engine	Raw biometric/behavioral data → Homomorphic transformation → Encrypted AI-ready input

(continued)

Layer	Main functionality	AI module(s) and components	Operational scope
Layer 2	Encrypted inference mapping with function preservation	Encrypted inference preservation module	Encrypted input → Semantic-preserving transformation → Encrypted model output
Layer 3	Adversarial robustness via gradient regularization	Adversarial training with gradient masking	Perturbed input data → Regularized loss function → Robust model parameters
Layer 4	Secure decision optimization via reinforcement learning	Deep Reinforcement Learning (DRL)	Encrypted observations → Policy gradient optimization → Optimal encrypted policy
Layer 5	Post-decision anomaly detection on encrypted logs	Multivariate Gaussian anomaly detection	Encrypted log vectors → Distributional deviation scoring → Intrusion alert flags
Layer 6	Immutable blockchain audit logging	Blockchain-based audit and compliance ledger	AI decision traces → Cryptographic hashing → Immutable audit trails

To address the threat of data poisoning in AI-driven surveillance, we propose a multi-layered secure learning framework in which adversarial robustness and data integrity monitoring are achieved through mathematically validated AI modules and secure infrastructure protocols. Data poisoning attacks, wherein adversaries inject carefully crafted inputs to compromise model behavior, not only distort predictions but also exacerbate systemic bias, particularly against vulnerable communities (Bruneault and Sabourin Laflamme 2021).

To resist such manipulations, the framework first applies adversarial training based on a minimax learning objective that confronts hostile perturbations during model optimization. The formulation enables the learning process to optimize parameters w against input perturbations δ drawn from a constrained set Δ. This is formally defined by:

$$\min_{w} \mathbb{E}_{(x,y)\sim\mathcal{D}}\left[\max_{\delta\in\Delta} \ell(f_w(x+\delta), y)\right]$$

where w, the AI model parameters, represent the learned decision-making weights, $\mathbb{E}_{(x,y)\sim\mathcal{D}}$, the expected loss over the data distribution, ensures robustness across diverse inputs, $\max_{\delta\in\Delta} \ell(f_w(x+\delta), y)$, the inner maximization, represents the worst-case adversarial perturbation, δ, the perturbation variable, manipulates inputs to induce misclassification, Δ, the adversarial constraint set, defines allowable perturbations. This function ensures that AI models are trained to recognize and resist adversarial data manipulations, significantly enhancing the robustness of AI-driven surveillance systems.

To compute the inner maximization term efficiently, the framework integrates iterative gradient-based perturbation using projected gradient descent (PGD), which generates adversarial examples as:

$$\delta^{(t+1)} = \Pi_\Delta\left(\delta^{(t)} + \alpha \cdot \text{sign}\left(\nabla_x \ell\left(f_w\left(x + \delta^{(t)}\right), y\right)\right)\right)$$

where $\delta^{(t)}$ adversarial perturbation at step t, α step size, Π_Δ projection operator onto Δ, $\nabla_x \ell$ gradient of the loss with respect to input. This function simulates worst-case attacks during training, making the model resilient under real-world adversarial exposure.

To detect poisoned data instances during or after training, the framework includes a statistical anomaly detector based on Kernel Density Estimation (KDE) over a trusted clean dataset D_{clean}. The anomaly flag $A(x)$ is raised when the probability density $p(x)$ of input x drops below a threshold τ:

$$A(x) = \mathbb{I}[p(x) < \tau], \quad \text{where} \quad p(x) = \frac{1}{nh}\sum_{i=1}^{n} K\left(\frac{x - x_i}{h}\right)$$

where $A(x)$ anomaly indicator, $p(x)$ estimated density, τ anomaly threshold, $K(\cdot)$ kernel function, h bandwidth, x_i training samples from the trusted dataset. This function flags low-density outliers indicative of poisoned data D_{clean}, reinforcing the integrity of the training process.

To further strengthen data integrity and authenticity in AI surveillance, blockchain-based data verification is introduced, preventing the injection of tampered or poisoned data into AI training pipelines. The data integrity hashing function is defined as:

$$H(D_i) = \text{SHA} - 256\left(D_i \middle\| t_i\right)$$

where $H(D_i)$, the cryptographic hash function, uniquely identifies dataset D_i, ensuring tamper resistance, SHA − 256, the secure hashing algorithm, prevents unauthorized modifications, D_i, the dataset entry, represents AI training data, t_i, the timestamp, ensures temporal consistency in data records. This function guarantees that AI training datasets are immutable and verifiable, preventing attackers from injecting poisoned data into AI surveillance systems.

This AI-driven model demonstrates a structured approach to securing the learning process of surveillance systems against adversarial manipulation, particularly data poisoning attacks. It integrates core machine learning mechanisms such as adversarial training with a minimax optimization objective and projected gradient descent to ensure that the model is not only trained on diverse inputs but also exposed to strategically crafted perturbations that simulate potential attacks. This process enables the system to develop internal resilience during training. In parallel, the model incorporates statistical anomaly detection through kernel density estimation, allowing it to identify and filter poisoned or out-of-distribution inputs that may evade initial defenses. Importantly, the model treats data authenticity as a foundational component: before data is admitted into the learning pipeline, it is verified using cryptographic hashing, anchoring each training instance with an immutable timestamped fingerprint. Through these combined elements, the section presents an AI framework that anticipates adversarial conditions across multiple levels training, validation, and input verification demonstrating how surveillance systems can maintain reliability and decision accuracy even when facing adversarial threats.

Framework: AI-Driven Secure Learning Framework Against Data Poisoning (AISLF-DP)

Layered AI Architecture for Adversarial Resilience and Data Integrity in Surveillance Systems

Layer	Main functionality	AI module(s) and components	Operational scope
Layer 1	Data integrity commitment prior to AI training	Blockchain-based data integrity verification module	Raw training data, timestamp → SHA-256 hash → Immutable data entry verification
Layer 2	Robust learning under adversarial perturbations	Adversarial training (minimax objective)	Training data → Minimax loss optimization → Robust model parameters
Layer 3	Generation of high-impact perturbations	Projected Gradient Descent (PGD)	Input sample, model gradient → PGD iterations → Adversarial examples
Layer 4	Detection of poisoned inputs via density estimation	Kernel density estimation anomaly detection module	Incoming data → Density-based scoring → Outlier rejection
Layer 5	Feedback-driven model refinement	Feedback-driven adaptive retraining engine	Anomaly flags, safe data → Filtered batch selection → Model update or retraining
Layer 6	Human-in-the-loop anomaly verification	Human validation and anomaly review interface	Flagged data instances → Expert validation → Confirmed anomaly labels or safe overrides

7.1.3 The Challenge of Anonymization

To address the limitations of traditional anonymization techniques in AI-driven data processing, a privacy-preserving AI anonymization framework is introduced, integrating Differential Privacy (DP) for robust anonymization and K-Anonymity with L-Diversity for enhanced data protection. While anonymization techniques aim to remove personally identifiable information (PII) to enable AI-driven analytics without compromising privacy, they are often vulnerable to re-identification attacks, particularly when anonymized datasets are cross-referenced with external sources (Saheb 2023). The proposed framework therefore integrates mathematically grounded constraints at the preprocessing level and differential privacy-aware training mechanisms at the core AI stage, enabling learning over anonymized data while strengthening resistance to identity and attribute inference attacks.

The framework initiates a formal data ingestion process grounded in the recognition of quasi-identifiers. These are structured attribute subsets $Q(x)$ that, when combined, present a risk of indirect re-identification. The mapping from each input vector to its corresponding quasi-identifier set is defined as:

$$Q : x \mapsto \{x_{i_1}, x_{i_2}, \ldots, x_{i_r}\}$$

where Q quasi-identifier mapping function, x input data instance, $x_{i_1}, x_{i_2}, \ldots, x_{i_r}$ selected quasi-identifying attributes from x. This function extracts structured features that form the basis for grouping, enabling the downstream enforcement of identity-obfuscating transformations.

To formalize anonymization at the dataset structure level, the model applies K-Anonymity constraints. Each equivalence class $Q(x)$ must satisfy the following cardinality condition:

$$|\{x \in D \mid Q(x) = Q(x_i)\}| \geq k$$

where $Q(x)$ is the quasi-identifier mapping, k is the anonymity parameter, D is the dataset. This ensures that each record shares identical identifiers with at least k others, mitigating re-identification risk.

To enhance intra-group diversity and prevent attribute disclosure, L-Diversity is enforced. Each equivalence class must contain multiple distinct sensitive values:

$$\forall Q(x) :| \{s \in S \mid s = \text{Sensitive}(x)\} | \geq l$$

where $Q(x)$ denotes an equivalence class, S the set of sensitive attributes, l the diversity threshold. This constraint ensures that even within anonymized groups, confidential fields remain semantically diverse.

The foundation of this model begins with differentially private learning, implemented through a noise-calibrated stochastic gradient descent (DP-SGD). This training mechanism ensures that the influence of any single data point on the model's parameters remains bounded. The privacy-preserving training objective is defined by the following mechanism:

$$\theta_{t+1} = \theta_t - \eta \left(\frac{1}{|B|} \sum_{x_i \in B} \text{clip}(\nabla_\theta \ell(f_\theta(x_i), y_i), C) + \mathcal{N}(0, \sigma^2 C^2 I) \right)$$

where θ_t model parameters at step t, η learning rate, B mini-batch, ℓ loss function, C clipping norm, $\mathcal{N}(0, \sigma^2 C^2 I)$ Gaussian noise calibrated to privacy budget ε. This function enforces output indistinguishability across neighboring datasets by introducing noise after bounding sensitivity.

Following the completion of differentially private training through DP-SGD, it is essential to validate that the calibrated noise levels introduced during optimization are sufficient to uphold the intended privacy guarantees. This verification is not performed heuristically but is grounded in a formal post-training audit of the sensitivity of model parameters with respect to individual data contributions. Such a check ensures that the observed gradient magnitudes remain within the clipping threshold used during training and that the applied noise variance aligns with the target privacy budget ε.

To express this formally, the post-training noise calibration condition is defined as:

$$\|\nabla_\theta \ell(f_\theta(x), y)\|_2 \leq C, \quad \text{with} \quad \sigma^2 \geq \frac{2\log(1.25/\delta)}{\varepsilon^2}$$

where $\nabla_\theta \ell(f_\theta(x), y)$ input gradient of the loss with respect to model parameters, $\|\cdot\|_2$ denotes the L2 norm, C gradient clipping threshold, σ^2 noise variance, ε privacy budget, δ privacy relaxation parameter. This function validates that the accumulated privacy-preserving perturbations are both bounded and correctly distributed, confirming that the trained model adheres to differential privacy constraints under the chosen budget ε.

Following post-training verification of differential privacy parameters, the framework extends its operational resilience through dynamic privacy–utility monitoring. This involves a multi-objective optimization formulation, in which the learned model is continuously recalibrated to balance accuracy and privacy preservation. The composite loss function is defined as:

$$\min_\theta \mathcal{L}_{\text{total}} = \lambda_1 \cdot \mathcal{L}_{\text{utility}}(f_\theta(x), y) + \lambda_2 \cdot \mathcal{L}_{\text{privacy}}(\sigma, \varepsilon)$$

where θ model parameters, $\mathcal{L}_{\text{total}}$ total objective function, $\mathcal{L}_{\text{utility}}$ predictive loss, $\mathcal{L}_{\text{privacy}}$ penalty based on noise and privacy budget, $f_\theta(x)$ model output, y ground truth, σ noise standard deviation, ε privacy budget, λ_1, λ_2 scalar trade-off weights. This function enables ongoing adaptation of the model by minimizing utility loss while penalizing privacy leakage, thus achieving equilibrium between performance and regulatory compliance.

The proposed AI-driven model demonstrates a structured approach to anonymization that integrates foundational data constraints with formal learning mechanisms to safeguard individual privacy throughout the data processing lifecycle. Beginning with the identification of quasi-identifiers, the framework ensures that potentially linkable attributes are isolated and structurally grouped according to established privacy constraints, such as kk-Anonymity and ll-Diversity. These preparatory stages form the basis for a privacy-aware learning process driven by differentially private stochastic gradient descent, which introduces calibrated noise into the model's parameter updates. The system extends beyond model training by incorporating post-training audits that verify compliance with predefined noise bounds and privacy budgets. Finally, it introduces a feedback mechanism that dynamically balances utility and privacy through multi-objective optimization, allowing the model to adapt in response to changing sensitivity thresholds or performance expectations. Altogether, the framework articulates a privacy-preserving pipeline that aligns data structuring, model training, and operational monitoring within a coherent AI lifecycle.

The challenge of effective anonymization underscores the need for robust data governance frameworks that prioritize privacy by design. Ensuring that data is truly anonymized and cannot be re-identified requires not only technical safeguards but also ethical oversight to monitor how data is collected, processed, and used.

Framework: AI-Powered Anonymization and Privacy-Preserving Learning Framework (AI-APPLF)

Layered AI Architecture for K-Anonymity, L-Diversity, DP Training, and Lifecycle Privacy Validation

Layer	Main functionality	AI module(s) and components	Operational scope
Layer 1	Quasi-identifier recognition and data ingestion	Quasi-identifier detection engine	Raw tabular input → Quasi-identifier schema mapping → Feature-aligned dataset
Layer 2	Structural grouping for identity obfuscation	K-Anonymity grouping module	Feature-aligned dataset → K-Anonymity constraint application → Identity-obfuscated dataset
Layer 3	Attribute diversity enforcement	L-diversity enforcement module	Grouped anonymized data → L-Diversity filtering → Attribute-balanced equivalence classes
Layer 4	Privacy-preserving model training	Differentially Private Stochastic Gradient Descent (DP-SGD)	Anonymized training data → DP-SGD optimization → Privacy-preserving model parameters
Layer 5	Post-training noise calibration verification	Differential privacy compliance audit engine	Model gradients, output sensitivity → Noise-audit compliance check → Verified privacy guarantees
Layer 6	Privacy–utility monitoring and recalibration	Dynamic privacy–utility optimization system	Model outputs, audit metrics → Privacy-utility scoring → Trade-off-aware adjustment or retraining

7.1.4 Balancing Security and Privacy

To address the ethical dilemma of balancing public safety with privacy rights in AI-driven surveillance, a Human-Centric AI Surveillance Framework (HC-AISF) is introduced. The framework combines Multi-Objective Optimization (MOO) for modeling the security–privacy trade-off with Explainable AI (XAI) for transparent and reviewable decision-making. Although AI surveillance may support public security, its deployment must be carefully governed to prevent disproportionate intrusions into personal privacy (Bruneault and Sabourin Laflamme 2021). The HC-AISF formalizes a dual-objective approach in which surveillance decisions are optimized across competing societal priorities while remaining interpretable, auditable, and accountable.

Before AI-based processing or optimization occurs, the framework incorporates a secure ingestion layer to protect the integrity and privacy of surveillance data. This layer transforms raw surveillance signals into identity-hardened feature representations suitable for downstream analytical modules. Rather than exposing personally identifiable attributes directly, the model applies a cryptographic hash function with time-indexed salting to reduce reversibility risks and strengthen resistance to linkage or collision-based attacks. Through this preprocessing stage, sensitive identity-bearing inputs are converted into protected representations that support subsequent AI analysis while limiting unnecessary exposure of personal information.

$$H(x_i) = \mathrm{SHA} - 256(x_i \parallel t_i)$$

where $H(x_i)$ hashed identity representation of the i-th input, SHA − 256 cryptographic hash function from the SHA-2 family, x_i original raw surveillance input signal, t_i timestamp associated with x_i. This function transforms identity-sensitive components into a fixed-length, non-reversible fingerprint that prevents back-calculation or inference of original input values, enabling secure transmission to downstream AI modules without compromising individual privacy.

To formally encode the trade-off between maximizing surveillance efficacy and minimizing privacy erosion, a multi-objective optimization function is introduced:

$$\max_{\mathrm{w}} [\lambda_1 S(\mathrm{w}) - \lambda_2 P(\mathrm{w})]$$

where S(w), the security utility function, quantifies the effectiveness of AI surveillance in detecting threats, P(w), the privacy cost function, measures the degree of intrusion into individual privacy, λ_1 and λ_2, the weighting parameters, balance security and privacy trade-offs based on societal values and legal constraints, w, the AI model parameters, regulate the optimization between security enforcement and privacy preservation. This function ensures that AI-driven surveillance optimally balances security effectiveness with privacy protection, preventing overreach while maintaining public safety.

To enhance transparency in AI surveillance decision-making, an Explainable AI (XAI)-driven fairness constraint is introduced using Shapley Additive Explanations (SHAP):

$$\phi_i = \sum_{S \subseteq N \setminus \{i\}} \frac{|S|!(|N| - |S| - 1)!}{|N|!} [f(S \cup \{i\}) - f(S)]$$

where ϕ_i, the Shapley value, quantifies the contribution of feature i to AI surveillance decisions, S, a subset of surveillance data features excluding i, represents the groups of features used for evaluation, N, the total set of features, defines the AI decision space, $f(S)$, the AI model output given feature subset S, ensures accountability in AI decision-making. This function ensures that AI surveillance models provide interpretable justifications for security-based decisions, preventing arbitrary or discriminatory actions that could infringe on individual freedoms.

To ensure that interpretability signals identified by SHAP analysis contribute meaningfully to model adaptation, we incorporate a fairness-aware parameter update rule that integrates fairness constraints directly into the model optimization process. Building upon previously optimized utility objectives, this mechanism combines gradients of the standard prediction loss with those derived from fairness auditing signals.

$$\theta_{t+1} = \theta_t - \eta\left(\nabla_\theta \mathcal{L}_{\text{utility}} + \lambda \nabla_\theta \mathcal{L}_{\text{fairness}}\right)$$

where θ_t model parameters at step t, η learning rate, $\nabla_\theta \mathcal{L}_{\text{utility}}$ gradient of the prediction loss, λ fairness trade-off coefficient, $\nabla_\theta \mathcal{L}_{\text{fairness}}$ gradient of the fairness penalty. This function ensures that the AI model updates its parameters by jointly minimizing prediction error and fairness violations derived from audit feedback.

To maintain the integrity of system updates in dynamic operational environments, the framework includes a feedback adaptation module. This module receives post-deployment audit signals from SHAP analyses and adjusts the trade-off weights λ_1, λ_2 or triggers retraining if fairness deviations exceed predefined tolerances. The feedback policy is defined via:

$$\lambda_j^{(t+1)} = \lambda_j^{(t)} - \eta \frac{\partial \mathcal{L}_{\text{audit}}}{\partial \lambda_j}$$

where λ_j trade-off coefficient for ($j = 1, 2$), η learning rate, $\mathcal{L}_{\text{audit}}$ fairness audit loss from XAI signals. This function enables continual recalibration of the system in response to real-time fairness and accountability metrics.

By employing multi-objective optimization with explainable AI and fairness adaptation, this AI-driven framework ensures that surveillance mechanisms are not only technically effective but also ethically sound and legally accountable. It encodes a dynamic feedback loop between decision transparency, performance optimization, and fairness assurance—a necessary triad for responsible surveillance governance.

Framework: Security–Privacy Balanced AI Surveillance Framework (SPB-AISF)

Layered AI Architecture for Secure and Transparent Surveillance Decision-Making

Layer	Main functionality	AI module(s) and components	Operational scope
Layer 1	Secure signal ingestion and identity hashing	Encrypted signal processing gateway	Input → Raw surveillance signals → Output → Identity-hashed and encrypted feature input
Layer 2	Policy threshold initialization	Governance parameter encoding engine	Input → Governance constraints, legal parameters → Output → Encoded trade-off coefficients

(continued)

Layer	Main functionality	AI module(s) and components	Operational scope
Layer 3	Security–privacy trade-off optimization	Multi-Objective Optimization (MOO)	Input → Encoded threat and privacy features → Output → Balanced surveillance policy function
Layer 4	Decision attribution and interpretability	Shapley Additive Explanations (SHAP)	Input → Feature vectors, model outputs → Output → SHAP-based feature attribution scores
Layer 5	Fairness audit feedback adjustment	Gradient-based fairness adaptation	Input → SHAP audit signals, legal boundaries → Output → Policy weight recalibration
Layer 6	Secure logging and human oversight	Blockchain-based decision audit interface	Input → Decision pathways, model outputs → Output → Auditable human-reviewed surveillance record

For instance, during the COVID-19 pandemic, AI-driven surveillance systems were used for contact tracing and public health monitoring. While these systems were effective in controlling the spread of the virus, they also raised concerns about the long-term implications of such surveillance on privacy (Saheb 2023). Once the pandemic subsided, questions arose about whether governments would continue using these systems for broader surveillance purposes, potentially leading to the normalization of mass surveillance practices.

Figure 7.1 outlines the key ethical dilemmas in AI-driven surveillance and security systems, focusing on Privacy and Data Security Concerns, the Challenge of Anonymization, and Balancing Security and Privacy. "Privacy and Data Security Concerns" addresses issues like mass data collection without consent, risks of long-term data storage, cyberattacks, and data poisoning. "The Challenge of Anonymization" highlights the limitations of anonymization and the need for robust data governance to prevent re-identification risks. "Balancing Security and Privacy" explores the tension between security justifications and privacy rights, using examples like pandemic surveillance to illustrate the implications of temporary vs. long-term surveillance measures. Together, these elements underscore the complex ethical landscape of deploying AI in surveillance and security.

Fig. 7.1 Ethical dilemmas in AI surveillance and security

Privacy and data security concerns in AI surveillance systems are critical ethical challenges that must be addressed to ensure the responsible use of these technologies. As AI continues to be integrated into public safety and governance systems, it is essential to establish robust data governance frameworks that protect individuals' privacy and ensure the security of personal data. Balancing the benefits of AI for public safety with the protection of privacy rights will be key to building public trust and ensuring the ethical deployment of AI surveillance technologies.

7.2 Bias and Discrimination

Artificial intelligence (AI) surveillance systems present profound ethical dilemmas related to bias and discrimination, which stem from how these systems are developed, trained, and deployed. Bias in AI surveillance technologies can lead to discriminatory outcomes that disproportionately affect marginalized communities, exacerbating social inequalities and eroding public trust in AI systems (Sargiotis 2025b).

7.2.1 Sources of Bias in AI Surveillance

Bias in AI surveillance systems typically originates from the datasets used during model training. Machine learning algorithms operate on large-scale structured representations of reality, and if these representations are skewed, either by overrepresenting dominant groups or by underrepresenting marginalized demographics, the learned decision functions may reproduce these imbalances (Fosch-Villaronga and Poulsen 2022; Sargiotis 2025b). The result is algorithmic behavior that may appear statistically optimal, yet remains socially discriminatory. To address this challenge, we propose a fairness-aware AI surveillance framework grounded in encoder-based representation learning and adversarial fairness optimization.

The process begins with a neural encoder function designed to transform the raw surveillance input x, such as facial imagery or motion vectors, into a compact latent representation z. This transformation enables downstream modules to process high-dimensional data with structural consistency and fairness-enforcing constraints.

$$z = \phi(x) = f_\theta(x)$$

where z, latent representation, xx, input data vector, $\phi(\cdot)$, encoder function, f_θ, neural encoder with parameters θ. This function transforms raw input features into a structured latent space suitable for fairness-aware training.

Then, to ensure demographic neutrality within the learned representation z, we impose a fairness regularization term via Kullback-Leibler divergence. This penalty enforces that the posterior $p(z \mid x)$ remains close to a fairness-informed prior $p(z)$, discouraging the encoding of sensitive information. The full representation learning objective is expressed as follows:

$$L_{\text{FRL}} = \mathbb{E}_{p(x)}\left[D_{\text{KL}}\left(p(z|x)\middle\|p(z)\right)\right] + \lambda\sum_{i=1}^{N}\ell(f_{\text{w}}(z_i), y_i)$$

where L_{FRL}, the fairness-aware loss function, ensures that learned representations are unbiased, $p(x)$, the input data distribution, represents the observed demographic distribution, $p(z|x)$, the conditional latent representation, encodes feature transformations ensuring fairness, $p(z)$, the prior distribution, regulates fairness constraints, $D_{\text{KL}}(p(z|x)\|p(z))$, the Kullback-Leibler (KL) divergence, penalizes deviations from a fair latent space, λ, the regularization coefficient, controls the balance between fairness and predictive accuracy, $\ell(f_{\text{w}}(z_i), y_i)$, the prediction loss function, ensures model effectiveness across demographic groups. This function ensures that AI-generated representations are unbiased, preserving equitable performance across all demographic subgroups.

The fairness-enforced representations z are then processed through a predictive function to yield surveillance-related decisions $\hat{y}$. This classifier operates exclusively on bias-reduced features and plays a pivotal role in linking representation learning with final model predictions.

$$\hat{y} = f_w(z)$$

where $\hat{y}$ predicted surveillance outcome, f_w, predictive classifier, z, fairness-aware latent representation. This function produces the model's primary output using representations that minimize demographic bias.

To ensure that no residual demographic information can be extracted from the model's internal or external behavior, we introduce an adversarial bias detection network g_v. Its purpose is to infer the protected attribute aa from the input xx, testing whether the model inadvertently encodes protected characteristics.

$$\hat{a} = g_v(x)$$

where $\hat{a}$, predicted protected attribute, g_v, adversarial debiasing network, x, input data. This function identifies hidden demographic signals and supports adversarial fairness enforcement.

The debiasing mechanism is embedded into a joint optimization framework where the classifier maximizes predictive accuracy, while the adversary attempts to reconstruct demographic attributes. The resulting adversarial loss is formulated to penalize demographic leakage:

$$L_{\text{adv}} = \sum_{i=1}^{N}\ell(f_{\text{w}}(x_i), y_i) - \gamma\sum_{i=1}^{N}\ell(g_{\text{v}}(x_i), a_i)$$

where L_{adv}, the adversarial fairness loss function, penalizes biased AI predictions, $f_w(x_i)$, the AI classifier, predicts the primary outcome y_i, $g_v(x_i)$, the adversarial debiasing network, predicts the protected attribute a_i, ensuring that biased information is not used in decision-making, γ, the fairness regularization coefficient, controls the strength of bias removal, $\ell\ (\cdot)$, the loss function, ensures optimization. This function enforces fairness by reducing the model's ability to retain or exploit protected demographic features.

To evaluate the fairness of decisions post-inference, a fairness audit function is applied. One canonical measure, demographic parity difference, quantifies the divergence in outcome distributions between protected groups:

$$\Delta_{\text{parity}} = | P(\hat{y} = 1 \mid a = 0) - P(\hat{y} = 1 \mid a = 1) |$$

where Δ_{parity}, demographic parity gap, $P(\hat{y} = 1 \mid a = 0)$, probability of positive outcome for group 0, $P(\hat{y} = 1 \mid a = 1)$, probability of positive outcome for group 1. This function quantifies the fairness gap in decision outcomes across sensitive demographic partitions.

Finally, to accommodate evolving demographic dynamics and to ensure temporal fairness consistency, the model undergoes continual retraining based on fairness drift. This is achieved through gradient-based update of encoder parameters using a total loss signal:

$$\theta \leftarrow \theta - \eta \nabla_{\theta} \mathcal{L}_{\text{total}}$$

where θ, encoder model parameters, η, learning rate, $\nabla_{\theta} \mathcal{L}_{\text{total}}$, gradient of total fairness loss. This function enables the encoder to adapt over time based on fairness feedback metrics and surveillance outcomes.

Through this architecture from input encoding and fairness representation learning to adversarial debiasing, fairness auditing, and continual adaptation, the proposed framework mathematically ensures that AI surveillance systems operate without embedding demographic discrimination, thus reinforcing public trust, ethical accountability, and policy-aligned governance. This issue is particularly concerning with facial recognition technologies, where biased training data has been shown to result in higher error rates for people with darker skin tones, women, and other marginalized populations (Saheb 2023). Studies have demonstrated that facial recognition systems often misidentify people of color at disproportionately higher rates compared to white individuals, leading to discriminatory outcomes in law enforcement and public safety applications. Such biases can result in wrongful arrests, increased surveillance of minority communities, and further marginalization of already vulnerable groups.

Framework: Fairness-Aware Surveillance Representation Framework (FASRF)

Layered AI Architecture for Debiasing Facial Recognition and Surveillance Systems

Layer	Main functionality	AI module(s) and components	Operational scope (Input → Processing → Output)
Layer 1	Input encoding of demographic and visual features	Neural encoder	Facial surveillance data → Representation encoding → Processed feature space
Layer 2	Representation fairness enforcement	Regularized representation learning	Encoded features → Fairness constraints applied → Demographic-invariant features
Layer 3	Outcome prediction from encoded features	Fairness-conscious classifier	Fair representations → Classification process → Surveillance decision
Layer 4	Adversarial detection of demographic leakage	Bias detection network	Input data → Protected attribute prediction → Exposure of bias patterns
Layer 5	Bias-gradient penalty enforcement	Adversarial fairness loss	Model outputs and protected predictions → Fairness regularization → Bias-mitigated outputs
Layer 6	Surveillance decision auditing	Equity audit and compliance reporting system	Surveillance results → Equity audit process → Fairness verification report
Layer 7	Continual adaptation with real-time fairness feedback	Continuous fairness monitoring and adaptive retraining engine	Performance monitoring → Fairness feedback loop → Model refinement and retraining

Bias in AI surveillance systems is not only rooted in data imbalance but is further exacerbated by the homogeneity of the teams developing these technologies. A lack of diversity in development teams can propagate implicit biases into the design, training, and deployment of AI models, leading to structural inequities in system outcomes (Bruneault and Sabourin Laflamme 2021). To address bias originating from both algorithmic behavior and dataset composition, we introduce a dual-module AI-driven bias mitigation framework. This architecture integrates (i) Fairness-Aware Reinforcement Learning (FARL) for decision-level bias reduction and (ii) Diverse Generative Adversarial Networks (D-GANs) for representative data synthesis.

The system begins by defining the surveillance environment as a Markov Decision Process (MDP), where the agent observes state vectors $s_t \in S$, selects surveillance actions $a_t \in A$, and transitions to the next state while receiving a fairness-aware reward signal. The agent's behavior is governed by a stochastic policy $\pi(a_t \mid s_t)$, which must be optimized not only for utility, but also for fairness across protected demographic attributes.

To model this trade-off mathematically, we define the reward function as:

$$R(s_t, a_t) = U(s_t, a_t) - \lambda \cdot D_{\mathrm{KL}}\left(P(\hat{Y}|A) \middle\| P(\hat{Y})\right)$$

where $R(s_t, a_t)$, the fairness-aware reward function, optimizes AI decision-making by balancing utility and fairness, $U(s_t, a_t)$, the utility function, quantifies the effectiveness of AI-driven surveillance decisions, λ, the fairness regularization parameter, controls the trade-off between performance and bias reduction, $D_{\mathrm{KL}}\left(P(\hat{Y}|A)\middle\|P(\hat{Y})\right)$, the Kullback-Leibler (KL) divergence, penalizes disparities in AI predictions across demographic groups. This function ensures that AI surveillance models are explicitly optimized to reduce bias, preventing discrimination against underrepresented populations.

Given this reward structure, the objective of the reinforcement learning module becomes to learn a policy that maximizes the expected cumulative discounted fairness-aware return:

$$\pi^* = \arg \underset{\pi}{max}\, \mathbb{E}_\pi \left[\sum_{t=0}^{T} \gamma^t R(s_t, a_t) \right]$$

where π^*, optimal policy, γ, discount factor, $R(s_t, a_t)$, fairness-aware reward from above. This function directs learning toward long-term policies that balance performance and equity.

However, optimizing policy fairness alone is insufficient if the training data itself is biased or demographically incomplete. To address representational gaps in the training dataset, we deploy a Diverse Generative Adversarial Network (D-GAN), designed to synthesize synthetic samples that enhance demographic variety and improve downstream generalization.

The D-GAN framework comprises a generator network $G(z)$ that produces synthetic data samples from noise z, and a discriminator $D(x)$ that distinguishes between real and generated samples. The training objective of the D-GAN is extended with a diversity regularization term using Jensen–Shannon divergence:

$$\underset{G}{min}\, \underset{D}{max}\, \mathbb{E}_{x\sim p_{\mathrm{data}}}[\log D(x)] + \mathbb{E}_{z\sim p_z}[\log(1 - D(G(z)))] + \beta \cdot \mathbb{E}_A \left[D_{\mathrm{JS}}\left(p^A_{\mathrm{data}} \middle\| p^A_{\mathrm{gen}}\right) \right]$$

where $G(z)$, the generator network, synthesizes new data samples, $D(x)$, the discriminator network, distinguishes between real and synthetic samples, pdatapdata, the real data distribution, represents existing AI training data, p_z, the noise prior, serves as the input for the generator, β, the diversity regularization parameter, controls demographic balance in generated data, $D_{\mathrm{JS}}\left(p^A_{\mathrm{data}}\middle\|p^A_{\mathrm{gen}}\right)$, the Jensen–Shannon divergence, ensures that AI-generated data distribution p^A_{gen} aligns with the real-world demographic distribution p^A_{data}. This function ensures that AI training datasets remain demographically diverse, reducing the risk of underrepresentation-induced bias.

The generated synthetic data from $G(z)$ is subsequently fed into the fairness-aware reinforcement learner, either augmenting the existing dataset or initializing the environment model. This establishes an end-to-end fairness reinforcement loop, wherein representative data and reward-level bias mitigation co-operate to produce equitable decision policies.

Following the generation of demographically diverse synthetic data samples via the generator G(z)G(z), it is critical to formally calibrate the alignment between the generated data distribution and the empirical demographic structure. To ensure this, we integrate a divergence regularization mechanism, utilizing the Jensen–Shannon (JS) divergence between the protected attribute distributions of the real and generated datasets. This step is pivotal for safeguarding demographic representativeness before the augmented data is reintroduced into the reinforcement learning framework.

$$D_{\mathrm{JS}}\left(\hat{p}_{\mathrm{data}}^{A} \middle\| \hat{p}_{\mathrm{gen}}^{A}\right)$$

where D_{JS}, Jensen–Shannon divergence, $\hat{p}_{\mathrm{data}}^{A}$, empirical demographic distribution over attribute A, $\hat{p}_{\mathrm{gen}}^{A}$, generator-induced demographic distribution. This function measures the divergence between the real and generated demographic distributions and is minimized to enforce population-level alignment within the augmented dataset.

Upon achieving demographic alignment in the synthesized data, the next stage involves retraining the fairness-aware reinforcement learning agent using this newly balanced dataset. The original policy optimization process, previously governed by the reward function $R(s_t, a_t)$, is now extended to integrate training with augmented trajectories. This retraining updates the parameters θθ of the policy model using a total loss function that captures both fairness and utility:

$$\theta \leftarrow \theta - \eta \nabla_{\theta} \mathcal{L}_{\mathrm{total}}^{\mathrm{aug}}$$

where θ, RL model parameters, η, learning rate, ∇_{θ}, gradient operator, $\mathcal{L}_{\mathrm{total}}^{\mathrm{aug}}$, total augmented loss incorporating fairness-aware rewards and policy regularization. This function adjusts the agent's decision-making policy using synthetic data, promoting generalization across previously underrepresented groups.

Following model retraining, it becomes essential to quantitatively audit the fairness of AI decisions across sensitive demographic groups. One widely accepted statistical fairness criterion is demographic parity, which evaluates whether the likelihood of a positive outcome differs between protected groups. This is formally expressed as:

$$\Delta_{\mathrm{parity}} = \mid P(\hat{y} = 1 \mid a = 0) - P(\hat{y} = 1 \mid a = 1) \mid$$

where Δ_{parity}, demographic parity gap, $P(\hat{y} = 1 \mid a = 0)$, probability of positive classification for group $a = 0$, $P(\hat{y} = 1 \mid a = 1)$, probability of positive classification for group $a = 1$. This function quantifies disparities in surveillance decisions and serves as a fairness auditing metric for validating model behavior.

By unifying fairness-aware reinforcement learning with demographically calibrated generative modeling, this dual-framework approach ensures that AI surveillance systems address bias both algorithmically and structurally. The reward function enforces equitable decision behavior, while the D-GAN enhances the demographic richness of training distributions. Together, they form an ethically grounded solution for developing inclusive and trustworthy AI in public safety contexts.

Framework: Fairness-Driven Reinforcement and Data Augmentation Architecture (FDR-DAA)

Layered AI Architecture for Bias Mitigation in Surveillance via *Reinforcement Learning and Diverse GANs*

Layer	Main functionality	AI module(s) and components	Operational scope (Input → Processing → Output)
Layer 1	Surveillance environment modeling	Markov Decision Process (MDP)	Environmental states, actions → State transition function → RL state space
Layer 2	Policy learning with fairness-aware rewards	Fairness-Aware Reinforcement Learning (FARL)	States, actions → Reward function with KL penalty → Updated fairness-optimized policy
Layer 3	Optimal policy generation	Stochastic policy optimization	Fairness-weighted returns → Policy gradient or value iteration → Optimal action policy
Layer 4	Synthetic data generation	Diverse Generative Adversarial Networks (D-GANs)	Noise vector, demographic condition → Generator-discriminator training → Demographically rich synthetic data
Layer 5	Data distribution calibration	JS divergence regularization	Real and generated data distributions → Diversity alignment → Balanced data representation
Layer 6	Augmented RL training	Reinforcement learning with data augmentation	Synthetic demographic data → Retraining RL policy → Bias-reduced decision agent
Layer 7	Bias auditing and metric evaluation	Fairness metric evaluation and reporting system	RL decisions, demographic groups → Fairness metrics (e.g., parity, equality) → Bias audit report
Layer 8	Continual adaptation and feedback integration	Adaptive policy feedback and optimization engine	Fairness feedback, model drift → Policy refinement loop → Long-term fairness adaptation

7.2.2 *Discriminatory Impacts of AI Surveillance*

The discriminatory effects of bias in AI surveillance systems are wide-ranging and have significant societal implications. In law enforcement, biased AI systems can lead to the over-policing of minority communities, reinforcing historical patterns of discrimination (Fosch-Villaronga and Poulsen 2022). Predictive policing algorithms, for example, often rely on historical crime data, which may reflect biases in how policing has been historically conducted. As a result, these algorithms may disproportionately target certain neighborhoods or demographic groups, leading to a cycle of over-policing and criminalization. To mitigate these effects, we introduce a fairness-constrained AI policing framework that integrates Causal Inference for Bias Detection and Fair Reinforcement Learning (Fair-RL) for equitable policy optimization.

The encoding of raw surveillance inputs into a compact and learnable representation forms the basis of the downstream fairness-aware inference pipeline. Given input data x, a parameterized neural encoder extracts the latent representation z through the transformation:

$$z = f_{\theta}(x)$$

where z, encoded latent representation, f_{θ}, neural encoder parameterized by θ, x, raw surveillance input. This function transforms high-dimensional observational data into a structured and trainable representation suitable for fair downstream predictions.

Building upon the latent encoding z, the prediction function aims to infer surveillance-relevant outcomes. However, to mitigate algorithmic bias, this predictive mechanism must be designed to maintain fairness across demographic groups. The fairness-aware classification function is denoted as:

$$\hat{y} = f_{w}(z)$$

where $\hat{y}$, predicted outcome, f_w, classifier parameterized by w, z, latent input representation. This function generates label predictions while serving as the foundation for subsequent fairness-regularized loss evaluation.

To operationalize these predictions in a sequential decision-making context, it is imperative to define the state upon which policy optimization depends. This is achieved by projecting the encoded features into a temporally aware state space representation:

$$s_t = \phi(z_t)$$

where s_t, Markovian state at time t, ϕ, state mapping function, z_t, latent features at time t. This function aligns latent representations with the structure required by reinforcement learning models to facilitate dynamic policy learning.

To enforce fairness at the prediction level, we define a fairness-constrained loss function that penalizes disparities across protected demographic groups. This function balances predictive accuracy with fairness regularization:

$$\mathcal{L}_{\text{fair}} = \sum_{i=1}^{N} \ell(f_w(z_i), y_i) + \lambda \cdot \mathbb{E}_{p(x)}\left[D_{\text{KL}}\left(P(\hat{Y} \mid A) \middle\| P(\hat{Y})\right)\right]$$

where $\mathcal{L}_{\text{fair}}$, total fairness-constrained loss, $f_w(z_i)$, prediction function, y_i, observed label, λ, regularization coefficient, D_{KL}, divergence between group-conditional and marginal predicted distributions. This function ensures that learned predictions remain unbiased across sensitive attributes such as race or location.

With the state s_t defined, the system proceeds to construct a probabilistic policy that governs the selection of enforcement actions. The policy function maps each state to a distribution over possible actions:

$$\pi_\theta(a_t \mid s_t)$$

where $\pi_\theta(a_t \mid s_t)$, probability of selecting action a_t in state s_t, θ, parameters of the stochastic policy, a_t, candidate enforcement action, s_t, current system state. This function enables principled exploration of surveillance strategies while preparing the foundation for optimization.

To refine this policy under fairness constraints, we integrate the fairness-augmented Proximal Policy Optimization (PPO) objective. This function jointly maximizes expected advantages while penalizing unfair predictions through KL divergence regularization:

$$J(\theta) = \mathbb{E}\left[min\left(\frac{\pi_\theta(a_t|s_t)}{\pi_{\theta_{\text{old}}}(\text{a}_t|s_t)}A_t, \text{clip}\left(\frac{\pi_\theta(a_t|s_t)}{\pi_{\theta_{\text{old}}}(a_t|s_t)}, 1-\epsilon, 1+\epsilon\right)A_t\right)\right]$$
$$-\gamma \cdot D_{\text{KL}}\left(P(\hat{Y}_t \mid A) \parallel P(\hat{Y}_t)\right)$$

where $J(\theta)$, the fairness-aware policy optimization function, regulates AI-driven policing decisions, $\pi_\theta(a_t| s_t)$, the AI policy, determines the probability of selecting enforcement action a_t given the state s_t, $\pi_{\theta_{\text{old}}}(a_t|s_t))$, the previous AI policy, represents past decision probabilities, A_t, the advantage function, quantifies the benefit of taking action a_t in state s_t, γ, the fairness regularization coefficient, penalizes biased predictions. This function ensures that AI policing policies dynamically adjust to balance fairness and law enforcement efficiency, preventing the disproportionate targeting of specific communities.

This AI-driven framework ensures that predictive policing systems remain unbiased and socially responsible. The causal fairness constraint detects and corrects bias in crime data analysis, while the Fair-RL function optimizes law enforcement policies to prevent systemic discrimination. This dual-framework approach mitigates over-policing, reinforces fairness in AI-driven surveillance, and fosters trust in AI-assisted public safety initiatives.

Framework: Fairness-Constrained Predictive Policing Architecture (FCPPA)

Layered AI Architecture for Bias Mitigation in AI-Driven Surveillance and Predictive Policing

Layer	Main functionality	AI module(s) and components	Operational scope (Input → Processing → Output)
Layer 1	Input encoding of surveillance features	Neural encoder	Raw surveillance data → Transformation process → Encoded feature space

(continued)

Layer	Main functionality	AI module(s) and components	Operational scope (Input → Processing → Output)
Layer 2	Fairness-constrained prediction modeling	Fairness-aware classifier	Encoded data → Fair prediction process → Bias-mitigated outputs
Layer 3	State definition for policy learning	State Mapper	Encoded features → State extraction process → Policy input states
Layer 4	Policy initialization and surveillance strategy	Stochastic policy model	Policy input states → Policy initialization → Action probabilities
Layer 5	Fairness-constrained policy optimization	Proximal Policy Optimization (PPO)	Action trajectories → Optimization with fairness control → Updated policy
Layer 6	Bias auditing and action disparity evaluation	Fairness disparity evaluation and reporting system	Predicted actions by group → Disparity measurement → Fairness audit report
Layer 7	Continual policy refinement and feedback loop	Continuous policy adaptation and fairness feedback engine	Fairness audit results → Policy refinement feedback → Adapted surveillance strategy

Moreover, biased AI surveillance systems can perpetuate discrimination in adjacent sectors, such as housing, employment, and public services. For example, surveillance systems deployed in public spaces may generate behavioral, demographic, or geospatial profiles that later inform decisions about resource allocation or service provision, disproportionately benefiting certain groups while disadvantaging others (Saheb 2023). These discriminatory effects undermine the principles of fairness and equality, contributing to the entrenchment of social inequalities. To address this risk, a fairness-constrained AI resource allocation framework is introduced, unifying Fairness-Aware Decision Optimization (FADO) with counterfactual fairness analysis.

To support fair allocation of public services and resources, the system begins by processing raw behavioral, demographic, and geospatial data through an encoding process. This transformation maps heterogeneous inputs into a structured representation suitable for downstream fairness-aware decision-making. The encoder function transforms the raw inputs into a latent representation, expressed as:

$$z_i = \sigma(W_e x_i + b_e)$$

where z_i, latent representation, W_e, encoder weight matrix, b_e, bias vector, x_i, input feature vector, σ, nonlinear activation function. This function performs a nonlinear feature transformation, converting the high-dimensional raw data into a structured, encoded form suitable for further analysis, enabling a robust foundation for fairness-aware resource allocation decisions.

Building on this, the encoded features are then passed through a prediction model, parameterized by appropriate weights and bias terms. This model outputs a probabilistic outcome, which estimates eligibility for services such as housing or public support. The process of generating these outcomes is captured by the following equation:

$$\hat{y}_i = \text{softmax}\left(W_p z_i + b_p\right)$$

where $\hat{y}_i$, predicted outcome probabilities, W_p, prediction model weights, b_p, bias vector, z_i, latent input. This equation converts the encoded features into a probability distribution over potential outcomes, thereby determining eligibility for resource allocation in a manner consistent with the fairness objectives defined in earlier stages of the model.

To ensure equitable decision-making, we introduce a fairness-constrained objective that augments the empirical risk with a demographic bias penalty using Kullback-Leibler (KL) divergence:

$$\min_{W_p} \frac{1}{N} \sum_{i=1}^{N} \mathcal{L}(\hat{y}_i, y_i) + \lambda D_{\mathrm{KL}}\left[P(\hat{Y} \mid A) \middle\| P(\hat{Y})\right]$$

where W_p, the AI model parameters, regulate resource allocation decisions, $\mathcal{L}$, prediction loss function, y_i, true label, λ, fairness regularization coefficient, D_{KL}, KL divergence, A, protected attribute. This function minimizes prediction error while penalizing allocation disparities across demographic groups.

To further ensure individual-level fairness, a counterfactual test is embedded to assess whether the AI's decision would remain stable under hypothetical changes in protected attributes, while keeping other features constant:

$$P\left(\hat{Y}_{A=a} | X\right) = P\left(\hat{Y}_{A=a'} | X\right)$$

where $\hat{Y}_{A=a}$, predicted outcome under attribute value a, $\hat{Y}_{A=a'}$, counterfactual outcome under altered attribute a', X, observed features. This ensures decisions are independent of protected characteristics and reveal hidden structural bias.

Together, this dual-function framework FADO for fairness-constrained optimization and counterfactual analysis for individual bias verification prevents systemic exclusion in resource allocation processes. The architecture promotes equitable outcomes in public service systems and reinforces social justice through AI-driven policy mechanisms.

Framework: Fairness-Constrained AI Resource Allocation Framework (FCRA-RF)

Layered AI Architecture for Equitable Decision-Making in Surveillance-Informed Public Services

Layer	Main functionality	AI module(s) and components	Operational scope
Layer 1	Input structuring and normalization	Data normalization and structuring engine	Raw behavioral, biometric, and geospatial data → Normalization and standardization → Structured input vectors

(continued)

Layer	Main functionality	AI module(s) and components	Operational scope
Layer 2	Latent feature encoding	Feedforward Neural Encoder (ReLU)	Structured input vectors → Feature transformation → Latent representations
Layer 3	Resource eligibility prediction	Softmax classifier	Latent representations → Predicted outcome probabilities → Resource eligibility decisions
Layer 4	Fairness-constrained optimization	Empirical Risk + KL divergence minimization	Prediction error and demographic data → Minimize prediction error and fairness disparity → Optimized fair allocation policy
Layer 5	Individual-level bias detection	Counterfactual fairness test	Predicted outcomes and protected attributes → Check for fairness under hypothetical changes → Validate individual fairness
Layer 6	Continuous equity monitoring	Long-term fairness tracking and disparity monitoring system	Allocation history and demographic outcomes → Update disparity metrics → Track long-term fairness
Layer 7	Feedback and policy adjustment	Adaptive policy adjustment and equity feedback engine	Monitoring results and citizen feedback → Refine policies → Updated model parameters and decision-making

7.2.3 Mitigating Bias and Discrimination in AI Surveillance

Addressing bias and discrimination in AI surveillance systems requires a multifaceted approach that includes both technical and ethical interventions. One key strategy is ensuring that the datasets used to train AI surveillance systems are diverse and representative of the populations they are intended to serve (Fosch-Villaronga and Poulsen 2022). This involves collecting and curating data that accurately reflects demographic, socioeconomic, and geographic diversity, as well as regularly auditing datasets for potential biases. To address these requirements, a Bias-Resilient AI Surveillance Framework is introduced, integrating Fair Representation Learning (FRL) for data diversity with Bias-Aware Dataset Auditing through Adversarial Debiasing Networks (ADNs). The Fair Representation Learning (FRL) objective function for dataset diversity is formulated as:

$$L_{\mathrm{FRL}} = \mathbb{E}_{p(x)}\left[D_{\mathrm{KL}}\left(p(z|x)\middle\|p(z)\right)\right] + \lambda\sum_{i=1}^{N}\ell(f_{\mathbf{w}}(z_i), y_i)$$

where L_{FRL}, the fairness-aware learning objective, ensures that learned representations remain unbiased, $p(x)$, the input data distribution, represents real-world demographic diversity, $p(z|x)$, the conditional latent representation, encodes fair feature transformations, $p(z)$, the prior distribution, regularizes fairness constraints,

$D_{KL}(p(z|x)\|p(z))$, the Kullback-Leibler (KL) divergence, penalizes deviations from unbiased data representations, v, the fairness regularization coefficient, balances between accuracy and fairness, $\ell(f_{\mathbf{w}}(z_i), y_i)$, the prediction loss function, ensures effective AI model training. This function ensures that AI models learn from demographically diverse and unbiased datasets, mitigating disparities in AI-driven surveillance.

To further systematically audit datasets for bias, an Adversarial Debiasing Network (ADN) for bias-aware dataset auditing is introduced:

$$L_{\mathrm{ADN}} = \sum_{i=1}^{N} \ell(f_{\mathrm{w}}(x_i), y_i) - \gamma \sum_{i=1}^{N} \ell(g_{\mathrm{v}}(x_i), a_i)$$

where L_{ADN}, the adversarial debiasing loss function, ensures that AI models do not learn biased patterns, $f_{\mathrm{w}}(x_i)$, the AI classifier, predicts the primary outcome y, $g_{\mathrm{v}}(x_i)$, the adversarial debiasing network, detects and removes patterns correlated with protected attributes a_i, γ, the fairness regularization coefficient, penalizes reliance on biased data. This function ensures that datasets used for AI surveillance undergo continuous bias auditing, allowing for systematic identification and correction of discriminatory data patterns.

Following the successful debiasing and fair representation learning steps, the model progresses to the prediction phase, where the learned representations are used to derive meaningful outcomes. In this phase, the Softmax Classifier is employed to map the learned features into interpretable predictions. This transition ensures that the predictions are based on the fair representations generated in earlier stages, aligning with the fairness constraints imposed throughout the process. The prediction function is given by:

$$\hat{y}_i = \mathrm{softmax}\left(W_p z_i + b_p\right)$$

where $\hat{y}_i$, predicted outcome probabilities, W_p, prediction model weight matrix, z_i, latent input from fair representations, b_p, bias vector. This function computes the predicted outcomes by transforming the encoded features into a probability distribution over the possible outcomes, ensuring the decision-making process remains fair and unbiased.

This AI-driven model demonstrates a structured approach to addressing bias and discrimination within AI surveillance systems, ensuring that decision-making processes remain both effective and equitable. The framework integrates advanced AI techniques, such as Fair Representation Learning (FRL) and Adversarial Debiasing Networks (ADN), to ensure that the datasets used for training are diverse, balanced, and free from discriminatory patterns. The FRL component enables the model to generate fair, latent representations of the input data, while the ADN function continuously monitors and corrects for bias within the dataset. Following these preprocessing steps, the model transitions to a prediction phase, where the Softmax Classifier converts the learned features into outcome probabilities, such as eligibility or resource allocation, based on the fair representations. The flow of this framework ensures that each decision point, from data collection to prediction, adheres to fairness constraints, mitigating the risk of bias while optimizing predictive accuracy. Ultimately, this AI model establishes a balanced approach to decision-making, promoting fairness in AI-driven systems while maintaining performance and utility across diverse populations.

Framework: Bias-Resilient AI Surveillance Framework (BR-ASF)

Layered AI Architecture for Bias-Aware Data Curation and Fairness in Surveillance Systems

Layer	Main functionality	AI module(s) and components	Operational scope
Layer 1	Data collection and preprocessing	Demographic and geospatial data preprocessing engine	Raw demographic, socioeconomic, and geographic data → Normalization and curation → Preprocessed input vectors
Layer 2	Fair representation learning	Fair representation learning	Input data → Optimization → Unbiased latent representation
Layer 3	Bias-aware dataset auditing	Adversarial debiasing network	Preprocessed data → Optimization → Bias-free data transformations
Layer 4	Outcome prediction	Softmax Classifier	Fair representations → Outcome prediction → Predicted outcomes
Layer 5	Continuous monitoring and bias auditing	Ongoing fairness audit and bias metric evaluation system	Predicted outcomes, feedback data → Bias metrics assessment → Updated fairness audits
Layer 6	Feedback and adjustment	Dynamic bias feedback and model adjustment engine	Bias reports, stakeholder feedback → Refinement of training and auditing processes

In AI surveillance systems, fairness metrics are crucial for evaluating how well the system performs across different demographic groups. These metrics enable practitioners to identify potential biases in the system's decision-making and take corrective action to mitigate disparities (Bruneault and Sabourin Laflamme 2021). In addition, transparency and explainability are essential for building trust in the system, ensuring that stakeholders can understand the rationale behind AI decisions. To achieve this, a fairness-aware AI performance evaluation framework is introduced, incorporating demographic fairness metrics for bias detection and Explainable AI (XAI) for decision transparency.

The first stage of the framework involves preprocessing the raw data, which includes demographic, socioeconomic, and geographic information. Proper preprocessing helps ensure that the data is well structured and normalized, laying a strong foundation for subsequent analysis in later AI layers.

Once the data has been preprocessed, the Fair Representation Learning (FRL) module is introduced. Its aim is to create a latent representation of the data that supports fairness by reducing bias in the input features. The FRL module optimizes this latent space so that the learned features become less sensitive to discriminatory patterns while preserving the essential decision-relevant characteristics of the data.

The fairness metric is designed to quantify disparities in the AI decision-making process by measuring how the system's outcomes differ across protected and non-protected demographic groups. The demographic disparity metric Δ_{disp} is introduced to capture this difference:

$$\Delta_{\text{disp}} = \mid P(\hat{Y} = 1 | A = 0) - P(\hat{Y} = 1 | A = 1) \mid$$

where Δ_{disp}, the demographic disparity metric, quantifies fairness in AI decision-making, $P(\hat{Y} = 1 | A = 0)$, the probability of a favorable outcome for the non-protected group, represents AI model predictions for one demographic, $P(\hat{Y} = 1 | A = 1)$, the probability of a favorable outcome for the protected group, ensures fairness across populations. This function ensures that AI surveillance systems do not disproportionately favor or disadvantage specific demographic groups, identifying disparities that require corrective action.

Once the fairness metric is in place, continuous bias detection and correction are essential to maintain system fairness over time. At this stage, the framework evaluates the model's outputs using the demographic fairness metric Δ_{disp} to monitor for any emerging disparities. The function Δ_{disp} quantifies the disparity in predictions between protected and non-protected groups, as discussed earlier. By comparing the model's outcomes across demographic groups, the framework identifies disparities that require corrective measures.

$$\text{Correction Trigger} = \begin{cases} \Delta_{\text{disp}} > \epsilon & \text{Apply corrective action} \\ \Delta_{\text{disp}} \leq \epsilon & \text{Continue monitoring} \end{cases}$$

where Δ_{disp} p, demographic disparity metric, ϵ, the predefined fairness threshold. This function ensures that corrective measures are applied when the disparity exceeds the fairness threshold, allowing the system to dynamically adapt and prevent systemic bias.

Transparency in AI decisions is vital for establishing accountability. Explainable AI (XAI) helps stakeholders understand the rationale behind decisions, which is crucial for trust. A key component of XAI is the Shapley Additive Explanation (SHAP), a method used to attribute contributions of each feature in the decision-making process. The Shapley value ϕ_i is computed for each feature ii to quantify its contribution to the AI's output. The Shapley value is given by:

$$\phi_i = \sum_{S \subseteq N \setminus \{i\}} \frac{\mid S \mid !(|N| - |S| - 1)!}{\mid N \mid !} [f(S \cup \{i\}) - f(S)]$$

where ϕ_i, the Shapley value, quantifies the contribution of feature i to AI surveillance decisions, S, a subset of AI input features excluding i, determines feature influence on model predictions, N, the total set of features, defines the AI decision space, $f(S)$, the AI model's output based on subset S, ensures interpretability in AI decision-making. This function ensures that stakeholders can trace AI decisions to their contributing factors, improving accountability and transparency.

The core of the AI system is the outcome prediction phase, where the AI model produces predictions based on the unbiased, fair features obtained in the earlier phases. Using the Softmax Classifier, the model assigns probabilities to each possible outcome based on the learned features. The prediction function is defined as:

$$\hat{y}_i = \text{softmax}\left(W_p z_i + b_p\right)$$

where $\hat{y}_i$, predicted outcome probabilities, W_p, the prediction model weight matrix, z_i, the latent input from fair representations, b_p, bias vector. This function converts the learned features into a probability distribution, generating predictions that inform decisions such as eligibility for housing, employment, or other public services.

This AI-driven framework ensures that AI surveillance systems are both fair and interpretable. The fairness metric function enables systematic detection of demographic disparities, while the XAI function provides transparent, justifiable AI decisions. This dual-framework approach fosters accountable AI-driven governance, ensuring equitable treatment across populations and reinforcing trust in AI-powered surveillance systems.

Framework: Fairness-Aware AI Performance Evaluation Framework (FAAI-PEF)

Layered AI Architecture for Bias Detection and Decision Transparency in AI Surveillance Systems

Layer	Main functionality	AI module(s) and components	Operational scope
Layer 1	Data collection and preprocessing	Demographic and socio-economic data normalization engine	Raw demographic, socioeconomic, and geographic data → Normalization and curation → Preprocessed input vectors
Layer 2	Fair representation learning	Fair Representation Learning (FRL)	Input data → Optimization → Unbiased latent representation
Layer 3	Bias detection and correction		Preprocessed data → Fairness metrics evaluation → Continuous bias identification and correction
Layer 4	Explainable AI (XAI) for feature attribution	Shapley Additive Explanations (SHAP)	Fair representations → Feature contribution calculations → Interpretability in decision-making
Layer 5	Outcome prediction	Softmax classifier	Fair representations → Outcome prediction → Predicted outcomes
Layer 6	Continuous monitoring and bias auditing	Real-time fairness monitoring and bias evaluation system	Predicted outcomes, feedback data → Bias metrics assessment → Updated fairness audits
Layer 7	Feedback and adjustment	Adaptive fairness refinement and model calibration engine	Bias reports, stakeholder feedback → Refinement of training and auditing processes

Another important strategy for mitigating bias is increasing diversity in the development of AI technologies. By involving individuals from diverse backgrounds in the design, development, and deployment of AI systems, organizations can reduce the risk of embedding biased assumptions into technology (Saheb 2023). This includes not only technical expertise but also ethical oversight to ensure that AI systems are aligned with societal values of fairness and equality. To mitigate bias in AI surveillance systems, a diversity-aware AI development framework is introduced, integrating diverse representation learning for inclusive model training and ethical AI oversight mechanisms for bias prevention. To mathematically model bias-aware learning from diverse data, a diverse representation learning objective function is introduced:

$$L_{\mathrm{DRL}} = \mathbb{E}_{p(x)}\left[D_{\mathrm{KL}}\left(p(z|x)\middle\|p(z)\right)\right] + \alpha\sum_{i=1}^{N}\ell(f_{\mathbf{w}}(z_i), y_i) + \beta H(p(z))$$

where L_{DRL}, the diverse representation learning objective, ensures equitable AI model training, $p(x)$, the input data distribution, represents demographic diversity, $p(z|x)$, the learned representation, encodes unbiased feature transformations, $p(z)$, the prior distribution, regulates fairness constraints, $D_{\mathrm{KL}}(p(z|x)\|p(z))$, the Kullback-Leibler (KL) divergence, prevents model overfitting to biased data distributions, α, the accuracy-fairness trade-off coefficient, balances prediction performance and diversity, $H(p(z))$, the entropy term, ensures diverse representation learning. This function ensures that AI models learn from diverse, representative datasets to prevent demographic bias in surveillance systems.

To further enhance ethical AI governance, an ethical oversight mechanism using fairness-constrained optimization is introduced:

$$\min_{\mathrm{w}}\sum_{i=1}^{N}\ell(f_{\mathrm{w}}(x_i), y_i) + \lambda \mid P(\hat{Y}|A=0) - P(\hat{Y}|A=1) \mid$$

where w, the AI model parameters, regulate ethical decision-making, $\ell(f_{\mathrm{w}}(x_i), y_i)$, the classification loss function, ensures predictive accuracy, λ, the fairness regularization coefficient, penalizes disparities in AI predictions, $P(\hat{Y}|A=0)$ and $P(\hat{Y}|A=1)$, the predicted probabilities of favorable outcomes for different demographic groups, quantify fairness. This function ensures that AI decision-making aligns with fairness constraints, reducing systemic bias while maintaining predictive reliability.

This AI-driven framework ensures that bias in AI surveillance systems is proactively mitigated at both the data and governance levels. The DRL function enhances demographic inclusivity in AI model learning, while the ethical oversight function ensures that AI-driven decision-making is systematically monitored for fairness and accountability. This dual-framework approach fosters the responsible and ethical development of AI surveillance technologies, ensuring alignment with societal values of fairness and equality.

Framework: Diversity-Aware AI Development Framework (DA-AI-DF)

Layered AI Architecture for Bias Detection and Ethical AI Governance in AI Surveillance Systems

Layer	Main functionality	AI module(s) and components	Operational scope
Layer 1	Data COLLECTION AND PREPROCESSING	Demographic and socio-economic data preprocessing system	Raw demographic, socioeconomic, and geographic data → Normalization and curation → Preprocessed input vectors
Layer 2	Diverse Representation Learning (DRL)	Diverse Representation Learning (DRL)	Input data → Optimization → Unbiased latent representation
Layer 3	Ethical oversight for decision-making	Fairness-constrained optimization	Fair representations → Apply fairness constraints → Ethical decision-making with fairness regularization
Layer 4	Continuous monitoring and bias auditing	Real-time fairness and bias auditing system	Predicted outcomes, feedback data → Bias metrics assessment → Updated fairness audits
Layer 5	Feedback and adjustment	Dynamic fairness feedback and model refinement engine	Bias reports, stakeholder feedback → Refinement of training and auditing processes

Figure 7.2 highlights the ethical concerns of bias and discrimination within AI surveillance systems. It is divided into three main sections: Sources of Bias in AI Surveillance, Discriminatory Impacts of AI Surveillance, and Mitigating Bias and Discrimination. "Sources of Bias" explores issues like biased training data due to group overrepresentation or underrepresentation and a lack of diversity among developers. "Discriminatory Impacts" addresses how biased AI affects law enforcement through over-policing minority communities and impacts public services by creating disparities in resource allocation. "Mitigating Bias" emphasizes the importance of using diverse and representative data, applying fairness metrics, enhancing transparency, and increasing diversity in AI development to reduce discriminatory outcomes and promote fairness.

Bias and discrimination in AI surveillance systems pose significant ethical challenges that must be addressed to ensure the fair and equitable use of these technologies. Without interventions to mitigate bias, AI surveillance systems risk perpetuating and amplifying social inequalities, particularly in law enforcement and public safety contexts. Addressing these challenges requires a commitment to improving data diversity, applying fairness metrics, and increasing diversity in AI development. By taking these steps, it is possible to create AI surveillance systems that are more inclusive, transparent, and fair.

7.3 Transparency and Accountability

The integration of AI into surveillance and security has brought about significant concerns regarding transparency and accountability. These two principles are essential in ensuring that AI systems are used responsibly, ethically, and in ways that do

Fig. 7.2 Bias and discrimination in AI surveillance

not infringe on human rights. However, AI systems, particularly in surveillance, often lack transparency due to the complexity of the algorithms and proprietary restrictions, raising questions about accountability when things go wrong (Sargiotis 2024a, 2025a).

7.3.1 Lack of Transparency

One of the primary challenges with AI surveillance systems is their opacity. Many AI systems, especially those based on machine learning and deep learning, function as "black boxes" (Saheb 2023). Their decision-making processes are often difficult to interpret, even for their creators. This lack of transparency makes it challenging for stakeholders, such as policymakers and the public, to understand how AI surveillance systems operate and how decisions are produced (Bruneault and Sabourin Laflamme 2021). For instance, in law enforcement, when AI systems are used for predictive policing, opacity makes it difficult to assess fairness or challenge algorithmic outputs in court. To address this problem, a transparent AI surveillance framework is introduced, integrating Explainable AI (XAI) for decision interpretability and causal inference for justifiable AI predictions. To mathematically model explainability in AI decision-making, a Shapley Additive Explanations (SHAP) feature attribution function is introduced:

$$\phi_i = \sum_{S \subseteq N \setminus \{i\}} \frac{|S|!(|N| - |S| - 1)!}{|N|!} [f(S \cup \{i\}) - f(S)]$$

where ϕ_i, the Shapley value, quantifies the contribution of feature i to AI surveillance decisions, S, a subset of AI input features excluding i, determines feature influence on model predictions, N, the total set of features, defines the AI decision space, $f(S)$, the AI model's prediction based on subset S, ensures accountability in AI decision-making. This function ensures that stakeholders can interpret how individual features influence AI surveillance decisions, increasing system transparency.

To further enhance justifiability in AI-driven law enforcement, a causal inference-based fair decision function is introduced:

$$P\left(\hat{Y}_{A=a}|X\right) = P\left(\hat{Y}_{A=a'}|X\right)$$

where $P\left(\hat{Y}_{A=a}|X\right)$, the probability of an AI decision given protected attribute $A = a$, represents the system's outcome for a specific demographic, $P\left(\hat{Y}_{A=a'}|X\right)$, the counterfactual probability, evaluates whether the AI decision would change if the protected attributes were different. This function ensures that AI-driven predictive policing and surveillance systems remain legally defensible by providing justifiable, bias-free decisions.

This AI-driven framework ensures that AI surveillance models operate transparently and fairly. The SHAP function provides stakeholders with interpretable explanations of AI decision processes, while the causal fairness function ensures that AI-driven law enforcement decisions remain justifiable and bias-free. This dual-framework approach strengthens trust in AI surveillance technologies, ensuring legal compliance, public accountability, and ethical decision-making in AI-driven governance.

Framework: Transparent AI Surveillance Framework (TAI-SSF)

Layered AI Architecture for Transparency and Accountability in AI Surveillance Systems

Layer	Main functionality	AI module(s) and components	Operational scope
Layer 1	Data collection and preprocessing	Demographic and geospatial data normalization engine	Raw demographic, socioeconomic, and geographic data → Normalization and curation → Preprocessed input vectors
Layer 2	Explainable AI (XAI) for Decision Interpretability	Shapley Additive Explanations (SHAP)	Input data → Feature attribution → Explainable decision-making
Layer 3	Causal inference for fair decision-making	Causal inference models	Predicted outcomes, protected attributes → Counterfactual analysis → Justifiable decisions

(continued)

Layer	Main functionality	AI module(s) and components	Operational scope
Layer 4	Continuous monitoring and transparency auditing	Continuous fairness and transparency auditing system	Predicted outcomes, feedback data → Fairness metrics evaluation → Continuous transparency checks
Layer 5	Feedback and adjustment	Adaptive transparency feedback and model adjustment engine	Bias reports, stakeholder feedback → Refinement of training and auditing processes

Moreover, the proprietary nature of many AI systems means that the underlying algorithms are not publicly available for scrutiny. Companies that develop these technologies often cite intellectual property concerns as reasons for not disclosing their algorithms. However, this lack of transparency undermines public trust in AI surveillance systems, as individuals cannot know whether these systems are making fair and unbiased decisions (Fosch-Villaronga and Poulsen 2022). Without transparency, it becomes nearly impossible to ensure that AI systems are aligned with ethical standards, especially when these systems are deployed in sensitive areas such as public safety and security. To address the lack of transparency in proprietary AI surveillance systems, a trustworthy AI surveillance framework is introduced, integrating Open Auditable AI (OAA) for Algorithmic Transparency and Differentially Private Federated Learning (DP-FL) for Ethical AI Deployment. To mathematically model auditable AI transparency, an Open Auditable AI (OAA) verification function is introduced:

$$V(A) = H(W_A) + \sigma$$

where $V(A)$, the verification function ensures that AI surveillance algorithms can be independently audited, $H(W_A)$, the cryptographic hash function, records AI model weights W_A on a decentralized ledger, preventing unauthorized modifications, σ, the verification signature, ensures that AI updates and modifications remain auditable. This function ensures that AI models remain publicly verifiable, even when developed by private companies, preventing algorithmic opacity and ensuring ethical compliance.

To further enhance ethical AI deployment while protecting intellectual property, a Differentially Private Federated Learning (DP-FL) objective function is introduced:

$$\min_{\mathrm{w}} \sum_{k=1}^{K} p_k F_k(\mathrm{w}) + \epsilon D_{\mathrm{KL}}\left(M(D) \middle\| M(D')\right)$$

where w, the AI model parameters, regulate decision-making, k, an individual client node, contributes to the decentralized AI learning process, p_k, the weighting factor, ensures data balance across training nodes, $F_k(\mathrm{w})$, the local loss function, optimizes AI models without centralized data access, ϵ, the privacy budget, controls the level of differential privacy, $D_{\mathrm{KL}}(M(D)\|M(D'))$, the Kullback-Leibler (KL) divergence, ensures that individual data points remain indistinguishable. This function ensures that AI models can be trained across multiple entities without exposing sensitive data, preserving privacy while allowing ethical oversight.

This framework ensures that proprietary AI surveillance systems remain verifiable, bias-free, and publicly accountable. The OAA function enables independent audits of AI decision-making, while the DP-FL function ensures that AI training remains decentralized, preventing unauthorized data exposure. This dual-framework approach fosters public trust in AI surveillance by ensuring algorithmic transparency, ethical governance, and privacy-preserving AI decision-making in critical areas such as public safety and security.

Framework: Trustworthy AI Surveillance Framework (TAI-SSF)

Layered AI Architecture for Algorithmic Transparency and Ethical AI Deployment

Layer	Main functionality	AI module(s) and components	Operational scope
Layer 1	Data collection and preprocessing	Demographic and socioeconomic data preprocessing engine	Raw demographic, socioeconomic, and geographic data → Normalization and curation → Preprocessed input vectors
Layer 2	Open Auditable AI (OAA) for algorithmic transparency	Open Auditable AI (OAA)	Input data → Cryptographic hashing → Ensuring algorithmic transparency and auditability
Layer 3	Differentially Private Federated Learning (DP-FL)	Differentially Private Federated Learning (DP-FL)	Decentralized model training → Privacy preservation → Ethical AI deployment without data leakage
Layer 4	Continuous monitoring and transparency auditing	Real-time transparency and fairness auditing system	Predicted outcomes, feedback data → Fairness and transparency metrics assessment → Ongoing audits for AI transparency and accountability
Layer 5	Feedback and adjustment	Continuous transparency feedback and model calibration engine	Bias reports, stakeholder feedback → Refinement of training and auditing processes for transparency and fairness

7.3.2 Accountability Gaps

The issue of accountability gaps in AI surveillance systems arises from opaque decision-making processes, which make it difficult to determine responsibility for errors such as facial recognition misidentification or bias in predictive policing. This lack of transparency raises ethical and legal concerns, especially in law enforcement and national security applications, where AI-driven decisions can directly affect individual rights and freedoms (Saheb 2023). Addressing these accountability challenges requires mechanisms for explainability, responsibility tracking, and adaptive risk assessment. A comprehensive framework for surveillance accountability can therefore combine Graph Neural Networks (GNNs) with Explainable AI (XAI) techniques. GNNs are well suited to modeling the complex interrelationships among system components, actors, data flows, and decision stages, while XAI methods support interpretability and traceability across the decision-making pipeline.

To quantify and track accountability, we introduce a graph-based AI accountability model, which structures AI decision processes as interconnected nodes in a graph, allowing responsibility attribution based on decision influence:

$$A_{t+1} = \sigma\left(W_A A_t + \sum_{i \in N(t)} W_N A_i + b \right)$$

where A_{t+1}, updated accountability state at time $t + 1$, W_A, weight matrix governing direct influence of previous decisions, A_t, previous accountability state, $N(t)$, set of neighboring nodes contributing to the accountability update, W_N, weight matrix for neighboring influences, A_i, accountability contribution from node i, b, bias term stabilizing updates, σ, activation function ensuring bounded outputs. This equation models the accountability propagation mechanism in an AI surveillance system. The accountability state at any time step is influenced by past accountability assignments and external influences from connected entities in the system (e.g., developers, organizations, and regulatory bodies). By structuring AI decision-making processes as interconnected nodes, GNNs allow responsibility tracking across the entire AI pipeline.

To enhance the explainability of decisions, we incorporate an explainable AI-based attribution function, leveraging Shapley values to quantify the contribution of each decision component:

$$S_i = \sum_{S \subseteq N \setminus \{i\}} \frac{|S|!(|N| - |S| - 1)!}{|N|!} (f(S \cup \{i\}) - f(S))$$

where S_i, Shapley value quantifying contribution of decision component i, S, subset of all decision components excluding i, N, total set of decision components, $f(S)$, AI decision output for subset S, $f(S \cup \{i\})$, AI decision output including i, factorial terms ensure fair distribution of responsibility across decision factors. This function ensures fairness in responsibility attribution by calculating how much each decision component (e.g., training data, algorithmic choice, and system deployment) contributes to the AI system's decision outcome. It allows policymakers and developers to pinpoint whether accountability gaps stem from flawed training data, algorithmic biases, or misapplications by end users.

To reinforce accountability tracking dynamically, a Policy Optimization Accountability Function using Proximal Policy Optimization (PPO) is introduced:

$$J(\theta) = \mathbb{E}\left[min\left(\frac{\pi_\theta(a_t|s_t)}{\pi_{\theta_{\text{old}}}(a_t|s_t)}A_t, \text{clip}\left(\frac{\pi_\theta(a_t|s_t)}{\pi_{\theta_{\text{old}}}(a_t|s_t)}, 1-\epsilon, 1+\epsilon\right)A_t\right)\right]$$

where $J(\theta)$, objective function for policy optimization, $\pi_\theta(a_t| s_t)$, probability of action a_t under policy θ, $\pi_{\theta_{\text{old}}}(a_t|s_t)$, probability of action under previous policy, A_t, advantage function evaluating action's improvement, ϵ, clipping parameter preventing excessive updates. This function optimizes AI system adjustments to minimize accountability gaps by dynamically adjusting the AI policy based on real-time performance assessments. By constraining accountability updates within safe bounds, the PPO-based mechanism ensures that responsibility shifts do not destabilize system fairness.

The proposed graph-based AI accountability model, integrated with explainable AI-based attribution function and policy optimization accountability function, creates a structured and interpretable framework for tracking responsibility in AI surveillance. The GNN-based accountability propagation mechanism enables structured responsibility tracking, while Shapley value-based attribution ensures transparent impact assessments. PPO-driven policy optimization refines accountability allocations dynamically. This AI-driven framework provides a method to address accountability gaps in AI surveillance systems, ensuring fair, transparent, and ethical decision-making.

Framework: AI-Driven Accountability Framework (AI-DAF)

Layered AI Architecture for Transparency, Responsibility, and Ethical AI Deployment in AI Surveillance Systems

Layer	Main functionality	AI module(s) and components	Operational scope
Layer 1	Data collection and preprocessing	Demographic and geospatial data normalization system	Raw demographic, socioeconomic, and geographic data → Normalization and curation → Preprocessed input vectors
Layer 2	Graph-based accountability tracking	Graph Neural Networks (GNNs)	Interconnected nodes in graph → Accountability propagation across the AI system → Responsibility attribution from decision influence
Layer 3	Explainability and attribution	Shapley Additive Explanations (SHAP)	Input features → Quantifying feature contribution to AI decisions → Ensuring transparency and accountability in decision-making
Layer 4	Policy optimization for accountability	Proximal Policy Optimization (PPO)	AI system adjustments → Real-time performance assessments → Optimizing accountability allocation while minimizing gaps
Layer 5	Continuous monitoring and accountability auditing	Real-Time accountability monitoring and auditing system	Predicted outcomes, feedback data → Accountability metrics assessment → Continuous auditing of responsibility across the system

(continued)

Layer	Main functionality	AI module(s) and components	Operational scope
Layer 6	Feedback and adjustment	Dynamic feedback and accountability refinement engine	Bias reports, stakeholder feedback → Refinement of training, auditing, and decision-making processes for fairness and accountability

The growing deployment of autonomous AI systems presents significant accountability challenges, particularly when these systems make decisions independently without human oversight. While AI-driven models are often designed to assist human decision-makers, fully autonomous systems operate without direct human intervention, creating ambiguity in responsibility attribution (Bruneault and Sabourin Laflamme 2021). If an autonomous AI system makes a mistake, it becomes difficult to determine whether liability lies with the developers, the deploying organization, system operators, or the governance bodies responsible for oversight. This lack of accountability raises ethical and legal concerns, particularly in high-stakes domains such as healthcare, law enforcement, and financial decision-making. To address these accountability challenges, a structured Reinforcement Learning-Based Explainable Accountability Model is introduced. This framework integrates Deep Reinforcement Learning (DRL) for adaptive decision-making, Explainable AI (XAI) for interpretability, and Bayesian Risk Assessment for uncertainty quantification in AI-driven decisions:

$$Q^*(s,a) = \mathbb{E}\left[r_t + \gamma \max_{a'} Q^*(s',a')\right]$$

where $Q^*(s,a)$, optimal action-value function guiding accountability optimization, r_t, immediate accountability reward function assessing decision reliability, γ, discount factor ensuring long-term responsibility tracking, $\max_{a'} Q^*(s',a')$, future accountability impact based on optimal next action selection. This function ensures that AI systems dynamically optimize accountability by reinforcing decisions that enhance transparency while penalizing actions leading to untraceable outcomes. The reinforcement learning model progressively refines the AI's ability to balance autonomy with accountability.

To quantify accountability risks, a Bayesian accountability risk function is incorporated, enabling uncertainty-aware decision assessments:

$$P(A|D) = \frac{P(D|A)P(A)}{P(D)}$$

where $P(A|D)$, probability of accountability failure given decision D, $P(D|A)$, likelihood of decision D occurring under accountability failure, $P(A)$, prior probability of accountability failure, $P(D)$, total probability of decision D occurring. This probabilistic model allows AI systems to estimate accountability risks dynamically, ensuring that high-risk decisions trigger precautionary measures, such as requiring human intervention.

To ensure interpretability, a Shapley value-based accountability attribution function is introduced, determining how much each decision factor contributes to accountability risks:

$$S_i = \sum_{S \subseteq N \setminus \{i\}} \frac{|S|!(|N| - |S| - 1)!}{|N|!} (f(S \cup \{i\}) - f(S))$$

where S_i, Shapley value quantifying contribution of decision component i, S, subset of all decision components excluding i, N, total set of decision components, $f(S)$, AI decision output for subset S, $f(S \cup \{i\}$, AI decision output including i, factorial terms ensure fair responsibility distribution.

This AI-driven framework provides a mathematically and ethically sound approach to addressing accountability gaps in autonomous AI systems. It ensures that AI models dynamically adjust accountability policies, estimate risk probabilities, and transparently distribute responsibility, mitigating the risk of untraceable harmful decisions.

Framework: Reinforcement Learning-Based Explainable Accountability Model (RL-XAM)

Layered AI Architecture for Transparency, Responsibility Tracking, and Ethical AI Deployment in Autonomous Systems

Layer	Main functionality	AI module(s) and components	Operational Scope
Layer 1	Data collection and preprocessing	Demographic and socioeconomic data preprocessing engine	Raw demographic, socioeconomic, and geographic data → Normalization and curation → Preprocessed input vectors
Layer 2	Reinforcement learning for accountability optimization	Deep Reinforcement Learning (DRL)	Dynamic decision-making → Accountability optimization → Reinforcement of decisions that enhance transparency and traceability
Layer 3	Bayesian risk assessment for accountability	Bayesian accountability risk function	Probability estimation → Uncertainty-aware decision assessments → Accountability risk quantification for dynamic decision-making
Layer 4	Explainable AI for responsibility attribution	Shapley Additive Explanations (SHAP)	Decision components → Fair responsibility attribution → Ensuring transparency in accountability decisions
Layer 5	Policy optimization for accountability	Proximal Policy Optimization (PPO)	AI policy adjustments → Real-time performance assessments → Optimizing accountability allocation with minimal gaps

(continued)

Layer	Main functionality	AI module(s) and components	Operational Scope
Layer 6	Continuous monitoring and accountability auditing	Continuous accountability monitoring and auditing system	Predicted outcomes, feedback data → Accountability metrics assessment → Continuous auditing of responsibility across the system
Layer 7	Feedback and adjustment	Adaptive accountability feedback and model refinement engine	Bias reports, stakeholder feedback → Refinement of training, auditing, and decision-making processes for fairness and accountability

7.3.3 *Addressing Transparency and Accountability*

Ensuring transparency in AI surveillance systems is a critical step toward mitigating accountability gaps. AI models, particularly in law enforcement and predictive policing, must be interpretable to prevent biases and unjustified decision-making. Explainable AI (XAI) techniques provide insights into how AI systems reach conclusions, allowing for greater scrutiny and oversight (Fosch-Villaronga and Poulsen 2022). Without clear explanations of AI-generated decisions, law enforcement agencies risk deploying opaque models that may reinforce biases, leading to wrongful targeting or discriminatory policing strategies. To ensure AI systems operate with accountability, transparency, and fairness, an AI-driven framework incorporating XAI-based interpretability mechanisms is necessary.

To enhance interpretability in AI surveillance, a Shapley value-based XAI framework is introduced, quantifying the contribution of each feature to an AI decision and supporting greater transparency in predictive policing and other surveillance applications:

$$S_i = \sum_{S \subseteq N \setminus \{i\}} \frac{|S|!(|N| - |S| - 1)!}{|N|!} \left(f(S \cup \{i\}) - f(S)\right)$$

where S_i, Shapley value quantifying contribution of feature i, S, subset of all input features excluding i, N, total set of features used in the AI model, $f(S)$, AI model's decision output using subset S, $f(S \cup \{i\})$, AI decision output including feature i, factorial terms ensure fair contribution attribution. This function ensures that AI-generated decisions are decomposed into individual feature contributions, providing transparency regarding the factors influencing an AI prediction. For example, in predictive policing, the model can reveal whether socioeconomic factors, geographic locations, or demographic variables disproportionately influence AI-generated risk scores.

To refine interpretability further, a Local Interpretable Model-Agnostic Explanations (LIME) function is incorporated, approximating AI decisions using locally interpretable surrogate models.

$$\hat{f}(x) = \arg \underset{g \in G}{min}\, L(f, g, \pi_x) + \Omega(g)$$

where $\hat{f}(x)$, interpretable approximation of the original AI function, g, locally interpretable model from hypothesis space G, $\underset{g \in G}{min}\, L(f, g, \pi_x)$, loss function measuring discrepancy between AI model f and surrogate model g, π_x, locality-aware weighting function, $\Omega(g)$, complexity penalty ensuring the surrogate model remains interpretable. This function ensures that AI predictions remain explainable even for complex black-box models. By approximating opaque AI decisions using simple interpretable models, stakeholders, including policymakers and law enforcement officers can understand the rationale behind AI-generated outcomes.

To ensure fairness and prevent biased decision-making, a fairness-constrained predictive function is introduced, ensuring that AI-generated decisions adhere to fairness constraints.

$$\underset{\theta}{min}\, \mathbb{E}[L(f_\theta(x), y) + \lambda \cdot D(f_\theta(x), Z)]$$

where θ, AI model parameters, $L(f_\theta(x), y)$, loss function measuring predictive accuracy, λ, fairness penalty coefficient, $D(f_\theta(x), Z)$, disparity function measuring bias in predictions against protected attributes Z. This function penalizes AI decisions that disproportionately affect protected demographic groups. By incorporating fairness constraints, AI models ensure that predictive policing algorithms do not systematically discriminate against certain populations.

The Shapley value-based XAI framework, integrated with the LIME-based interpretability function and the fairness-constrained predictive function, creates a mathematically rigorous and ethically robust AI-driven transparency framework. This model enables law enforcement agencies and policymakers to scrutinize AI-generated decisions, ensuring that predictive policing models rely on justifiable, unbiased, and explainable data. The integration of XAI-based explanations and fairness constraints guarantees that AI-driven surveillance operates with transparency and accountability, mitigating risks associated with opaque and potentially discriminatory decision-making.

Framework: Shapley Value-Based XAI Framework (SVB-XAI)

Layered AI Architecture for Transparency, Fairness, and Accountability in AI Surveillance

Layer	Main functionality	AI module(s) and components	Operational scope
Layer 1	Data collection and preprocessing	Demographic and geospatial data preprocessing engine	Raw demographic, socioeconomic, and geographic data → Normalization and curation → Preprocessed input vectors

(continued)

Layer	Main functionality	AI module(s) and components	Operational scope
Layer 2	Shapley value-based accountability attribution	Shapley Additive Explanations (SHAP)	Feature attribution → Quantifying contribution of features to accountability risks → Ensuring transparency in decision-making
Layer 3	Local Interpretable Model-Agnostic Explanations (LIME)	Local Interpretable Model-Agnostic Explanations (LIME)	AI decision output → Surrogate model approximation → Ensuring interpretability of complex black-box models
Layer 4	Fairness-constrained predictive function	Fairness-constrained predictive function	Predicted outcomes → Ensuring fairness in predictive decisions → Penalizing biased predictions against protected demographic groups
Layer 5	Continuous monitoring and transparency auditing	Continuous accountability and fairness auditing system	Predicted outcomes, feedback data → Accountability metrics evaluation → Continuous auditing of decision-making processes for transparency and fairness
Layer 6	Feedback and adjustment	Dynamic feedback and model refinement engine	Bias reports, stakeholder feedback → Refinement of training, auditing, and decision-making processes to ensure fairness and transparency

Second, accountability in AI systems requires the establishment of clear legal and regulatory frameworks that define responsibility for AI-driven decisions. Without explicit policies governing AI accountability, organizations deploying AI systems may evade liability, leading to unchecked decision-making processes. Governments and regulatory bodies must develop policies that ensure organizations remain accountable for the consequences of AI driven actions (Saheb 2023). Key regulatory measures should include mandatory audits, compliance checks, and avenues for individuals to challenge AI decisions that negatively impact them. However, ensuring effective oversight necessitates an AI-driven framework that integrates automated auditing, compliance verification, and challenge mechanisms to enhance regulatory enforcement. An AI-driven model to support regulatory compliance in AI governance is the automated AI audit model, which employs Graph Neural Networks (GNNs) for compliance tracking, federated learning for decentralized auditing, and Reinforcement Learning (RL) for optimizing accountability enforcement. The compliance verification function is defined as:

$$C_t = \sigma\left(W_C C_{t-1} + \sum_{i \in N(t)} W_N C_i + b\right)$$

where C_t, compliance state at time t, W_C, weight matrix governing historical compliance trends, C_{t-1}, compliance state from the previous audit cycle, $N(t)$, set of regulatory entities monitoring compliance, W_N, weight matrix governing regulatory influence, C_i, compliance contribution from regulatory node i, b, bias term stabilizing compliance updates, σ, activation function ensuring smooth compliance evaluations. This function tracks the compliance state of AI systems dynamically over time, enabling continuous assessment of whether organizations adhere to regulatory mandates. By structuring AI compliance data as interconnected nodes in a GNN-based accountability graph, regulatory bodies can trace patterns of non-compliance and enforce corrective measures effectively.

To facilitate decentralized auditing without compromising privacy, federated learning-based AI auditing is introduced, allowing multiple regulatory entities to audit AI systems collaboratively while preserving sensitive data. The federated AI auditing function is defined as:

$$\theta^* = \arg \min_{\theta} \sum_{k=1}^{K} (L_k(\theta) + R(\theta))$$

where θ^*, optimized AI auditing model parameters, K, number of distributed audit entities, $L_k(\theta)$, local audit loss function for entity k, $R(\theta)$, regularization term ensuring stable model convergence. This function enables AI auditing models to be trained across multiple regulatory entities without requiring centralized access to AI decision logs. By leveraging federated learning, organizations can ensure compliance with AI regulations while preserving the confidentiality of proprietary AI models.

To enhance AI accountability enforcement, reinforcement learning-based compliance optimization is introduced, ensuring that AI governance policies dynamically adapt to evolving regulatory requirements. The Compliance optimization function is defined as:

$$J(\theta) = \mathbb{E}\left[min\left(\frac{\pi_\theta(a_t|s_t)}{\pi_{\theta_{\text{old}}}(a_t|s_t)}A_t, \text{clip}\left(\frac{\pi_\theta(a_t|s_t)}{\pi_{\theta_{\text{old}}}(a_t|s_t)}, 1-\epsilon, 1+\epsilon\right)A_t\right)\right]$$

where $J(\theta)$, compliance optimization function, $\pi_\theta(a_t| s_t)$, probability of compliance action a_t under policy θ, $\pi_{\theta_{\text{old}}}(a_t|s_t)$, probability under previous policy, A_t, advantage function assessing improvement in compliance performance, ϵ, clipping parameter preventing excessive updates. This function enables AI regulatory frameworks to optimize policy enforcement dynamically. By leveraging Proximal Policy Optimization (PPO), compliance mechanisms ensure that AI systems are continuously updated to align with evolving regulations.

The automated AI audit model, integrating GNN-based compliance tracking, federated AI auditing, and reinforcement learning-based compliance optimization, ensures that AI accountability mechanisms operate effectively within legal and regulatory frameworks. The compliance verification function tracks organizational adherence to regulatory mandates, while federated AI auditing ensures privacy-preserving oversight across multiple regulatory entities. Finally, the reinforcement learning-based compliance optimization dynamically refines AI accountability enforcement. This AI-driven framework provides an approach to ensuring that organizations deploying AI systems remain fully accountable for AI-driven decisions, fostering responsible AI governance.

Framework: Automated AI Audit Model (AAAM)

Layered AI Architecture for Compliance Tracking, Decentralized Auditing, and Accountability Optimization

Layer	Main functionality	AI module (s) and components	Operational scope
Layer 1	Data collection and preprocessing	Demographic and socioeconomic data normalization engine	Raw demographic, socioeconomic, and geographic data → Normalization and curation → Preprocessed input vectors
Layer 2	Graph-based compliance tracking	Graph Neural Networks (GNNs)	Compliance tracking → Dynamic updates of compliance states → Accountability propagation across AI systems
Layer 3	Federated learning for decentralized auditing	Federated learning	Decentralized model training → Privacy-preserving audits → Regulatory compliance checks without centralized access
Layer 4	Reinforcement learning-based compliance optimization	Reinforcement Learning (RL)	Dynamic policy optimization → Accountability enforcement → Adjustments based on real-time regulatory performance
Layer 5	Continuous auditing and compliance verification	Continuous compliance monitoring and verification system	Audit feedback, compliance data → Continuous compliance assessment → Adjustments based on auditing results
Layer 6	Feedback and adjustment	Adaptive feedback and accountability adjustment engine	Bias reports, stakeholder feedback → Refinement of auditing, compliance, and decision-making processes for dynamic accountability enforcement

Finally, building public trust in AI surveillance systems requires a commitment to transparency, ensuring that these technologies are developed and deployed in ways that uphold fairness, accountability, and respect for individual rights. Public skepticism often arises when AI systems operate as "black boxes," making decisions without clear explanations or public scrutiny. Greater openness in AI development, including stakeholder participation from diverse backgrounds, fosters inclusivity and ensures that AI models are aligned with societal values (Bruneault and Sabourin Laflamme 2021). To institutionalize this transparency, an AI-driven framework integrating Explainable AI (XAI), federated learning for decentralized stakeholder involvement, and Multi-Agent Reinforcement Learning (MARL) for collaborative decision-making is introduced. A key component of this framework is the stakeholder-centric explainability function, which ensures that AI decisions are interpretable for all affected groups, including policymakers, civil rights organizations, and the public. The stakeholder-centric explainability function is defined as:

$$E_t = \sigma\left(W_E E_{t-1} + \sum_{i \in S} W_S E_i + b\right)$$

where E_t, explainability state at time t, W_E, weight matrix controlling past explainability trends, E_{t-1}, previous explainability state, S, set of stakeholder groups involved in AI system evaluation, W_S, weight matrix governing stakeholder influence, E_i, explainability contribution from stakeholder i, b, bias term ensuring stability, σ, activation function maintaining bounded outputs. This function models how stakeholder involvement influences the transparency of AI decisions over time. By structuring public engagement within an AI-driven accountability framework, organizations ensure that diverse perspectives are integrated into the explainability of AI surveillance models.

To ensure inclusive participation in AI system development while protecting data privacy, federated learning for stakeholder collaboration is introduced. This mechanism allows multiple stakeholder groups to review and refine AI models collectively without centralized data exposure. The federated stakeholder learning function is defined as:

$$\theta^* = \arg \min_{\theta} \sum_{k=1}^{K} (L_k(\theta) + R(\theta))$$

where θ^*, optimized AI model parameters incorporating stakeholder feedback, K, number of participating stakeholder entities, $L_k(\theta)$, local model refinement function for stakeholder k, $R(\theta)$, regularization term preventing overfitting to biased perspectives. This function ensures that multiple stakeholder groups such as civil rights organizations, policymakers, and independent auditors can collaboratively refine AI models without compromising data privacy. By distributing AI learning across multiple independent entities, federated learning prevents any single organization from unilaterally shaping AI development.

To further enhance fairness and accountability in AI deployment, Multi-Agent Reinforcement Learning (MARL) for collaborative governance is introduced, optimizing AI policies based on feedback from multiple stakeholders. The collaborative AI governance function is defined as:

$$J(\theta) = \mathbb{E}\left[\sum_{i=1}^{N} \sum_{t=1}^{T} \gamma^t R_i(s_t, a_t)\right]$$

where $J(\theta)$, policy optimization function for AI governance, N, total number of stakeholder agents contributing feedback, T, decision-making horizon, γ^t, discount factor prioritizing long-term fairness, $R_i(s_t, a_t)$, reward function assessing governance improvements from stakeholder i at state s_t and action a_t. This function ensures that AI governance policies dynamically adapt to stakeholder feedback. By incorporating Multi-Agent Reinforcement Learning (MARL), AI decision-making is continuously refined based on diverse input, preventing biases from dominating AI policy evolution.

The stakeholder-centric explainability function, integrated with federated learning for stakeholder collaboration and multi-agent reinforcement learning for collaborative governance, provides a rigorous AI-driven framework for fostering public trust in AI surveillance. The explainability function ensures transparency by structuring stakeholder involvement, the federated learning mechanism enables decentralized and privacy-preserving AI refinement, and the collaborative AI governance function optimizes AI policies dynamically based on multi-stakeholder feedback. Together, these mechanisms ensure that AI surveillance operates with fairness, transparency, and accountability, strengthening public confidence in AI-driven decision-making.

Framework: Stakeholder-Centric Explainability Framework (SCEF)

Layered AI Architecture for Transparency and Collaborative Governance

Layer	Main functionality	AI module(s) and components	Operational scope
Layer 1	Data collection and preprocessing	Demographic and socioeconomic data normalization engine	Input → Raw demographic, socioeconomic, and geographic data → Normalization and curation → Preprocessed input vectors
Layer 2	Transparency and explainability	Transparent decision-making and explainability module	Input → Stakeholder feedback, transparency metrics → Output → Transparent AI decisions with stakeholder input
Layer 3	Federated Learning for Stakeholder Collaboration	Federated Stakeholder Learning Function (FL)	Input → Stakeholder feedback, decentralized model learning → Output → Collaborative AI model refinement without centralized data exposure
Layer 4	Collaborative AI governance	Multi-Agent Reinforcement Learning (MARL)	Input → Stakeholder feedback, decision states → Output → Optimized AI governance policies based on multi-stakeholder input
Layer 5	Governance and oversight	Stakeholder governance and accountability system	Input → Stakeholder feedback, governance models → Output → Enhanced accountability and regulation of AI decisions
Layer 6	Feedback and evaluation	Continuous feedback and governance refinement engine	Input → Continuous feedback from stakeholders → Output → Refined governance policies and transparency models

Fig. 7.3 Transparency and accountability in AI surveillance

Figure 7.3 illustrates the critical issues of transparency and accountability in AI surveillance. It is divided into three main sections: Lack of Transparency, Accountability Gaps, and Addressing Transparency and Accountability. "Lack of Transparency" highlights the challenges of opaque "black box" systems and proprietary restrictions, which limit public understanding and trust. "Accountability Gaps" addresses the difficulty in assigning responsibility for AI decisions, especially when autonomous systems operate without human oversight, leading to ambiguity in responsibility. "Addressing Transparency and Accountability" emphasizes solutions such as Explainable AI (XAI) to enhance decision transparency, the establishment of legal and regulatory frameworks to define accountability, and the importance of public trust through open development and stakeholder involvement.

Transparency and accountability are crucial for the ethical deployment of AI surveillance systems. The current lack of transparency in these systems creates significant ethical concerns, particularly in terms of accountability for the decisions made by AI. Addressing these challenges will require a combination of technical solutions, such as explainable AI, and regulatory measures to ensure that organizations using AI systems are held accountable for their actions.

7.4 Surveillance and Human Rights

AI-driven surveillance systems have a profound impact on human rights, particularly regarding privacy, freedom of expression, and freedom of assembly. These systems, often deployed by governments or corporations, can intrude on individuals' private lives, monitor public spaces, and track personal data without explicit consent. The use of AI in surveillance raises ethical concerns about how these technologies might undermine fundamental human rights and the broader implications for democratic societies (Sargiotis 2024a, b).

7.4.1 Privacy Violations

The most pressing human rights concern related to AI surveillance systems is their impact on privacy. Technologies such as facial recognition, biometric scanning, and predictive monitoring collect vast amounts of personal data, often without individuals' knowledge or consent (Saheb 2023). This pervasive surveillance allows AI systems to track, record, and analyze human behavior in both public and private spaces, directly undermining the right to privacy, a fundamental human right enshrined in international legal frameworks such as the Universal Declaration of Human Rights (UDHR) and the International Covenant on Civil and Political Rights (ICCPR) (Bruneault and Sabourin Laflamme 2021). To address these concerns, an AI-driven framework integrating differential privacy, federated learning for decentralized data protection, and adversarial debiasing for privacy-aware AI decision-making is introduced. A central mechanism within this framework is the differential privacy protection function, which ensures that AI surveillance systems process data while minimizing the risk of re-identifying individuals:

$$P(A|D) - P(A|D') \leq \epsilon$$

where $P(A|D)$, probability of AI decision AA given dataset D, $P(A|D')$, probability of the same decision given a slightly different dataset D', ϵ, privacy budget parameter controlling data leakage risk. This function ensures that AI-generated decisions remain stable even when individual data points change, preventing privacy breaches. By injecting noise into AI computations, differential privacy techniques preserve anonymity, making it difficult to trace AI decisions back to specific individuals.

To further safeguard personal data, federated learning for decentralized privacy preservation is introduced, ensuring that AI surveillance models are trained without centralizing sensitive user data:

$$\theta^* = \arg \min_{\theta} \sum_{k=1}^{K} (L_k(\theta) + \mathrm{R}(\theta))$$

where θ^*, optimized AI model parameters ensuring privacy, K, number of distributed data sources, $L_k(\theta)$, local AI model loss function for entity k, $R(\theta)$, regularization term preventing overfitting. This function enables AI models to be trained across multiple decentralized nodes, preventing the need for centralized data collection. By keeping sensitive user information localized while still improving AI accuracy, federated learning minimizes privacy risks.

To prevent AI surveillance models from disproportionately targeting individuals based on biased data, adversarial debiasing for privacy-aware AI decision-making is incorporated:

$$\min_{\theta} \mathbb{E}[L(f_\theta(x), y) + \lambda D(f_\theta(x), Z)]$$

where θ, AI model parameters, $L(f_\theta(x), y)$, primary learning objective ensuring prediction accuracy, λ, privacy-aware debiasing penalty coefficient, $D(f_\theta(x), Z)$, disparity function assessing bias against protected attributes Z. This function ensures that AI decisions do not disproportionately affect specific groups by reducing biases in the training data. By incorporating adversarial fairness constraints, AI models mitigate privacy risks while preventing discriminatory outcomes.

The differential privacy protection function, combined with federated privacy-preserving learning and adversarial privacy-preserving debiasing, provides a rigorous AI-driven framework for safeguarding privacy in AI surveillance. The differential privacy mechanism ensures AI decisions remain anonymized, the federated learning approach prevents centralized data collection risks, and the adversarial debiasing function mitigates privacy-related biases. Together, these AI-driven mechanisms establish an approach to AI surveillance, ensuring that personal data protection remains a fundamental priority in AI governance.

In countries where AI surveillance is deployed extensively, the lack of transparency about how data is collected, stored, and used intensifies privacy concerns. Individuals are often unaware of the extent to which their personal data is being used, making it difficult for them to exercise control over their information. This type of invasive surveillance has been criticized for creating a "surveillance society," where individuals' actions are constantly monitored and scrutinized by state and corporate actors.

Framework: Privacy-Preserving AI Surveillance Framework (PPAISF II)

Layered AI Architecture for Privacy Protection in AI Surveillance

Layer	Main functionality	AI module(s) and components	Operational scope
Layer 1	Data collection and preprocessing	Biometric and behavioral data preprocessing engine	Input → Raw biometric, facial recognition, and behavioral data → Normalization and curation → Preprocessed input vectors
Layer 2	Differential privacy protection	Differential privacy	Input → Raw data, privacy budget → Processing → Noise injection to ensure privacy and prevent re-identification → Output → Anonymized AI decisions
Layer 3	Federated learning for decentralized privacy preservation	Federated Learning (FL)	Input → Decentralized data sources → Processing → Local AI model training → Output → Optimized global AI model without centralized data collection
Layer 4	Adversarial debiasing for privacy-aware AI decision-making	Adversarial debiasing	Input → Biased data → Processing → Debiasing AI models to reduce privacy-related biases → Output → Fair and privacy-preserving AI decisions
Layer 5	Governance and oversight	Privacy governance and compliance system	Input → Privacy regulations, ethical guidelines → Processing → Policy development and enforcement → Output → Governance for AI surveillance privacy practices

(continued)

Layer	Main functionality	AI module(s) and components	Operational scope
Layer 6	Feedback and evaluation	Privacy impact feedback and evaluation system	Input → Privacy and fairness audits → Processing → Evaluating privacy-preserving models → Output → Refined privacy measures and improved AI practices

7.4.2 *Chilling Effects on Freedom of Expression and Assembly*

Beyond privacy, AI surveillance systems, while designed for security and efficiency, also pose significant risks to fundamental human rights, particularly freedom of expression and freedom of assembly. When individuals know they are being monitored, they may self-censor their speech or refrain from participating in public demonstrations out of fear of being tracked, identified, or targeted by law enforcement (Fosch-Villaronga and Poulsen 2022). This chilling effect suppresses democratic participation and discourages political activism, ultimately undermining societal freedoms. To counteract these risks, an AI-driven framework integrating Differential Privacy, Adversarial Obfuscation Techniques, and Multi-Agent Reinforcement Learning (MARL) for privacy-aware crowd analysis is introduced. A core component of this framework is the privacy-preserving feature masking function, which ensures that AI models process surveillance data without violating individuals' rights to anonymity in public spaces:

$$\tilde{X} = X + \mathcal{N}(0, \sigma^2)$$

where $\tilde{X}$, privacy-protected input data, X, original surveillance data, $\mathcal{N}(0, \sigma^2)$, Gaussian noise function with variance σ^2 ensuring anonymization. This function prevents AI surveillance models from directly identifying individuals in public spaces by introducing controlled noise into facial recognition and movement-tracking algorithms. By ensuring that identifiable features are masked before AI analysis, it protects individuals' rights to participate in public discourse without fear of surveillance-based retaliation.

To reinforce protection against AI-driven behavioral profiling, adversarial obfuscation for de-identification in surveillance systems is introduced:

$$\min_{\theta} \mathbb{E}[L(f_\theta(X), y) - \lambda D(f_\theta(X), Z)]$$

where θ, AI model parameters, $L(f_\theta(X), y)$, predictive loss function, λ, adversarial regularization coefficient controlling de-identification, $D(f_\theta(X), Z)$, disparity function reducing identifiable features from protected attributes Z. This function ensures that AI models cannot reconstruct identifiable details from blurred or obfuscated images, preventing AI-based tracking from being used to suppress protests or monitor political activism unfairly. By incorporating adversarial learning techniques, AI surveillance systems can maintain security functions while respecting the right to freedom of expression and assembly.

To further ensure that AI surveillance does not disproportionately target specific individuals or groups, Multi-Agent Reinforcement Learning (MARL) for privacy-aware crowd analysis is introduced, optimizing AI decision-making while preserving public freedoms:

$$J(\theta) = \mathbb{E}\left[\sum_{i=1}^{N}\sum_{t=1}^{T}\gamma^{t}R_{i}(s_{t}, a_{t})\right]$$

where $J(\theta)$, policy optimization function for privacy-aware AI surveillance, N, total number of agents analyzing crowd behavior, T, decision horizon for real-time AI monitoring, γ^t, discount factor controlling long-term fairness, $R_i(s_t, a_t)$, reward function measuring ethical compliance of AI decision-making in state s_t under action a_t. This function ensures that AI surveillance optimizes security decisions without suppressing public freedoms. By dynamically adjusting AI policies based on real-time assessments of privacy risks, MARL-based systems ensure that individuals are not unfairly targeted for participating in lawful assemblies.

The privacy-preserving feature masking function, combined with adversarial obfuscation for de-identification and multi-agent reinforcement learning for privacy-aware crowd analysis, provides a mathematically rigorous AI-driven framework for protecting freedom of expression and assembly in AI surveillance. The feature masking mechanism ensures anonymized data processing, the adversarial obfuscation function prevents AI models from reconstructing identifiable details, and the MARL-based privacy-aware analysis function optimizes AI policies to balance security with democratic rights. Together, these AI-driven mechanisms ensure that AI surveillance systems do not become tools of repression, allowing individuals to express themselves and assemble freely without fear of undue monitoring or targeting.

Framework: Privacy-Preserving AI Surveillance Framework (PPAISF)

Layered AI Architecture for Protecting Freedom of Expression and Assembly

Layer	Main functionality	AI module(s) and components	Operational scope
Layer 1	Data collection and preprocessing	Raw surveillance data preprocessing engine	Input → Raw surveillance data → Preprocessing for anonymization → Preprocessed data ready for AI models
Layer 2	Privacy-preserving feature masking	Privacy-protective data masking module	Input → Original surveillance data → Data masking for privacy protection → Output → Privacy-protected data ready for analysis

(continued)

Layer	Main functionality	AI module(s) and components	Operational scope
Layer 3	Adversarial obfuscation for de-identification	Adversarial debiasing	Input → Biased surveillance data → Processing to remove identifiable features → Output → Anonymized data without identifiable characteristics
Layer 4	Multi-agent reinforcement learning for privacy-aware crowd analysis	Multi-Agent Reinforcement Learning (MARL)	Input → Crowd behavior data → Optimization of decision-making policies → Output → Optimized surveillance decisions that balance privacy and security
Layer 5	Governance and oversight	Privacy compliance and governance system	Input → Privacy regulations, ethical guidelines → Development and enforcement of policies → Output → Governance mechanisms ensuring privacy compliance
Layer 6	Feedback and evaluation	Privacy impact feedback and risk evaluation system	Input → Feedback on privacy measures → Evaluation of privacy risks and impacts → Output → Refined policies and enhanced AI systems

In authoritarian regimes, AI surveillance is frequently deployed as a tool to suppress dissent and monitor political opponents. Governments utilize AI-driven systems to track individuals' online activities, monitor communications, and predict potential involvement in political activism (Saheb 2023). This type of surveillance directly infringes on the rights to freedom of expression and assembly, forcing individuals to weigh the risks of speaking out against their government or participating in protests. AI-powered tracking mechanisms can generate detailed behavioral profiles, enabling state authorities to preemptively suppress opposition. To counteract these repressive applications, an AI-driven framework integrating privacy-preserving AI, federated learning for decentralized resistance, and adversarial detection mechanisms is introduced. A key element of this framework is the privacy-preserving anonymization function, which protects users' digital identities from AI surveillance tools that seek to link online activity with real-world identities:

$$X' = X + \mathcal{N}(0, \sigma^2)$$

where X', anonymized online activity data, X, original digital activity, $\mathcal{N}(0, \sigma^2)$, Gaussian noise ensuring privacy-preserving obfuscation. This function prevents AI tracking systems from accurately identifying users based on online behavior, ensuring that individuals can express political opinions and engage in activism without fear of state retaliation. By introducing controlled noise into digital identifiers, AI systems are unable to create deterministic mappings between online and real-world identities.

To further enhance resistance against AI-powered political suppression, federated learning for decentralized digital privacy is introduced, allowing dissidents to communicate securely without centralized surveillance risks:

$$\theta^* = \arg \min_{\theta} \sum_{k=1}^{K} (L_k(\theta) + R(\theta))$$

where θ^*, optimized privacy-preserving communication model parameters, K, number of decentralized nodes protecting dissident communications, $L_k(\theta)$, local AI model loss function ensuring encryption, $R(\theta)$, regularization term preventing overfitting to adversarial monitoring attempts. This function enables decentralized AI training across independent nodes, ensuring that encrypted communication models remain robust against surveillance efforts. By distributing data processing across multiple private networks, activists and independent media can communicate without exposing centralized logs that authoritarian regimes could exploit.

To counteract predictive AI models used to anticipate political dissent, adversarial detection for activist protection is introduced, preventing AI surveillance from accurately forecasting opposition movements:

$$\min_{\theta} \mathbb{E}[L(f_\theta(X), y) - \lambda D(f_\theta(X), Z)]$$

where θ, AI model parameters, $L(f_\theta(X), y)$, predictive loss function ensuring accurate detection of AI-based repression attempts, λ, adversarial countermeasure coefficient, $D(f_\theta(X), Z)$, disparity function reducing authoritarian model accuracy for activist profiling. This function ensures that predictive surveillance models used by authoritarian governments fail to accurately anticipate political activism. By deploying adversarial countermeasures, activists can prevent AI from systematically identifying, tracking, and suppressing dissent.

The privacy-preserving anonymization function, integrated with federated digital privacy protection and adversarial detection for activist protection, provides a mathematically rigorous AI-driven framework for resisting authoritarian AI surveillance. The anonymization function disrupts deterministic AI tracking, the federated learning model decentralizes digital privacy protection, and the adversarial detection function counters predictive suppression models. Together, these mechanisms provide activists, journalists, and civil rights defenders with a robust AI-driven defense against authoritarian surveillance, ensuring that fundamental rights to freedom of expression and assembly are preserved in digital spaces.

Framework: Authoritarian AI Surveillance Resistance Framework (AASRF)

Layered AI Architecture for Protecting Freedom of Expression and Assembly

Layer	Main functionality	AI module(s) and components	Operational scope
Layer 1	Data collection and preprocessing	Digital activity data preprocessing engine	Input → Raw digital activity data → Preprocessing and anonymization → Anonymized data ready for AI models

(continued)

Layer	Main functionality	AI module(s) and components	Operational scope
Layer 2	Privacy-preserving anonymization function	Privacy-preserving data anonymization module	Input → Original digital activity data → Noise injection for privacy protection → Output → Privacy-protected data ready for analysis
Layer 3	Federated learning for decentralized digital privacy	Federated Learning (FL)	Input → Decentralized nodes protecting communications → Local model training → Global model optimization without central data collection
Layer 4	Adversarial detection for activist protection	Adversarial debiasing	Input → Biased data (with identifiable features) → Processing to de-bias and prevent profiling → Output → Anonymized data without identifiable characteristics
Layer 5	Governance and oversight	Privacy compliance and ethical governance system	Input → Privacy regulations, ethical guidelines → Policy development → Enforcement of privacy and ethical practices in AI systems
Layer 6	Feedback and evaluation	Privacy impact assessment and feedback loop	Input → Feedback on privacy-preserving mechanisms → Evaluation of privacy risks and impacts → Output → Refined policies and enhanced AI systems

7.4.3 Discriminatory Impacts on Marginalized Communities

AI surveillance systems disproportionately impact marginalized communities, reinforcing systemic discrimination by amplifying biases embedded in historical data. The deployment of AI in racial and ethnic minority communities often results in over-policing and heightened surveillance, exacerbating social inequalities rather than addressing public safety concerns (Fosch-Villaronga and Poulsen 2022). AI-driven decision-making mechanisms tend to reflect societal biases, leading to discriminatory enforcement patterns that disproportionately target already vulnerable groups. To counteract these biases, an AI-driven framework integrating fairness-constrained learning, bias-aware representation learning, and reinforcement learning for equitable surveillance optimization is introduced. A fundamental component of this framework is the fairness-constrained decision function, which ensures that AI surveillance systems do not disproportionately target marginalized communities based on biased training data:

$$\min_{\theta} \mathbb{E}[L(f_\theta(x), y) + \lambda D(f_\theta(x), Z)]$$

where θ, AI model parameters, $L(f_\theta(x), y)$, primary learning objective ensuring prediction accuracy, λ, fairness penalty coefficient regulating bias mitigation, $D(f_\theta(x), Z)$, disparity function measuring bias against protected attributes Z. This function introduces explicit fairness constraints, penalizing AI models that produce discriminatory outcomes based on race, ethnicity, or socioeconomic status. By embedding fairness constraints directly into AI learning objectives, the function ensures that AI decision-making remains equitable across different demographic groups.

To prevent bias from being embedded into AI feature representations, bias-aware representation learning is introduced, optimizing AI feature extraction mechanisms to reduce discriminatory influences from historical data:

$$\mathcal{L}_{\text{total}} = \mathcal{L}_{\text{pred}} + \alpha \mathcal{L}_{\text{bias}}$$

where $\mathcal{L}_{\text{total}}$, total loss function incorporating both prediction accuracy and bias mitigation, $\mathcal{L}_{\text{pred}}$, prediction loss ensuring AI effectiveness, α, bias mitigation coefficient regulating trade-off, $\mathcal{L}_{\text{bias}}$, bias reduction loss function minimizing the influence of protected attributes. This function ensures that AI-generated features are adjusted to minimize bias, preventing racial and socioeconomic disparities from influencing AI decision-making processes. By actively reducing discriminatory correlations in AI feature spaces, the function ensures that AI systems do not reinforce systemic inequalities.

To dynamically optimize AI surveillance enforcement patterns, reinforcement learning for equitable surveillance optimization is introduced, ensuring that AI policies adapt over time to minimize discriminatory enforcement patterns:

$$J(\theta) = \mathbb{E}\left[\sum_{t=1}^{T} \gamma^t R(s_t, a_t)\right]$$

where $J(\theta)$, policy optimization function ensuring equitable AI enforcement, T, decision horizon for AI surveillance adjustments, γ^t, discount factor controlling long-term fairness, $R(s_t, a_t)$, reward function penalizing discriminatory surveillance actions in state s_t under action a_t. This function ensures that AI surveillance systems continuously refine their enforcement policies based on fairness constraints. By integrating reinforcement learning mechanisms, the function dynamically adjusts AI surveillance strategies to reduce racial profiling and biased enforcement patterns over time.

The fairness-constrained decision function, integrated with bias-aware representation learning and reinforcement learning for equitable surveillance optimization, provides a mathematically rigorous AI-driven framework for ensuring fairness in AI surveillance systems. The fairness-constrained function directly penalizes biased AI decision-making, the bias-aware learning function eliminates discriminatory influences from AI feature spaces, and the reinforcement learning optimization function dynamically refines AI surveillance policies to reduce systemic disparities over time.

Together, these mechanisms ensure that AI surveillance does not exacerbate social inequalities, fostering a more just and equitable application of AI-driven law enforcement.

Framework: Fairness-Constrained AI Surveillance Framework (FC-AISF)

Layered AI Architecture for Fairness in AI Surveillance Systems

Layer	Main functionality	AI module(s) and components	Operational scope
Layer 1	Data collection and preprocessing	Raw surveillance data preprocessing engine	Input → Raw surveillance data → Preprocessing → Data ready for AI models
Layer 2	Fairness-constrained decision function	Fairness-constrained learning	Input → Surveillance data → Learning with fairness penalty → Output → Fairer AI decisions
Layer 3	Bias-aware representation learning	Bias-aware representation learning	Input → Raw features → Feature learning with bias reduction → Output → Fairer, unbiased features
Layer 4	Reinforcement learning for equitable surveillance optimization	Reinforcement learning for equitable surveillance optimization	Input → Surveillance policy data → RL optimization → Output → Optimized, non-biased enforcement strategies
Layer 5	Governance and oversight	Ethical governance and compliance system	Input → Ethical guidelines and policies → Governance mechanisms → Output → Ethical compliance and oversight
Layer 6	Feedback and evaluation	Fairness impact feedback and performance evaluation system	Input → Evaluation feedback → Assessment of fairness → Output → Improved fairness and performance

For example, predictive policing algorithms are designed to anticipate crime by analyzing historical data; however, these models may disproportionately target minority communities, contributing to higher levels of surveillance, intervention, or arrest (Bruneault and Sabourin Laflamme 2021). Because AI models learn from past crime reports, which may already reflect systemic biases, they risk reproducing discriminatory policing patterns that conflict with the principle of equality before the law, as articulated in international human rights frameworks such as the Universal Declaration of Human Rights (UDHR) and the International Covenant on Civil and Political Rights (ICCPR). Addressing this issue requires an AI-driven framework integrating causal inference for bias detection, algorithmic debiasing techniques, and dynamic reinforcement learning for fair predictive policing. a key component of this framework is the causal Fairness Assessment Function, which identifies whether disparities in predictive policing outcomes stem from historical biases rather than objective risk factors:

$$\Delta_{\text{fair}} = P(Y|X,Z) - P(Y|X)$$

where Δ_{fair}, fairness disparity metric quantifying bias, $P(Y|X, Z)$, probability of an AI decision Y given feature set X and protected attribute Z, $P(Y|X)$, probability of the same decision without considering Z. This function measures whether AI models disproportionately associate crime predictions with race, ethnicity, or socioeconomic status. If the disparity Δ_{fair} is significant, it indicates bias in predictive policing algorithms, necessitating correction.

To mitigate these biases, algorithmic debiasing for fair predictive modeling is introduced, ensuring that AI models do not overfit to historically biased data patterns:

$$\min_{\theta} \mathbb{E}[L(f_{\theta}(X), Y) + \lambda D_{\text{fair}}(f_{\theta}(X), Z)]$$

where θ, AI model parameters, $L(f_{\theta}(X), Y)$, predictive loss function ensuring model accuracy, λ, fairness penalty coefficient regulating bias mitigation, $D_{\text{fair}}(f_{\theta}(X), Z)$, disparity function penalizing biased associations with protected attributes Z. This function penalizes predictive models that exhibit racial, ethnic, or economic biases, ensuring that predictive policing decisions are driven by legitimate crime factors rather than biased historical patterns.

To dynamically adjust predictive policing strategies based on fairness constraints, dynamic reinforcement learning for fair AI decision-making is introduced:

$$J(\theta) = \mathbb{E}\left[\sum_{t=1}^{T} \gamma^{t} R_{\text{fair}}(s_t, a_t)\right]$$

where $J(\theta)$, policy optimization function ensuring fairness in predictive policing, T, decision horizon, γ^{t}, discount factor prioritizing long-term fairness, $R_{\text{fair}}(s_t, a_t)$, reward function penalizing biased enforcement actions in state s_t under action a_t. This function enables AI-driven predictive policing models to adjust dynamically to ensure fairness over time, correcting for historical biases while optimizing crime prediction accuracy without disproportionate enforcement on marginalized communities.

The causal fairness assessment function, combined with algorithmic debiasing for fair predictive modeling and dynamic reinforcement learning for fair AI decision-making, establishes a rigorous AI-driven framework for ensuring fairness in predictive policing. The causal fairness function detects bias in predictive models, the algorithmic debiasing function removes discriminatory patterns from AI decision-making, and the reinforcement learning optimization function dynamically adjusts AI policing policies to promote equitable law enforcement. Together, these mechanisms ensure that AI-driven predictive policing models uphold the principle of equality before the law, rather than reinforcing systemic biases.

Framework: Fairness-Constrained Predictive Policing Framework (FCPP)

Layered AI Architecture for Fair Predictive Policing

Layer	Main functionality	AI module(s) and components	Operational scope
Layer 1	Data collection and preprocessing	Historical crime data preprocessing engine	Input → Raw historical crime data → Preprocessing to remove irrelevant data → Data ready for AI analysis
Layer 2	Causal fairness assessment function	Causal inference for bias detection	Input → Data with protected attributes → Causal analysis to identify fairness disparities → Output → Fairness disparity metric that quantifies bias
Layer 3	Algorithmic debiasing for fair predictive modeling	Algorithmic debiasing	Input → Biased historical data → Debiasing model to adjust for fairness → Output → Fairer predictive model that is not biased toward specific demographics
Layer 4	Dynamic reinforcement Learning for fair AI decision-making	Reinforcement learning for fair AI decision-making	Input → Surveillance data → RL optimization with fairness penalties → Output → Optimized decision-making model that minimizes biased enforcement
Layer 5	Governance and oversight	Ethical governance and compliance system	Input → Ethical guidelines, privacy regulations → Policy development and enforcement → Output → Governance mechanisms to ensure compliance with fairness and equality
Layer 6	Feedback and evaluation	Fairness and performance feedback system	Input → Performance and fairness evaluation → Feedback to improve AI systems → Output → Refined fairness measures in AI models

7.4.4 Legal and Ethical Challenges

The rapid advancement of AI surveillance technologies has outpaced existing legal and ethical frameworks, creating significant gaps in the protection of privacy, freedom of expression, and freedom of assembly. While international human rights law provides a foundational safeguard for these rights, many legal systems lack specific AI governance mechanisms, particularly in countries with weak privacy laws and inadequate regulatory oversight (Saheb 2023; Sargiotis 2024a, b). This allows AI surveillance systems to be deployed without sufficient transparency or accountability, exacerbating risks of misuse and human rights violations. To bridge this gap, an AI-driven framework integrating automated legal compliance auditing, adaptive regulatory policy optimization, and ethical risk assessment for AI governance is introduced. A key element of this framework is the AI legal compliance verification function, which assesses whether AI surveillance systems adhere to evolving international legal and ethical standards:

$$C_t = \sigma\left(W_C C_{t-1} + \sum_{i \in N(t)} W_N C_i + b\right)$$

where C_t, compliance state at time t, W_C, weight matrix regulating legal adaptation, C_{t-1}, prior compliance state, $N(t)$, set of regulatory bodies overseeing AI deployment, W_N, weight matrix adjusting for jurisdictional differences, C_i, legal compliance input from governing entity i, b, bias term ensuring stability, σ, activation function maintaining bounded compliance updates. This function enables AI-driven systems to automate legal compliance verification by continuously integrating updates from international, national, and regional legal frameworks, ensuring that AI surveillance technologies remain aligned with human rights principles.

To optimize legal policy enforcement dynamically, adaptive regulatory policy optimization is introduced, allowing governments and regulatory bodies to update AI governance strategies in real time:

$$J(\theta) = \mathbb{E}\left[\sum_{t=1}^{T} \gamma^t R_{\text{policy}}(s_t, a_t)\right]$$

where $J(\theta)$, policy optimization function for AI governance, T, regulatory enforcement timeline, γ^t, discount factor controlling long-term legal adherence, $R_{\text{policy}}(s_t, a_t)$, reward function ensuring regulatory compliance at state s_t under action a_t. This function ensures that AI governance strategies remain flexible, allowing policymakers to adjust regulatory policies in response to emerging AI threats while maintaining alignment with human rights protections.

To assess ethical risks dynamically, ethical risk scoring for AI governance is introduced, quantifying potential legal and ethical violations in AI surveillance applications:

$$R_{\text{ethics}} = P(V|D) \cdot S(V)$$

where R_{ethics}, ethical risk score, $P(V|D)$, probability of a violation V given AI decision D, $S(V)$, severity of violation assessed based on international legal standards. This function enables AI ethics auditors and policymakers to quantify risks associated with AI surveillance, ensuring that high-risk AI applications are subject to stricter oversight.

The AI legal compliance verification function, integrated with adaptive regulatory policy optimization and ethical risk assessment for AI governance, provides a mathematically rigorous AI-driven framework for ensuring that AI surveillance technologies comply with evolving legal and ethical standards. The compliance verification function dynamically aligns AI systems with international human rights frameworks, the regulatory policy optimization function adapts legal enforcement mechanisms to emerging AI risks, and the ethical risk assessment function quantifies potential violations in AI surveillance applications. Together, these AI-driven mechanisms close the legal and ethical gap in AI governance, ensuring that AI surveillance systems operate within the boundaries of human rights law while maintaining transparency and accountability.

Framework: AI Surveillance Legal and Ethical Compliance Framework (AISLECF)

Layered AI Architecture for Legal and Ethical Governance in AI Surveillance

Layer	Main functionality	AI module(s) and components	Operational scope
Layer 1	Data collection and preprocessing	Raw surveillance data preprocessing engine	Input → Raw surveillance data → Preprocessing → Data ready for AI analysis
Layer 2	AI legal compliance verification function	Automated legal compliance auditing	Input → Legal frameworks and updates → AI model adaptation → Output → Compliance state verification in AI systems
Layer 3	Adaptive regulatory policy optimization	Reinforcement learning for policy optimization	Input → Regulatory enforcement data → Policy optimization via reinforcement learning → Output → Adjusted and optimized legal policies
Layer 4	Ethical risk scoring for AI governance	Ethical risk assessment for AI governance	Input → AI decision-making → Ethical violation probability and severity → Output → Ethical risk score
Layer 5	Governance and oversight	Legal and ethical governance system	Input → Legal and ethical guidelines → Policy development → Output → Governance mechanisms ensuring compliance with human rights standards
Layer 6	Feedback and evaluation	Ethical compliance feedback and system refinement engine	Input → Feedback on AI system performance → Evaluation of ethical compliance → Output → Refined AI systems with updated policies

As AI surveillance systems become more pervasive, governments and international organizations must implement new regulations that uphold human rights while preventing the misuse of AI technologies (Fosch-Villaronga and Poulsen 2022). To achieve this, legal frameworks must mandate transparency in AI deployments, restrict excessive data collection, and establish mechanisms for individuals to contest AI-driven decisions. Robust data protection laws are also essential to ensure that personal data collected by AI systems remains secure and is not exploited for purposes beyond its original intent. However, given the complexity and rapid evolution of AI, a dynamic AI-driven framework is required to enforce these regulations effectively. This framework integrates automated transparency auditing, privacy-preserving AI governance, and reinforcement learning for regulatory adaptation. A critical element of this framework is the AI transparency auditing function, which supports verification that AI surveillance systems provide appropriate disclosure regarding their decision-making processes and data collection practices:

$$T_t = \sigma\left(W_T T_{t-1} + \sum_{i \in N(t)} W_N T_i + b\right)$$

where T_t, transparency compliance state at time t, W_T, weight matrix regulating transparency trends, T_{t-1}, previous transparency score, $N(t)$, set of regulatory bodies monitoring AI deployment, W_N, weight matrix for compliance influence, T_i, transparency feedback from monitoring entity i, b, bias term ensuring stability, σ, activation function ensuring smooth updates. This function ensures that AI surveillance systems are regularly audited for transparency, tracking whether organizations disclose their AI models' decision-making processes and data-handling practices.

To prevent unauthorized personal data collection, privacy-preserving AI governance is introduced, ensuring that AI surveillance adheres to strict data protection regulations:

$$X' = X + \mathcal{N}(0, \sigma^2)$$

where X', anonymized data processed by AI, X, original collected data, $\mathcal{N}(0, \sigma^2)$, controlled noise function ensuring differential privacy. This function ensures that AI models process data without compromising individual privacy, introducing mathematically rigorous anonymization techniques that prevent unauthorized tracking and profiling.

To dynamically refine regulatory enforcement, reinforcement learning for regulatory adaptation is introduced, ensuring that legal policies evolve based on real-world AI compliance trends:

$$J(\theta) = \mathbb{E}\left[\sum_{t=1}^{T} \gamma^t R_{\text{reg}}(s_t, a_t)\right]$$

where $J(\theta)$, regulatory adaptation policy function, T, regulatory adjustment timeline, γ^t, discount factor prioritizing long-term governance stability, $R_{\text{reg}}(s_t, a_t)$, reward function incentivizing AI compliance at state s_t under action a_t. This function enables regulatory bodies to dynamically adjust AI governance policies based on real-time AI compliance assessments, ensuring that legal frameworks remain responsive to emerging AI challenges.

The AI transparency auditing function, combined with privacy-preserving AI governance and reinforcement learning for regulatory adaptation, establishes a rigorous AI-driven framework for ensuring responsible AI surveillance governance. The transparency auditing mechanism enforces accountability in AI decision-making, the privacy-preserving function safeguards personal data, and the regulatory adaptation model dynamically refines AI policies in response to emerging risks. Together, these mechanisms create a comprehensive AI governance system that ensures AI surveillance systems respect fundamental human rights while preventing misuse and unauthorized data collection.

Framework: AI Surveillance Legal and Ethical Governance Framework (AISLEG)

Layered AI Architecture for Legal and Ethical Governance in AI Surveillance

Layer	Main functionality	AI module(s) and components	Operational scope
Layer 1	Data collection and preprocessing	Raw surveillance data preprocessing engine	Input → Raw surveillance data → Preprocessing → Data ready for AI models
Layer 2	AI transparency auditing function	Automated transparency auditing	Input → Regulatory feedback → Audit AI systems for transparency → Output → Transparency compliance tracking
Layer 3	Privacy-preserving AI governance	Privacy-preserving AI governance	Input → Raw data → Apply privacy-preserving techniques → Output → Anonymized data for AI processing
Layer 4	Reinforcement Learning for Regulatory Adaptation	Reinforcement Learning for Regulatory Adaptation	Input → AI compliance feedback → Policy optimization → Output → Updated regulatory policies for AI governance
Layer 5	Governance and oversight	Ethical and legal governance system	Input → Ethical and legal frameworks → Policy development → Output → Governance mechanisms ensuring AI compliance
Layer 6	Feedback and evaluation	Compliance feedback and governance refinement engine	Input → Feedback on AI performance → Evaluate compliance and risks → Output → Refined AI governance policies

AI surveillance systems pose significant risks to human rights, particularly concerning privacy, freedom of expression, and freedom of assembly. These systems can lead to widespread privacy violations, chilling effects on free speech, and discriminatory practices that disproportionately impact marginalized communities. To mitigate these risks, there is a need for stronger legal and ethical frameworks that regulate the use of AI surveillance technologies and protect individuals' fundamental human rights.

Figure 7.4 illustrates the ethical implications of AI-driven surveillance on human rights, focusing on privacy violations, chilling effects on freedom of expression and assembly, discriminatory impacts on marginalized communities, and the legal and ethical challenges surrounding these technologies. Privacy concerns highlight data collection without consent, lack of transparency, and the rise of a "surveillance society." Freedom of expression and assembly are threatened by self-censorship, suppression of dissent, and monitored public spaces. Marginalized communities face

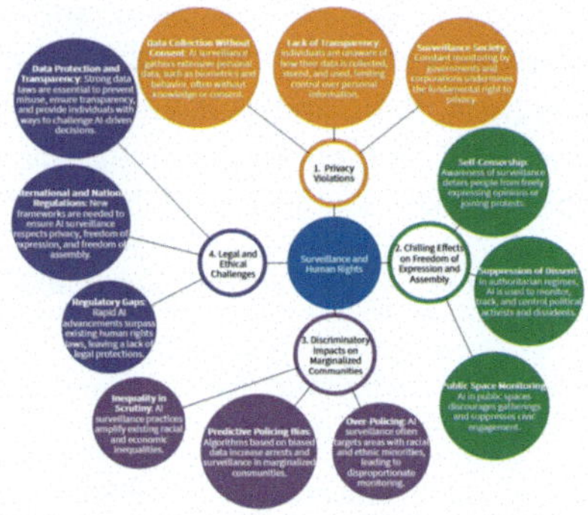

Fig. 7.4 Surveillance and human rights: ethical concerns in AI integration

discriminatory impacts through predictive policing bias and over-policing. Legal and ethical challenges stress the need for regulatory frameworks, transparency, and data protection to prevent AI misuse and ensure accountability.

The ethical dilemmas surrounding AI surveillance and security are complex and multifaceted. While AI offers significant opportunities to enhance public safety and improve security, it also presents substantial risks to privacy, human rights, and social justice. Addressing these ethical challenges requires a careful balance between technological innovation and the protection of individual freedoms, with a particular focus on transparency, accountability, fairness, and the responsible use of data.

Limitations and Governance Implications

The surveillance governance frameworks developed in this chapter address one of the most contested domains of urban AI: the use of intelligent systems for security, monitoring, risk assessment, and public-order management. AI-Driven Privacy and Security in Surveillance (AI-PSS), Privacy-Preserving AI Surveillance Framework (PPAISF and PPAISF II), Privacy-Conscious Adaptive Surveillance Intelligence Framework (PCASIF), Secure Encrypted AI Surveillance Framework (SEAISF), Fairness-Constrained Predictive Policing Architecture (FCPPA), Fairness-Constrained AI Surveillance Framework (FC-AISF), Reinforcement Learning-Based Explainable Accountability Model (RL-XAM), Shapley Value-Based XAI Framework (SVB-XAI), Automated AI Audit Model (AAAM), Stakeholder-Centric Explainability Framework (SCEF), AI Surveillance Legal and Ethical Compliance Framework (AISLECF), and AI Surveillance Legal and Ethical Governance Framework (AISLEG) provide mechanisms for proportionality, privacy preservation, anti-bias safeguards, legal compliance, independent audit, and stakeholder-centric explainability. Their governance value is considerable, but their normative limits must be stated with particular clarity. Surveillance technologies operate close to fundamental rights, including privacy, freedom of movement, freedom of assembly, freedom of expression, and protection from discriminatory state power. For this reason, no privacy-preserving architecture, fairness constraint, encryption protocol, explainability model, or audit function can by itself legitimate surveillance practices that lack necessity, proportionality, democratic authorization, and enforceable legal safeguards. The governance implication is that AI surveillance must remain exceptional, bounded, reviewable, and institutionally accountable. Technical safeguards may reduce specific risks, but the authority to deploy, restrict, suspend, or prohibit surveillance systems must remain subject to public law, independent oversight, judicial review, civic scrutiny, and the protection of human rights, especially for communities historically exposed to over-policing, political monitoring, or discriminatory security practices.

7.5 Conclusion

This chapter offers a profound exploration of the ethical dilemmas inherent in AI-driven surveillance, focusing on the tension between enhancing public safety and protecting fundamental human rights, particularly privacy, freedom of expression, and assembly. Through a multi-layered and mathematically rigorous framework, it integrates key privacy-preserving mechanisms such as Differential Privacy (DP) and Federated Learning (FL) with advanced ethical principles like fairness, transparency, and accountability.

The chapter methodically addresses the challenges posed by the mass collection of personal data, offering solutions that ensure individuals' privacy is safeguarded while still enabling the AI systems to function effectively. It highlights the importance of developing AI models that not only perform optimally but also adhere to stringent privacy regulations, emphasizing the role of differential privacy in anonymizing data and federated learning in decentralizing data storage to prevent unauthorized access.

A significant portion of the analysis focuses on bias and discrimination in AI surveillance systems, with the chapter outlining how biased datasets can perpetuate social inequalities. It proposes frameworks for fairness-aware AI models, including adversarial debiasing and representation learning, to mitigate such biases and ensure equitable outcomes across diverse demographic groups. Furthermore, the chapter demonstrates how transparency can be achieved through explainable AI (XAI) mechanisms, such as Shapley Additive Explanations (SHAP), to make AI decisions interpretable and auditable.

The chapter culminates in a call for a human-centric equilibrium in AI surveillance, advocating for the continuous recalibration of AI models through Multi-Objective Optimization (MOO) to balance the security benefits with the protection of individual freedoms. The frameworks presented here not only equip AI systems to safeguard privacy and fairness but also empower human oversight mechanisms, ensuring that surveillance technologies do not become tools of repression but serve the greater good of public safety without infringing on rights.

This work lays a comprehensive foundation for the responsible, ethical, and transparent deployment of AI in surveillance systems, offering a blueprint for policymakers, technologists, and scholars who seek to navigate the complex ethical landscape of AI governance. By blending rigorous computational models with robust ethical guidelines, this chapter provides a pathway toward an AI-driven surveillance infrastructure that is both secure and respectful of human rights.

References

Al-kfairy, M., Mustafa, D., Kshetri, N., Insiew, M., & Alfandi, O. (2024). Ethical challenges and solutions of generative AI: An interdisciplinary perspective. *Informatics*. https://doi.org/10.3390/informatics11030058

Bruneault, F., & Sabourin Laflamme, A. (2021). AI Ethics: How can information ethics provide a framework to avoid usual conceptual pitfalls? *AI & Society*, 36, 757–766. https://link.springer.com/article/10.1007/s00146-020-01077-w

Fosch-Villaronga, E., & Poulsen, A. (2022). Diversity and inclusion in artificial intelligence. In B. Custers & E. Fosch-Villaronga (Eds.), Law and Artificial Intelligence. https://doi.org/10.1007/978-94-6265-523-2_6

Saheb, T. (2023). Ethically contentious aspects of artificial intelligence surveillance: A social science perspective. *AI and Ethics*, 3, 369–379. https://doi.org/10.1007/s43681-022-00196-y

Sargiotis, D. (2024a). Ethical AI in Information Technology: Navigating Bias, Privacy, Transparency, and Accountability. *Advances in Machine Learning & Artificial Intelligence*, 5(3), 01–14. https://doi.org/10.33140/AMLAI.05.03.03

Sargiotis, D. (2024b). Advancing Civil Engineering with AI and Machine Learning: From Structural Health to Sustainable Development. *International Journal of Architecture and Planning*, 4(2), 54–81. https://doi.org/10.51483/IJARP.4.2.2024.54-81

Sargiotis, D. (2025a). Transforming civil engineering with AI and machine learning: Innovations, applications, and future directions. *International Journal of Research Publication and Reviews*, 6(1), 3780–3805. https://doi.org/10.55248/gengpi.6.0125.0528

Sargiotis, D. (2025b). Fostering ethical and inclusive AI: A human-centric paradigm for social impact. *International Journal of Research Publication and Reviews*, 6(1), 3754–3779. https://doi.org/10.55248/gengpi.6.0125.0529

Conclusion

This book has articulated a profound reorientation of the smart city: from a technologically optimized machine to a morally aware civic system. In tracing the ethical architecture of Artificial Intelligence (AI) within urban environments, this book asserts that the intelligence of a city cannot be measured solely by its capacity to calculate, predict, or automate but by its ability to uphold justice, protect autonomy, and foster inclusive participation. The future of AI-powered urbanism must be built not only on code and computation but on conscience.

Through the deployment of mathematically rigorous models including Adversarial Debiasing, Fairness-Constrained Reinforcement Learning, Interpretable Policy Optimization, and Federated Differential Privacy the book has shown that ethical AI is not an abstraction. It is an implementable, auditable, and computable framework for ensuring equity, transparency, and accountability across all layers of smart city governance. AI systems, when designed with integrity, become instruments of repair: they can redress historical biases, distribute public services with spatial fairness, and empower marginalized communities through intelligible, participatory decision-making.

At the core of this ethical vision lies a call for algorithmic justice—a normative stance in which the deployment of AI in cities is bound by rules of explanation, consent, accessibility, and accountability. Participatory Explainable AI (P-XAI), SHAP-based attribution tools, Policy-Justification Functions, and Dispute Resolution Modules embedded within Multi-Agent Reinforcement Learning architectures transform AI from a black box into a site of democratic negotiation. This book has thus advanced not only a technical paradigm but a civic ethic: where code serves citizenship and automation respects agency.

The book concludes with a call to redesign the governance of AI itself. Through frameworks such as the Risk-Aware Urban Regulation Framework (RAURF), the Ethically Constrained Nash Equilibrium, and Blockchain-Auditable Consent Models, the city becomes a place where algorithmic power is regulated, reasoned,

D. Sargiotis, *AI with Integrity*, https://doi.org/10.1007/978-3-032-31133-7

and reciprocally governed. Urban intelligence must be transparent not only to engineers and policymakers but to all citizens who inhabit its decisions. Only then can AI serve its highest urban purpose not as an instrument of surveillance or efficiency alone, but as an ethical co-architect of futures that are inclusive, sustainable, and human-centered.